Ben Richardson
SUGAR: REFINED POWER IN A GLOBAL REGIME

Simon Rushton and Owain David Williams (*editors*)
PARTNERSHIPS AND FOUNDATIONS IN GLOBAL HEALTH GOVERNANCE

Marc Schelhase
GLOBALIZATION, REGIONALIZATION AND BUSINESS
Conflict, Convergence and Influence

Herman M. Schwartz and Leonard Seabrooke (*editors*)
THE POLITICS OF HOUSING BOOMS AND BUSTS

Leonard Seabrooke
US POWER IN INTERNATIONAL FINANCE
The Victory of Dividends

Timothy J. Sinclair and Kenneth P. Thomas (*editors*)
STRUCTURE AND AGENCY IN INTERNATIONAL CAPITAL MOBILITY

J.P. Singh (*editor*)
INTERNATIONAL CULTURAL POLICIES AND POWER

Susanne Soederberg, Georg Menz and Philip G. Cerny (*editors*)
INTERNALIZING GLOBALIZATION
The Rise of Neoliberalism and the Decline of National Varieties of Capitalism

Kenneth P Thomas
INVESTMENT INCENTIVES AND THE GLOBAL COMPETITION FOR CAPITAL

Helen Thompson
CHINA AND THE MORTGAGING OF AMERICA
Economic Interdependence and Domestic Politics

Ritu Vij (*editor*)
GLOBALIZATION AND WELFARE
A Critical Reader

Matthew Watson
THE POLITICAL ECONOMY OF INTERNATIONAL CAPITAL MOBILITY

Owen Worth and Phoebe Moore
GLOBALIZATION AND THE 'NEW' SEMI-PERIPHERIES

Xu Yi-chong and Gawdat Bahgat (*editors*)
THE POLITICAL ECONOMY OF SOVEREIGN WEALTH FUNDS

International Political Economy Series
Series Standing Order ISBN 978-0–333–71708–0 (hardback)
Series Standing Order ISBN 978-0–333–71110–1 (paperback)

You can receive future titles in this series as they are published by placing a standing order. Please contact your bookseller or, in case of difficulty, write to us at the address below with your name and address, the title of the series and one of the ISBNs quoted above.

Customer Services Department, Macmillan Distribution Ltd, Houndmills, Basingstoke, Hampshire RG21 6XS, England

---

Global Commodity Governance

State Responses to Sustainable Forest and Fisheries Certification

Fred Gale
Senior Lecturer, School of Government, University of Tasmania, Australia

Marcus Haward
Associate Professor, School of Government, University of Tasmania, Australia

First published 2011 by
PALGRAVE MACMILLAN

Palgrave Macmillan in the UK is an imprint of Macmillan Publishers Limited, registered in England, company number 785998, of Houndmills, Basingstoke, Hampshire RG21 6XS.

Palgrave Macmillan in the US is a division of St Martin's Press LLC, 175 Fifth Avenue, New York, NY 10010.

Palgrave Macmillan is the global academic imprint of the above companies and has companies and representatives throughout the world.

Palgrave® and Macmillan® are registered trademarks in the United States, the United Kingdom, Europe and other countries

ISBN 978–0–230–51663–2   hardback

A catalogue record for this book is available from the British Library.

A catalog record for this book is available from the Library of Congress.

10   9   8   7   6   5   4   3   2   1
20   19   18   17   16   15   14   13   12   11

Transferred to Digital Printing in 2014

Contents

List of Figures and Tables

Figures

Tables

Acknowledgements

We are indebted to many people who have contributed their time and insights to this study. In particular, we would like to acknowledge the Australian Research Council's Discovery Grant Program for funding the research. This ensured the project's completion by enabling us to undertake field research in each of our three comparator countries.

We would also like to acknowledge the contributions and insights of Tim Cadman, Anthony Charles, Roxanne Comeaux, Simon Counsell, Phyllis Dale, Jim Farrell, Mike Fullerton, Mike Garforth, Alistair Hobday, Duncan Leadbitter, Rich Lincoln, Tavis Potts, Keith Sainsbury, Hannah Scrase, Leonie Van Der Maesen, Martin Von Mirbach and Peter Wilson. These individuals took time out of their busy schedules to meet with us and discuss the FSC, MSC and evolution of forest and fisheries certification in their respective jurisdictions.

Marcus Haward thanks Jon Sumby for his invaluable research support on fisheries certification. Fred Gale thanks Ben Atkins for his tireless efforts in tracking down studies and information on forest certification.

Books take a long time to research and write and this book is no exception. Families often bear a hidden burden of the research endeavour. Fred Gale sincerely thanks his wife, Beverly and son, Evan for their patience and understanding and constant support. Marcus Haward sincerely thanks his wife, Anne and his son, Charles for their ongoing support in the research and writing of this book.

List of Abbreviations

ACF	Australian Conservation Foundation
AFMA	Australian Fisheries Management Authority
AFPA	American Forestry and Paper Association
AFS	Australian Forestry Standard
AHP	Analytical Hierarchy Process
ASI	Accreditation Services International
ASIC	Australian Seafood Industry Council
ATFS	American Tree Farm System
BCEN	British Columbia Environmental Network
C&I	Criteria and Indicators
CAR	Corrective Action Requests
CCFM	Canadian Council of Forest Ministers
CCWA	Conservation Council of Western Australia
CFMEU	Construction, Forestry, Mining and Energy Union
CFP	Common Fisheries Policy
CLA	Country Landowners Association
CoC	Chain of Custody
COFI	Confederation of Forest Industries, British Columbia
CPET	Central Point for Expertise on Timber
CPPA	Canadian Pulp and Paper Association
CSA	Canadian Standards Association
CSFCC	Canadian Sustainable Forestry Certification Coalition
CSIRO	Commonwealth Scientific and Industrial Research Organisation
CSO	Civil Society Organisation
DAFF	Department of Agriculture, Fisheries and Forestry
DEFRA	Department of Environment, Food and Rural Affairs
DFID	Department for International Development
DFO	Department of Fisheries and Oceans
ECSO	Environmental Civil Society Organisation

EEZ	**Exclusive Economic Zone**
EMS	**Environmental Management System**
ENGO	Environmental Non-Governmental Organisation
EU	**European Union**
FAO	**United Nations Food and Agriculture Organization**
FFIC	Forestry and Forest Industries Council of Tasmania
FIAT	Forest Industries Association of Tasmania
FICGB	Forest Industry Council of Great Britain
FLEGT	Forest Law Enforcement, Governance and Trade
FLO	**Fairtrade Labelling Organisation International**
FoE	Friends of the Earth
FPAC	Forest Products Association of Canada
FRA	Forest Resources Assessment
FSC	Forest Stewardship Council
FSC-AC	Forest Stewardship Council *Asociación Civil*
FSC-GD	Forest Stewardship Council-Global Development
FSC-IC	Forest Stewardship Council-International Centre
GATT	General Agreement on Tariffs and Trade
GBCA	Green Building Council of Australia
HACCP	Hazard Analysis Critical Control Point
HCVF	High Conservation Value Forests
HVP	Hancock Victoria Plantations
ICA	International Commodity Agreement
IFF	Intergovernmental Forum on Forests
IFMA	International Forest Monitoring Agency
IFOAM	International Federation of Organic Agricultural Movements
IPC	Integrated Programme for Commodities
IPF	Intergovernmental Panel on Forests
ISO	International Organization for Standardization
ITTO	International Tropical Timber Organization
IUCN	World Conservation Union (formerly the International Union for the Conservation of Nature)
IUU	Illegal, Unregulated and Unreported Fishing
IWA	Industrial, Wood and Allied Workers of Canada

LEED	Leadership in Energy and Environmental Development
LOSC	United Nations Law of the Sea Convention
MAC	Management Advisory Committee
MCFFA	Ministerial Council on Forestry, Fisheries and Aquaculture
MCPFE	Ministerial Council for the Protection of Forests in Europe
MIS	Managed Investment Scheme
MSC	Marine Stewardship Council
NAFI	National Association of Forest Industries
NFFO	National Federation of Fishermen's Organisations
NSMD	Non-State, Market Driven
OCS	Offshore Constitutional Settlement
OPEC	Organization of Petroleum Exporting Countries
PCC	Pacific Certification Council
PEFC	Programme for the Endorsement of Forest Certification
PIMC	Primary Industries Ministerial Council
PPMs	Product and Processing Methods
RCEN	Canadian Environmental Network
RFA	Regional Forest Agreement
SCS	Scientific Certification Systems
SFI	Sustainable Forestry Initiative
SGS	Sociéte Générale de Surveillance
SSSI	Site of Special Scientific Interest
TAB	Technical Advisory Board
TAC	Total Allowable Catch
TCA	Timber Communities Australia
TGA	Timber Growers Association
TTF	Tropical Timber Federation
TWS	The Wilderness Society
UK	United Kingdom
UKWAS	United Kingdom Woodland Assurance Standard
UNCED	United Nations Conference on Environment and Development
UNCLOS	United National Convention on the Law of the Sea
UNCTAD	United Nations Conference on Trade and Development

UNECE	United Nations Economic Commission for Europe
VAFI	Victoria Association of Forest Industries
WARP	Woodworkers Alliance for Rainforest Protection
WCL	Wildlife and Countryside Link
WCWC	Western Canada Wilderness Committee
WSSD	World Summit on Sustainable Development
WTO	World Trade Organization
WWF	World Wide Fund for Nature (formerly World Wildlife Fund)

1
Commodity Governance in a Globalising World

Commodities play a vital role in world trade. Every day, huge quantities of oil, iron, gold, silver, tin, copper, coffee, tea, cocoa, sugar, rice, wheat, corn, fish, cotton, wool and timber and other products are transported around the world. The trade in commodities links a vast number of producers and consumers together via lengthy and complicated chains. Whether engaged in extraction, transportation, refining, transformation, packaging, wholesaling, retailing or marketing, or credit or consumption, we are all enmeshed in these trade arrangements and share responsibility for their consequences.

Commodity production is now truly globalised. In the past 50 years, revolutions in transport and communications coupled with runaway global demand and the liberalisation of trade, investment and finance relations have turbocharged the trade component of world economic output. Whereas total world economic output increased by three per cent between 1974 and 2007, the concomitant growth in world trade was five per cent (Cohn 2009, 167). Technological development has raised the capacity of merchant fleets and innovation, such as containerisation, has meant that ever-more goods are transported at ever-lower cost. Technological developments have interacted with increased trade liberalisation via the General Agreement on Tariffs and Trade (GATT). Despite widespread disappointment at the US Senate's refusal to ratify the 1948 Havana Charter which thwarted the establishment of a potentially powerful International Trade Organization, the GATT facilitated the pursuit of a remarkably successful post-war trade liberalisation agenda via successive 'rounds' of negotiations. GATT negotiations during the 1960s and 1970s resulted in significant, if not necessarily equitable, reductions in tariffs: for example, the Kennedy Round reduced the tariffs on industrial goods by an average of 35 per cent (Cohn 2009, 174).

The production, transformation and exchange of commodities encompass much more than just the economy. These activities have profound, frequently negative, impacts on people and the planet. When commodity

1

production respects community rights, pays a living wage and maintains biodiversity, it makes a net contribution to the global economic welfare. Unfortunately, however, in countries around the world, commodity production exploits communities and workers while simultaneously degrading or destroying the environment. The fine furniture in our living rooms, the carpets on our floors and the clothes on our backs are nothing more than assemblages of commodities. In the absence of appropriate regulation, there is every possibility that these embody illegal and unsustainably managed materials manufactured from sweatshop labour that infringed the rights of indigenous and local communities (Dauvergne 2008).

That economic activity can have highly detrimental environmental and social consequences is hardly revelatory. Indeed, all countries have national laws in place to mitigate potential damage. The early twentieth century saw a steady improvement in conditions as unions fought to reduce working hours, improve pay and conditions and outlaw child and prison labour. Women, too, struggled to turn around millennia of discrimination at home and at work, although much remains to the done. Towards the end of the twentieth century, and even as some of the original gains were being rolled back under the neoliberal revolution, attention focused on the natural environment. Increasingly concerned about the detrimental ecological consequences of economic activity, countries introduced planning, pollution, environmental assessment and common-pool resource laws to improve outcomes.

These national regulatory efforts have not been effective. They were not harmonised across states, leading over time to a competitive 'race to the bottom' in labour and environmental standards. Even more problematic, however, is the gap between the policies legislated by states and their implementation on the ground. The rise of environmental regulation coincided with a decline in state capacity as governments retrenched after the 1973–4 recession induced by the Organization of Petroleum Exporting Countries (OPEC) oil price rise. The 'hollowing out' of the state accelerated in the 1980s as neoliberal ideas gained ascendancy, reducing the capacity of governments to exercise the required oversight over a broad swathe of policy sectors (Rhodes 1994).

Regulatory efforts were ineffective even where state capacity was not curtailed. As globalisation deepened and widened, variations in national standards appeared that international organisations were powerless to address. Shoes made by companies like Nike, Adidas and Reebok under allegedly sweatshop conditions in places like Indonesia were sold in air-conditioned malls in London at fashion-designer prices (Bartley 2003, forthcoming). Environmentally polluting production in the Maquiladora region of Mexico was effectively imported into the US despite the North American Agreement on Environmental Cooperation and its associated Commission (Markell and Knox 2003). And even though a powerful new international public agency was set up to monitor and police world trade after 1995, the World Trade Organization (WTO) refused to act on environmental and social issues fearing that the inclusion of 'process and production methods' (PPMs) in trade law would be the slippery slope to protectionism (Gale 1998).

With the state increasingly ineffective in regulating global production chains, concerned groups and individuals began to focus on internal, market-based governance arrangements in the shape of voluntary codes of conduct, quality and environmental management systems (EMS) and certification and labelling schemes. Such 'Non-State, Market-Driven' (NSMD) schemes (Cashore, Auld and Newsom 2004) sought to overcome the twin deficiencies of lack of state capacity and bounded jurisdiction. Today, there are a huge number of such global governance arrangements. These include Responsible Care, a code of conduct for the continual improvement of social and environmental practices in the chemical industry; the Fairtrade Labelling Organizations International (FLO), a civil society initiative of 24 companies that sets standards for the use of the FAIRTRADE ecolabel; and the International Organization for Standardization's Environmental Management Series (ISO 14000), which sets out the requirements for companies to 'plan-do-improve' with respect to their environmental practices. What unites these different initiatives is that they are all voluntarily embraced by the companies that use them, have been initiated by business and/or civil society actors, and utilise the market to encourage adoption, implementation and compliance. Many studies now highlight the important contribution these 'new governance' arrangements are making to regulating global production (Meidinger, Elliott and Oesten 2003; Bartley 2003; Cashore, Auld and Newsom 2004; Pattberg 2007; Tollefson, Gale and Haley 2008; Gulbrandsen 2009; Cadman 2009; Lister 2009; Gulbrandsen 2010). While many authors view the diverse set of certification arrangements as evidence of a single 'new governance' phenomenon, in this book, we focus on differences. We are interested in differences between schemes, differences between governments and differential governmental responses to certification schemes. Our aim is to understand why some governments responded more favourably to some schemes than others and why, over time, governments reversed their positions and became supportive of schemes they previously excoriated. Our double comparison of schemes and states offers an important lens to understand the global political economy of certification, potentially making both scheme proponents and governments more self-reflective about their actions.

Commodity governance in historical perspective

The crucial role that commodities play in economic growth and development means that governments and businesses have used every conceivable means to access them. No commodity is more important today than oil, a fact that partly explains why the US decided to overthrow tyranny in Iraq while turning a blind eye to the equally vicious rule of dictators in Zimbabwe and Burma/Myanmar (Stokes 2009). The complex amalgam of public–private interests that lay behind the US invasion of Iraq is redolent of the colonial era, a major motivation for which was to secure the flow of resources from colonies for the benefit of the 'motherland'. Under the British variant, the colonial idea of the 'white man's burden' rationalised the invasion of vast areas of the world as a benevolent civilising mission that brought religion and 'enlightenment' to the ignorant (Dossa 2007). The private sector was a full partner in this colonial example of 'global commodity governance'. In return for monopoly rights over territory and resources, the Hudson's Bay and East India companies organised the trade in furs, timber, fish, spices and cotton, creating links between colonial outposts and London, the centre of imperial power.

Imperialism, then, is a well-established form of global commodity governance. Its inherent features are the capture and rule of other lands as a consequence of a set of complex religious, economic and humanitarian impulses. Despite the expressed nobleness of purpose, practical colonialism involved the exploitation of natural and human resources in the colonies for the benefit of the imperial centre and to offset the associated costs. Dutch, French, German, Italian and later-American imperialisms may have explained their invasions in high-sounding tones, but whatever the rationale, colonisation cost money and colonies were expected to pay their way (Weatherby 2010). Japanese imperialism is instructive. A country with few natural resources, especially oil, Japan feared that it would forever remain insecure if it could not industrialise and catch up with the West. To industrialise, however, the country required access to natural resources, and the flow of these became ever-more uncertain as tensions mounted in the 1930s between Allied and Axis powers. Japan launched the Pacific War not only to obtain 'first-mover' advantage against a more powerful enemy but also to secure access to regional resources including oil in the Dutch East Indies, now modern-day Indonesia (Beeson 1999).

Imperialism is no longer considered a legitimate means to secure access to commodities, a fact recently illustrated by the depth of the Bush administration's duplicity over the merits of invading Iraq. Colonialism disintegrated in the aftermath of World War II for a set of complex reasons (Weatherby 2010; Payne and Nasser 2009). The rise of nationalism meant that colonialism increasingly had to be imposed by military force; this, in turn, increased its costs and further undermined it as a legitimating ideology. In addition, public and private elites also realised that a core colonial objective of accessing commodities could be achieved by another means: free trade. Viewed by Western elites as egalitarian, voluntary and non-political and seen as crucial for 'Third World' growth and development, 'free trade' became a central plank of post-war development thinking. There was much to do too because, at the time, very little trade was free. The 1950s were an era of high tariffs and it was only successive rounds of GATT negotiations, especially the 1960s Kennedy Round and the 1970s Tokyo Round that saw substantial reductions (Cohn 2009). Even today, significant trade anomalies remain with industrialised countries wedded to high subsidies and tariffs on agricultural commodities that developing countries have a competitive advantage in producing.

Despite the departure of colonial administrators, therefore, the domination of the now former colonies continued through the structure of global markets. The essence of what became known as 'neo-colonialism' was developing-country dependence on the production of a handful of unprocessed commodities that were exported to markets in the industrialised 'core' to earn valuable foreign exchange to buy needed imports. Analysts such as Raul Prebisch argued that poor countries, locked in a long-term 'secular' decline in their terms of trade, constantly had to produce a greater volume of commodities to secure the same volume of industrialised imports (Peet and Hartwick 1999). The reason was that commodity prices declined over time relative to industrial goods due to the interaction of supply and demand. On the supply side, a net expansion in fields sown, oceans fished and mines dug increased the volume of commodities in peripheral countries putting downward pressure on prices. More commodities were also being produced because of efficiency gains due to the spread of new, more productive technologies. However, the substantial increase in supply occurring in developing countries interacted with a decline in the rate of growth of industrialised-country demand since there was only so much coffee, tea, sugar, cocoa and other goods that rich people could consume. Moreover, the incessant search for cheaper alternatives invariably resulted in the substitution of artificial for natural materials – as when nylon, rayon and polyester fabrics displaced cotton.

In the 1960s, the endpoint of a secular decline in the terms of trade was identified as a condition of 'dependency'. Operating within broadly neo-Marxist frameworks, writers such as Andre Gunder Frank and Immanuel Wallerstein divided the world into a small number of very large regions (Peet and Hartwick 1999). They argued that a rich, industrial 'core' exploited a poor, rural 'periphery' via trade, investment and financial relations. For some *dependendistas*, industrialisation would only occur if developing countries cut off relations with the core and embraced Import Substitution Industrialisation (ISI). It was the mirror opposite of this policy of autarchy, Export-Oriented Industrialisation, that eventually triumphed, however. Building on the inter-war experience of Japan, strong business-corporatist states in Asia such as Korea, Taiwan, Singapore and Hong Kong tied their national development strategies to the export of basic products to industrialised countries.

In the 1970s, and in a context of crisis induced by OPEC, developing countries attempted to use their numerical superiority in some United Nation's forums to demand a New International Economic Order. Impressed by the power of the OPEC in restructuring oil prices to the benefit of oil-producing countries, leaders in the South sought to exploit their 'commodity power' and pushed for the negotiation of international commodity agreements (ICAs). The United Nations Conference on Trade and Development (UNCTAD) was given responsibility for this initiative and set up the Integrated Programme for Commodities (IPC) in 1976 to investigate which commodities would be appropriate and to initiate negotiations (Chimni 1986). Commodity power turned out, however, to be a chimera. At the end of the decade, the election of Margaret Thatcher in Britain and Ronald Reagan in the US signalled the arrival of a new era of deregulation, privatisation and individualism which was summarised in a single word: neoliberalism (Harvey 2005).

With its eyes firmly fixed on the virtues of the 'free' market, neoliberalism undercut the theoretical premise for ICAs because they involved 'interfering' in markets to stabilise commodity prices, usually by manipulating a buffer stock. The hostility of the Reagan Administration to the UN in general and UNCTAD in particular meant that little progress was achieved within the IPC (Taylor 2003). By the mid-1980s the ideology of the free market was in the ascendancy and industrialised countries turned their attention to expanding the power and mandate of the GATT to secure a more liberalised global trade system. Negotiations on the Uruguay Round commenced in Montivideo in 1986 and ended eight years later with the establishment of the WTO, a much more powerful institution than the GATT and with a much wider mandate. Developing countries remained suspicious of the WTO however; not only had industrialised countries presented them with take-it-or-leave-it positions on agriculture and intellectual property rights but the promised benefits that were to flow from the agreement failed to materialise, perpetuating the organisation's legitimacy crisis (Esty 2002).

Efforts to reform the global trade system to address the trade distortions embedded in the WTO agreements have ended in dismal failure. In 1999, President Clinton launched the Millennium Round in Seattle, Washington, but demonstrations outside and remonstrations inside blocked progress. Following the 'Battle in Seattle', a new attempt at negotiations occurred in Doha, Qatar, in 2001, with agreement reached on the Doha Development Agenda (DDA). The DDA was to focus on securing the benefits of trade liberalisation for developing countries, especially with respect to commodity production. This required, in turn, that the Europeans and the Americans agree to a compromise deal on agricultural subsidies and tariff and non-tariff barriers that was acceptable to the South led by the G-20 countries.[1] While the US and the European Union (EU) eventually managed to agree on the parameters of a deal, the compromise ultimately proved unacceptable to developing countries. At Cancun, in 2003, the WTO Ministerial meeting ended in disarray with developing countries refusing to agree on an agenda that did not meet their needs. All subsequent attempts to revive the talks have ended in failure (Balaam and Veseth 2008).

Commodity governance and the environment

The post-war period has not merely been a North/South struggle over how to govern the international trading system. Beginning with Rachel Carson's book, *Silent Spring*, Americans and then the rest of the world have become increasingly aware of the consequences of unsustainable industrialisation for the natural environment. Initial concern focused on pollution with lead poisoning in Minamata, Japan, and toxic waste in Love Canal, sensitising people to the interconnections between the abiotic and biotic elements of 'ecosystems', a word that began to enter public consciousness. In 1972, the first major international conference on the environment was held in Stockholm, Sweden. Called the United Nations Conference on the Human Environment, its major achievements were the establishment of the United Nations Environment Program and a US-sponsored resolution calling on the International Whaling Commission to introduce a moratorium on whaling to conserve stocks.

The 1980s saw a further significant development of public awareness of the importance of the environment. In 1980, the International Union for the Conservation of Nature (IUCN) launched the World Conservation Strategy that broadened the focus from protecting endangered species to safeguarding ecosystems and bioregions (Redclift 1987). However, the decade is most memorable for the publication in 1987 of the Brundtland Commission's report, *Our Common Future*, which constituted a heroic attempt to reconcile environmental protection with industrialisation via the bridging term 'sustainable development'. Defined as 'development that meets the needs of the present without compromising the ability of future generations to meet their own needs', sustainable development embraced economic growth and environmental protection as important contributors to improving overall human welfare (WCED 1987).

Implicit in the Brundtland Commission's notion of sustainability is the image of a 'three-legged stool'. Transplanted to the world of policy, decisions only 'stand up' when they are based on a thorough evaluation of each of the three 'legs' – economic, social, environmental. The implications of sustainable development required policies, programmes and projects to be fully assessed to take account not only of their economic impacts but also of their social and environmental consequences. Yet, despite the compelling image, the systematic integration of environmental, social and economic concerns has proven extremely difficult to achieve at any level – local, national, regional or global. At the local and national levels, representation based on territorial constituencies empowers interests concerned with the maintenance of national security and economic accumulation. An absence of effective countervailing institutions diminishes the capacity of social and environmental interests to influence decision making. These local and national political ecological deficiencies in how the public interest is determined are transported to the regional and international level via plurilateral and multilateral mechanisms (Gale and Cadman, forthcoming). Despite the increasing activism of civil society groups, only officials from states are recognised as having the right to negotiate in regional and international organisations. Since national governments have been elected to ensure national security and economic growth, these interests are over-represented in international forums too. Again, the absence of countervailing institutions means that deals are done which turn out to be socially and environmentally sub-optimal. It is this structural domination of regional and international forums by economic and security interests that sparked civil society interest in alternative arrangements to effectively integrate the environmental, social and economic components of sustainability.

The recent failure of the United Nations Summit on Climate Change at Copenhagen highlights, once again, the many dysfunctional elements of multilateralism (Doelle 2010; Dimitrov 2010). In the forestry and fisheries sectors, international negotiations over the past three decades have been based on two inviolable principles: state sovereignty and economic accumulation. In both sectors, intergovernmental negotiations dating back to the 1970s have failed miserably to halt the destruction of forests or preserve the productive capacity of fisheries. In the forestry sector, a litany of international organisations has had responsibility for improving forest planning and management. Negotiations have occurred at the United Nations Food and Agricultural Organization (FAO), especially its Committee on Forestry (FAO-COFO) since 1947; at the International Tropical Timber Organization (ITTO) since 1983; at the Tropical Forestry Action Program since 1985; at the Preparatory Committees of the United Nations Conference on Environment and Development (UNCED) between 1990 and 1992; and at the post-UNCED forestry institutions of the Intergovernmental Panel on Forests (IPF) (1995–7), the Intergovernmental Forum on Forests (IFF) (1997–2000), and the United Nations Forum on Forests (2000–present) (Humphreys 1996, 2006; Gale 1998). Despite all the talk, FAO's own data show that rates of deforestation and degradation continue apace, with forests located in Brazil, Burma/Myanmar, Cameroon, Ecuador, Ghana, Indonesia, and Papua New Guinea singled out as especially threatened (FAO 2006).

The 1990s brought the problem of inadequate controls over fishing into sharp relief. Well-publicised stock collapses such as the Northern cod fishery of Newfoundland drew attention to failures in existing management arrangements. While the problems caused by non-compliance with fisheries regulations are not new, the post-Brundtland focus on sustainability has increased scrutiny over the performance of international and regional bodies and states. At the same time the sustainability agenda has been challenged by traditional freedom to fish and the lack of effective enforcement of and compliance with management measures such that according to the FAO many of the world's fisheries resources are fully if not over-exploited (FAO 2000).

Non-state commodity governance

It is in the shadow of intergovernmental failure that business and civil society actors have sought to create alternative governance institutions to negotiate the meaning of and implement sustainable development. One important initiative to emerge in the past two decades is voluntary certification, an arrangement that consists of the development of a sustainable management standard, the auditing of companies to determine if they meet the standard, the certifying of those companies that do and the appending of a 'label' to signal this achievement in the marketplace. Environmental and social certification holds out the promise that consumers can engage in socially and environmentally responsible consumption. However, this promise depends on the rigour of standard used to assess management practices and the validity of the truth claims signalled by the label. Lots can go wrong, giving rise to legitimate concerns about 'greenwashing' – the marketing of products as sustainable that, in fact, continue to do environmental and social damage.

Analytic attention has focused on the source of authority and legitimacy of these new global governance instruments. For Cashore, Auld and Newsom (2004), these derive from the marketplace rather than from the Westphalian principle of state sovereignty. Voluntary certification schemes are NSMD governance arrangements that derive their authority and legitimacy from their acceptance by market actors and a standard that is overly rigorous is rejected by business as too costly. From this perspective, market actors drive certification schemes and do so for a variety of reasons including fear of more onerous governmental regulation, the search for a competitive advantage and to achieve corporate social responsibility. This argument can be turned around. If a standard is overly lax and endorses environmentally and socially damaging business-as-usual practices, it will fail to garner the support of social and environmental actors. Thus voluntary certification schemes are institutions that facilitate bargaining about the standards to apply and the arrangements for auditing firms, tracking products, labelling goods and adjudicating complaints. The authority and legitimacy of such schemes derive as much from the quality of the internal governance arrangements in place as from their market acceptability (Cadman 2009).

Of the many certification schemes in operation today, two stand out as being particularly well institutionalised: the Forest Stewardship Council

(FSC) and the Marine Stewardship Council (MSC). The FSC has been a pioneer in the field of environmental and social certification. Its roots lie in a set of initiatives in the late 1980s by Friends of the Earth-UK (FoE-UK), the Ecological Trading Company and the Woodworkers Alliance for Rainforest Protection (WARP) (Gale 1998; see also Chapter 3). FoE-UK's proposal, which was presented to a meeting of the International Tropical Timber Organization, caused consternation among developing country governments and the tropical timber industry that demanded it be withdrawn.[2] Rather than accept defeat, however, environmentalists argued it could be a practical lever to transform the tropical timber industry, a view later backed by the London Environmental Economics Centre (LEEC 1993). Following extensive worldwide negotiations, the FSC was founded in 1993 in Toronto, Canada. In the ensuing years, it has evolved into an important global forest governance organisation certifying over 130 million hectares of forests as well managed. In 1996, it was joined by a sister organisation on the fisheries side, the MSC. The architects of the MSC were the global environmental organisation, the World Wide Fund for Nature International (WWF-International) and the food multinational, Unilever. The MSC has certified numerous fisheries around the world at various scales, including the New Zealand Hoki fishery and the Bering Sea Pollock fishery.

Despite both organisations carrying the 'stewardship' appellation and being enthusiastically backed by WWF-International, the institutional structure of the FSC and the MSC differ markedly. The FSC is a membership association: members join one of three 'chambers' depending on whether they represent economic, social or environmental interests. The FSC is governed by a General Assembly of members which meets every three years to consider all aspects of its operation. In contrast, the MSC is a foundation managed by a self-appointed board of trustees. The MSC's board has ultimate responsibility for the MSC's operation, although it receives input from two other bodies: a Technical Advisory Board and a Stakeholder Council. These different organisational arrangements are a product of different histories and have given rise to rather different experiences. While the FSC's complex representative arrangements have created implementation difficulties, the legitimacy of its decision-making processes has rarely been called into question. In contrast, the MSC has operated more efficiently, but has encountered ongoing criticism concerning the adequacy of its standard and the rigour with which it is implemented. While both organisations have evolved in terms of their organisational structures to make them 'fit for purpose', the MSC has found itself undertaking much more significant governance changes in response to ongoing criticism from environmental and social stakeholders.

The FSC and the MSC are examples of non-state global governance arrangements – institutions that seek rule-making authority over designated sectors of an increasingly integrated global economy. Other examples of the phenomenon include the organic movement, whose many national bodies are organised under the International Federation of Organic Agricultural Movements (IFOAM). Formed in 1972, IFOAM's aim is to coordinate a disparate organic movement that has evolved independently at different times and in different countries (Raynolds 2000). In 2005, it established an agreed set of principles of organic agriculture (covering health, ecology, fairness and care) and sought to embed these principles in organic production around the world through a system of accreditation of national organic schemes. Thus, for example, the Australian Certified Organic, Canadian Organic Growers and the UK's Biodynamic Agricultural Association are all accredited to IFOAM providing a guarantee they are legitimately practising organic agriculture. In addition to organics, other sectors where non-state global schemes have emerged or are being mooted include fair trade (under FLO), fair labour (under the Fair Labor Association), tourism (under the Sustainable Tourism Certification Network) and mining (under the Mining Certification Evaluation Project).

Non-state governance schemes have not only emerged in different sectors but competing schemes also exist within sectors. Organics is a case in point, where several schemes often exist at the national level and compete with each other for clientele. In the forest sector, the FSC's establishment quickly gave rise to the formation of several other national competitor schemes such as the US Sustainable Forestry Initiative (SFI), Canada's Canadian Standards Association scheme (CSA) and Australia's Australian Forestry Standard (AFS).[3] Conversely, in the fisheries sector, national competitor schemes to the MSC's have not emerged, although more locally developed, alternative methods of assessing and labelling fisheries are in use.

Approach and methodology

The emergence of non-state global governance schemes in a variety of sectors poses a range of intriguing empirical and normative questions about how governments have responded and should respond to them. At the most basic level, governments may choose a policy of support, neutrality or opposition to a non-state global governance scheme. What determines the position they adopt? And if they later alter that position, what factors led them to change their stance? Several important studies of non-state global governance have reflected on these questions (Elliott 1999; Bartley 2003; Meidinger, Elliott and Oesten 2003; Cashore, Auld and Newsom 2004; Gulbrandsen 2005; Pattberg 2007; Gulbrandsen 2009; Lister 2009; Gulbrandsen 2010). We build on these and other studies to develop an analytic approach to understanding state responses to certification schemes in the forestry and fisheries sectors. We adopt a critical, ecological political economy approach that builds on the policy network, commodity chain and sustainability literatures. From the policy network literature (e.g. Atkinson and Coleman 1989; Howlett and Rayner 2001), we adopt the view of the disaggregated state that behaves differently depending on the structure of the policy networks in operation in a given sector. This approach enables us to examine how the structure of a policy network (bureaucratic, clientelistic, triadic or pluralistic) shapes the state's response to specific certification schemes. The commodity chain literature enables us to focus on the flow of products from producers through intermediaries to retailers and final consumers and on the power relations within the chain. From the sustainability literature, we adopt the view that forestry and fisheries management raises not merely technical issues but paradigmatic issues as well. We view the concept of 'sustainable management' of forests and fisheries as contested within the applied sciences depending on whether managers operate within a sustained yield or ecosystem-based discourse.

In the book, we focus on how governments have responded to the FSC and MSC schemes. Our core argument is that a state's response to a specific certification scheme depends on the structure of the policy network responding to the scheme and on the rigour of the scheme's principles, criteria and indicators. We adopt the comparative method and examine state responses in three countries – Australia, Canada and the UK – to two certification schemes in the forestry (FSC) and fisheries (MSC) sectors. Our three comparator countries have been selected to minimise the impact of confounding variables related to differences in levels of development, political structure and culture. Australia, Canada and the UK are similar in several important respects. Notably, all three are developed, stable, constitutional democracies with well-established political parties representing centre-right, centre and Green perspectives. While it could be argued that Australia and Canada are formal federations while the UK is a unitary state, we argue that all are engaged in managing 'multilevel governance' arrangements. In Australia and Canada, this takes a conventional federal form, while in the UK it links sub-national, devolved regions (England, Scotland, Wales and Northern Ireland) with the supranational governance arrangements of the EU. Thus the three countries have all wrestled with the same difficulties when it comes to natural resource management related to jurisdictional authority and policy competence.

We have chosen our case study countries because they differ in one important respect: in terms of their economic dependence on the forestry and fisheries industries. Canada is a major producer of forest and fish products and is particularly dependent on the forestry industry. Timber products play a vital role in the Canadian economy and the country is heavily dependent on export markets in the US and Europe. Large forestry and fisheries lobbies exist and have exercised an important political influence at the regional and national levels. In contrast, the forestry and fisheries industries in the UK are small and marginal to its economy which relies on heavy industry and finance. While each sector has sought to exercise influence over policy, this has been offset by other actors within and outside the state. Also, the UK is a net importer of forestry and fisheries products and this marks its commodity chain out as very different from Canada. Australia constitutes the intermediate case. At the national level, neither forestry nor fishing is especially important compared to mining and tourism, for example. Australia has a trade deficit in forest products, exporting large quantities of raw materials to Asian markets and importing fine paper and furniture in return. While not especially important to the national economy, the forestry and fisheries industries make a substantial contribution to some regions, and this creates considerable Commonwealth-state tension over jurisdictional competence and policy substance.

Overview of the book

To provide an account of state responses to certification schemes, we have adopted the following structure. In Chapter 2, and following an account of the historical importance of different forms of global commodity governance, we set out our analytic framework in detail. This builds on a range of diverse literatures and aims to integrate structure, agency and discursive elements that are relevant to any specific analysis of political causation. Despite a certain eclecticism, the framework falls broadly within the emerging critical ecological political economy tradition that treats actors as gaining power and legitimacy due to their structural location in the system of market-capitalism, requiring countervailing power to be exercised if environmentally and socially sustainable outcomes are to be achieved. Following the presentation of our analytic framework, in Chapter 3 we provide a detailed historical overview of the FSC and the MSC, including an account of their respective governance arrangements, standards, certification systems and evolution.

In Chapter 4 we set out the policy context of our three comparator countries, Australia, Canada and the UK. A close reading of this chapter will orient the reader to the subtle differences between the operation of Australian and Canadian federalism, to the difficulties associated with multilevel governance in all three jurisdictions and to the core policymaking bodies in all three. This chapter sets the context for a detailed analysis in Chapter 5 (Australia), Chapter 6 (Canada) and Chapter 7 (UK) of the evolution of forest and fisheries certification and the response of the state to its emergence, consolidation and institutionalisation. In Chapter 8 we undertake a comparative analysis of the three case studies across the two certification systems, highlighting not only how states responded initially to each scheme but also the dynamic nature of the adjustments they made over the course of the past two decades as each scheme evolved both locally and globally. In Chapter 9, the Conclusion, we summarise our argument and critically assess the ability of our theoretical framework to account for state responses to non-state global governance schemes.

2
Global Commodities, Sustainable Governance: An Analytic Framework

A major challenge confronting the global community is the restructuring of production and consumption relations to meet the requirements of environmental and social sustainability. Two dimensions to this challenge stand out. The first is to reduce the overall scale of production measured in terms of throughput of energy and materials, while simultaneously closing the gap between under- and over-consumers. The second is to ensure the goods that are consumed are appropriately sourced, transformed, transported and disposed of according to the principles of sustainability. The emerging literature on sustainable consumption addresses the first issue related to scale effects. The literature pits analysts who view the capitalist-market system as fundamental to the solution via the system's role in stimulating innovation and efficiency (e.g. Huber 2000) against proponents of autochthonous communities that eschew exchange relations in favour of self-sufficiency (e.g. Goldsmith 1993). In between lie the environmental nationalists who, building on a conception of the 'steady-state' (Daly 1973), view governments as playing a crucial role in regulating production and consumption relations in the interests of sustainability.

Non-state global governance schemes that rely on codes of conduct, voluntary guidelines and certification, auditing and labelling systems are ill equipped to tackle the scale effects arising from an expansionist market-capitalist system. The market-capitalist system currently requires ever-increasing material throughput to sustain itself and other institutional innovations will be required to address this sustainability challenge. For non-state global governance schemes to be fully effective, therefore, they will need to work in tandem with approaches that address the issue of scale. Otherwise, the beneficial effects of substitution that occur when consumers switch to purchasing certified products may be swamped by the detrimental effects of a net overall increase in material throughput. Considered thus, non-state global governance schemes are a necessary but not sufficient means to achieve global ecological sustainability.

The basic idea underpinning many non-state global governance schemes is not new. A need for high-quality information has always existed in the marketplace. This is because when goods are exchanged, one party can come out the poorer if the product or service they receive turns out to be defective in some manner. In the past, personal reputation provided a guarantee that a purchased product conformed to pre-agreed specifications. In the early twentieth century, corporate branding arose to serve a similar function. Consumers gave preference to branded over unbranded products when the brands become associated with certain desirable qualities such as value-for-money, reliability, money-back guarantee, good after-sales service and so forth.

An emerging problem with market exchange relations is that a segment of consumers increasingly desire to purchase products that are socially and environmentally 'sustainable'. They seek assurances, for example, that products are not produced by child labour, are not contributing to biodiversity loss, are 'carbon neutral', are legal and do not damage their health. Whereas in the past, consumers were motivated mainly by concerns over price, quality and availability, today a growing segment that is engaged in discretionary purchases is motivated by social and environmental values. Appropriately designed non-state global governance schemes that certify and label products as meeting a high standard of production are assisting these consumers to meet these preferences as they go about their daily lives.

Globalisation and commodities

Governance schemes that employ certification and labelling have taken on added significance in the post–World War II era as a consequence of globalisation. Broadly defined as the 'recognition of the world as a single place' (Scholte 2001), globalisation features not only a new image of social interconnectedness but is fundamentally related to a widening and deepening of production, consumption, investment and financial networks that challenge the traditional regulatory capacity of states. One consequence of economic globalisation has been a significant increase in interconnectedness with the welfare of many actors fundamentally dependent on the actions of others many thousands of miles away. The recent global financial crisis (GFC) has provided a dramatic illustration of such interdependence. High-risk mortgage loans made in the US to the so-called Ninjas (individuals with no income, no jobs or assets) were repackaged as residential mortgage backed securities (RMBS) by investment banks and rated as low risk by the top credit rating agencies. When the US housing market plummeted, a domestic crisis in the US was globalised as the financial contagion spread across the world to embrace not only the banking system but even small municipalities, universities and charities (Baily, Litan and Johnson 2008; Wade 2009).

The interconnectedness of global production and consumption relations extends well beyond financial markets. A consumer purchasing a good or service represents the final link in a long supply chain composed of primary, secondary and tertiary producers and associated service providers. A wooden table sold in a British store may have been manufactured in China from wood sourced in Indonesia logged by a company financed by an international banking syndicate. The fish purchased in a supermarket may have been caught by trawlers off Australia before being frozen and transported half way around the world within a mere 24 hours. The clothes on our backs and the rugs under our feet likely derive from fibres sourced on one continent, processed on a second and manufactured in a third, sometimes in sweatshop conditions using child labour and at a very high environmental cost. This 'distancing' of production and consumption relations (Princen 2002) creates a set of social and environmental risks that governments acting alone or in concert have been powerless to regulate effectively. The steady stream of news reports about toxic products, illegally traded goods, abusive labour practices and production-related pollution and biodiversity loss testifies both to the importance of PPMs as well as the inability of governments to tackle them. It is here that non-state global governance in the shape of certification and labelling schemes is making a difference with sectoral schemes emerging to secure greater sustainability with respect to commodity flows.

Commodity production has expanded enormously in the latter half of the twentieth century and now touches almost every part of the world. To take just one example, coffee production expanded from about 81 million bags in 1980–1 to almost 130 million bags in 2008–9 (Table 2.1), an increase of approximately 60 per cent. This expansion is due to several factors including an increase in hectares sown, new market entrants, improved production techniques and better transportation infrastructure. However, notwithstanding the steady general trend upward in coffee production in the past 40 years, Table 2.1 reveals another ubiquitous feature of commodity markets: the tendency for production volumes and prices to fluctuate wildly from one year to the next. In the agricultural sector, this 'boom-and-bust' cycle is often due to natural factors related to weather (frosts, droughts, storms) and disease (pests, viruses, fungi). The large dip in coffee production in the 1995–6 season identified in Table 2.1 was due to an early, hard frost in 1994 that caused widespread plant damage in Brazil, significantly affecting the following season's output (Talbot 2004).

Of the huge number of commodities entering into world trade currently, we focus on two of enormous importance: wood and fish. While both are widely traded, they differ significantly from each other in terms of biology (inanimate/animate), ecosystem structure (terrestrial/aquatic) and usage (non-edible/edible). The political economy of production and consumption of wood and fish is also very different, with many more actors intervening in the supply chain for wood than for fish. To better appreciate the differences between these commodities and the trade that occurs within their

Table 2.1   Coffee production trends, 1980–2009

Year	Production ('000 bags)	% Increase/decrease
1980/81	80,730	
1985/86	90,173	+11.7%
1990/91	93,253	+3.4%
1995/96	86,979	−6.7%
2000/01	113,033	+30%
2005/06	110,181	−2.5%
2006/07	127,908	+16.1%
2007/08	118,086	−7.7%
2008/09	128,790	+9.1%

Source: Adapted from the International Coffee Organisation 2009.

supply chains, we now turn to a more detailed analysis of each of them, commencing first with wood.

Forests and timber

There are several important international data sources on forests and wood. The FAO produces the *Forest Resources Assessment* (FRA), a detailed statistical analysis of the status of the world's forests. Initiated in 1948, the early reports were relatively unsophisticated, but considerable improvement has occurred since 1980, when a more systematic approach to analysis has been taken. The most recent FRA report for the period 1990 to 2005 is the most sophisticated analysis yet and, unlike previous reports, attempts to assess the sustainability of forest management across all three of its dimensions: economic, environmental and social (FAO 2006). We use data from the 2005 FRA in discussing forest types and the distribution and decline in global forest cover. FAO also produces an annual *Forest Products Yearbook*, the most recent of which was published in 2009 for the period 2003–7 (FAO 2009a). The yearbook provides data on the structure of the global forest industry, including the trade in different types of forest products. We use data from the yearbook in our analysis of the trade in forest products, supplemented by data from other sources where appropriate.[1]

There are several different ways to categorise forests. Three different types can be distinguished based on latitude. In high latitudes north of 50 degrees, where temperatures are cold, precipitation occurs mostly as snow and soils are thin, boreal forests grow that are composed mainly of coniferous tree species such as pine, fir and spruce (UCMP 2009). Extensive regions of boreal forests are found across northern Russia, Canada and Scandinavia providing habitat for moose, bear, lynx and hawks (UCMP 2009). In contrast, tropical forests are found between the Tropic of Capricorn and Tropic of Cancer. These forests evidence high levels of biodiversity (FAO 2009a, xx) and contain a huge number of species of trees, many of which are of especial commercial value like mahogany, teak and ebony. Temperate forests occur in the latitudes between these two extremes and are dominated by broadleaved trees such as eucalypts in Australia, maples in Canada and oaks in the UK.

Another way to classify forests is by the degree to which they have been disturbed by humans. FAO uses a five-fold classification to distinguish forests on this basis. The FRA 2005 distinguishes between primary, modified natural and semi-natural forests on the one hand and productive and protective plantations on the other. Primary forests are defined as 'Forest/other wooded land of native species, where there are no clearly visible indications of human activities and the ecological processes are not significantly disturbed' (FAO 2006, 171). In contrast, modified natural forests are defined as 'Forest/other wooded land of naturally regenerated native species where there are clearly visible indications of human activities' (FAO 2006, 171). Semi-natural forests are distinguished from modified natural forests and pure plantations. Unlike modified natural forests, managers have used a range of silvicultural techniques (planting, seeding, fertilising, thinning) to deliberately alter regeneration processes. However, unlike plantations, exotic trees species are not planted and/or planting is not done in straight rows. Finally, FAO distinguishes two types of plantations based on purpose. Productive plantations are defined as 'Forest/other wooded land of introduced species and in some cases native species, established through planting or seeding, mainly for production of wood or nonwood goods' (FAO 2006, 171). These differ from protective plantations which are planted to provide ecosystem services for the protection of soils, water and biodiversity. According to FAO, over half of the world's forests is now modified natural forest (about 53%), with primary forest (about 36%), semi-natural forest (7%), productive plantations (3%) and protective plantations (1%) making up the balance (FAO 2006, 27).

Forests are unevenly distributed between and within countries around the world. The top ten countries with the greatest forest area are set out in Table 2.2. It can be seen that the Russian Federation contains a significant portion of the world's forests, much of it boreal forest in the Siberian region. Similarly, Brazil, Canada and the US are very important forested countries, with the Brazilian Amazon being renowned for its tropical forests, and Canada and the US for important stands of temperate forests.

Forests are valued for a diversity of purposes. These include the role they play in protecting against soil erosion and flooding, in providing habitat for animals and biodiversity and as recreational locations for ramblers, joggers, mountain bikers, hikers and so forth. One important, often primary, role for forests is the production of wood products for domestic and global markets. A huge number of wood products are in use around the world, which can be divided into six broad categories. These are (i) roundwood,

(ii) charcoal (including woodchips), (iii) sawnwood, (iv) wood-based panels, (v) pulp (including recovered paper) and (vi) paper and paperboard (FAO 2009a, xx–xi). Roundwood is the basic category from which all other products are derived. It comprises wood 'removed from forests and from trees outside the forest, including wood recovered from natural, felling and logging losses during the period, calendar year or forest year' (FAO 2009a, xx).

Wood removed from forests is used for two basic purposes: to burn for energy and as raw material for manufacturing products. A large percentage of the wood removed from forests in some regions is to meet immediate energy needs. Table 2.3 shows that of just over 3 billion cubic metres (m³) of roundwood removals in 2005, 1.2 billion m³ or approximately 40 per cent was for fuelwood. Table 2.3 also reveals how regions differ in terms of

Table 2.2 Top ten countries by forest area

Country	Million ha	% of total
Russian Federation	809	20
Brazil	478	12
Canada	310	8
United States	303	8
China	197	5
Australia	164	4
Democratic Republic of the Congo	134	3
Indonesia	88	2
Peru	69	2
India	68	2
All Others	1,333	34
Total	3,953	100

Source: UNEI (United Nations Environmental Indicators) 2010.

Table 2.3 Wood removals, 2005

Region	Industrial roundwood	Fuelwood		Total removals
	Million m³	Million m³	% of total	Million m³
Africa	79	591	88	670
Asia	174	189	52	362
Europe	543	139	20	681
North/Central America	725	112	13	837
Oceania	54	10	15	64
South America	225	173	44	398
World	1,799	1,214	49	3,013

Source: FAO 2006, 90.

industrial roundwood and fuelwood removals. In Africa, close to 90 per cent of total roundwood removals is for fuelwood purposes, whereas in the North and Central American region, the figure is only 13 per cent and for all of Europe, only 20 per cent. These statistics reflect the broader socio-economic divide in the global system between developed and developing countries, with people in developed countries having access to electricity (from oil, coal and nuclear generators) while people in developing countries continue to rely on the collection of wood to meet their direct energy requirements.

Total removals of roundwood have remained largely static since 1990 at 2005 levels of about 3 billion m³. The global figure disguises some significant regional differences, however. Removals in Africa have increased from 499 million to 661 million m³, or by about 32 per cent; in contrast, removals in Asia have declined from 454 to 362 million m³, or approximately 20 per cent, mainly as a consequence of China's ban on timber harvesting (FAO 2006, 89). Modest declines have occurred also in Europe and South America. The structure of the regional increases and decreases is also revealing, with fuelwood removals growing substantially in Africa from 445 to 585 million m³, or about 85 per cent of the total increase in African removals. In contrast, fuelwood removals in South America dropped dramatically from 302 million m³ in 1990 to 173 million m³ in 2005 (FAO 2006, 90).

The levelling off in total global roundwood removals contrasts with a significant increase in the value of the trade in wood products over the past 15 years. EarthTrends reports that total exports of wood products increased by 38 per cent from over US$147 billion in 1995 to US$203 billion in 2006. While the value of North America's exports over the period stabilized at around US$46 billion, Europe's and South America's exports almost doubled from US$69 billion to US$112 billion and from US$6 billion to US$10 billion respectively. The overall balance in terms of wood product exports remains heavily weighted in favour of developed countries. Of a total export value of US$203 billion, US$166 billion or 82 per cent was exported by developed countries, with developing countries contributing only US$38 billion or 18 per cent of the total (EarthTrends 2009).

Fisheries and fish

The FAO regularly publishes an assessment of the world's fisheries: the *State of World Fisheries and Aquaculture* (SOFIA). This report provides a detailed snapshot of the world's fisheries and provides a lens on emerging issues through coverage of 'selected issues' and 'special studies' in each biennial issue. FAO produces SOFIA reports in each even year – the most recent report (2008) was published in 2009, reporting data to end of 2006. The 2006 and 2008 SOFIA reports have addressed ecolabelling in special studies or as a selected issue, reinforcing the salience of this topic. The 2006 report has a section dealing with 'impact of market based standards and labels on international fish trade' (FAO 2007: 88–93). The 2008 SOFIA report included as one of its 'selected issues' the topic 'private and public standards and certification schemes: synergy or competition' (FAO 2009b: 95–104).

The world's capture fisheries production is concentrated in four regions: the Northwest Pacific, the Southeast Pacific, the Western Central Pacific and Northeast Atlantic. Marine catches from these areas comprise over two-thirds of the world catch (FAO 2007: 32). Fish and fishing activities are central to the world's population.

The FAO states that 'capture fisheries and aquaculture supplied the world with about 110 million tonnes of food fish in 2006. ... Of this total, aquaculture accounted for 47 percent' (FAO 2009b, 3). Global aquaculture production has grown at about 9 per cent per year since 1970 (FAO 2007), with much of this increased production coming from China.[2] Aquaculture operations will continue to be an important component in global fisheries supply, with potential impacts on capture fisheries. These impacts include targeting stocks for aquaculture feed. Ranching of high-value species such as southern bluefin tuna in Australia indicates the potential and impacts of developing aquaculture operations, with good returns from high-end markets but with feed stock derived from industrial fishing of small pelagic species. Demand for food fish is estimated at 110–20 million tonnes in 2010, based on current population growth rates. FAO estimated that production of marine capture fisheries and aquaculture was 144 million tonnes in 2006 with 76 per cent of this production for human consumption. The remaining demand will be met by aquaculture operations.

Fish trade is significant, opening up new avenues for regulation well away from the problematic issues of illegal, unregulated or unreported (IUU) capture operations.

Table 2.4 Marine and capture fisheries: Top ten producer countries, 2006

Country	Production (Million tonnes)
China	17.1
Peru	7.0
United States	4.9
Indonesia	4.8
Japan	4.2
Chile	4.2
India	3.9
Russian Federation	3.3
Thailand	2.8
Philippines	2.3

Source: FAO SOFIA 2008 (2009b, 11).

Fish and fishery products are highly traded, with more than 37 percent (live weight equivalent) of total production entering international trade as various food and feed products. World exports of fish and fishery products reached US$85.9 billion in 2006. In real terms (adjusted for inflation), exports of fish and fishery products increased by 32.1 percent in the period 2000–06. Exports of fish for human consumption have increased by 57 percent since 1996.

(FAO 2009a: 8)

The scope of fish trade increases the salience of 'port' and 'market' states – places where fish are unloaded or sold – to provide surveillance of catches. Catch documentation and certification is increasing in key fisheries (Haward 2004).

Table 2.5   Capture fisheries: Principal fishing areas, 2006

Area	Production (Million tonnes)
Northwest Pacific	21.6
Southeast Pacific	12.0
West Central Pacific	11.2
Northeast Atlantic	9.1
Eastern Indian Ocean	5.8
Western Indian Ocean	4.5
Eastern Central Atlantic	3.3
Northeast Pacific	3.3
Southwest Atlantic	2.4
Northwest Atlantic	2.2

Source: FAO SOFIA 2008 (2009a, 11).

Environmental consequences of commodity globalisation

Globalised commodity production generates a host of positive and negative consequences. Market production puts a premium on the specialisation and division of labour within firms, countries and across the globe as companies compete to produce goods and services at low cost to secure sales and grow market share. Competition induces technological and managerial innovation and, by deterring collusion, protects consumers from defective, monopolised production. However, market capitalism also puts a premium on private property, viewing it as a *sine qua non* of distributed production systems. The operation of market-capitalism over time tends to concentrate wealth at the individual, regional and global levels, generates envy by the have-nots and, via advertising, deliberately stimulates the desire for ever more material consumption, which becomes a proxy for social success.

For much of the post-war period, the negative social effects of market capitalism were regulated by the state, which was empowered to distribute the benefits of production more widely through the nation. Greater social equity was achieved via progressive tax polices that raised the revenue needed to finance the welfare state. Efforts to extend these arrangements to regulate the system's negative environmental consequences commenced in the early 1970s with the establishment of national environment agencies but were cut short following the crisis of Keynesianism and the rise of neoliberalism in the 1980s (Harvey 2005). While the immediate causes of the welfare state's crisis lay in fiscal problems exacerbated by the Organisation for Petroleum Exporting Countries' massive oil price increases of 1974 (O'Connor 1973), the more fundamental issue was the expansive nature of the market-capitalist system, the rise of multinational corporations (MNCs), the promotion of foreign direct investment (FDI) and financial liberalisation. In effect, globalisation undermined the conditions for the welfare state's continuation. As globalisation proceeded, states increasingly found it necessary to deregulate in order to compete as attractive locations for increasingly mobile capital. Meanwhile, efforts to regulate various aspects of the market-capitalist system have shifted to the international level and to the negotiation of 'regimes', usually presided over by an international organisation (Keohane 1984; Young 1989; Hasenclever et al. 1997).

The negative social consequences of globalising market capitalism are most evident in the dramatic divide in life chances between individuals born in rich and poor countries. Both the World Bank and the United Nations Development Program have developed indices to measure the extent of this gap. The World Bank divides countries into four categories based on the per capita gross national income (World Bank 2009a). High Income Economies are those with a per capita GNI per annum of more than $11,901 and include all three countries in our study: Australia, Canada and the UK. Low Income Economies, in contrast, are those with a per capita GNI of less than $975 per annum. Much of Africa falls into this category including large African nations such as Kenya, Tanzania and Zimbabwe (World Bank 2009). In these Low Income Economies, poverty is endemic and individuals lack access to basic health, education, sanitation and transportation infrastructure.

The negative social consequences of market capitalism reappear with respect to the environment. Although ecological modernisation theorists point out that not all consequences of market-led development are bad for the environment – rising national income is associated with declining population growth rates, a reduction in some types of pollution and an increased concern in environmental issues – significant cause for concern remains. To the noted positive environmental effects of rising national incomes can also be added a host of negative environmental effects associated with the increased scale of production, the spread of a consumerist culture, and the widening and deepening of unsustainable trade, investment and financial relations (Dauvergne and Clapp 2005).

In the two commodity sectors we investigate in this book, the negative environmental consequences of market- and state-led development have been evident for decades. For example, FAO notes that over half (52%) of the world's capture fish stock is fully exploited, 'and therefore producing catches that were at or close to maximum sustainable yield with no room for further expansion' (FAO 2009a, 7). Further, the organisation observes that 'In 2007, about 28 percent of stocks were either overexploited (19 percent), depleted (8 percent) or recovering from depletion (1 percent) and thus yielding less than their maximum potential owing to excess fishing pressure' (FAO 2009b, 7). At the same time FAO estimates that 1 per cent of the main stocks are recovering slowly from depletion (FAO 2007, 29). In a stark warning FAO noted: 'As stated before in *The State of World Fisheries and Aquaculture*, the maximum wild capture fisheries potential from the world's oceans has probably been reached, and a more closely controlled approach to fisheries management is required, particularly for some highly migratory, straddling and other fishery resources that are exploited solely or partially in the high seas' (FAO 2009b, 7–8).

While there can be marked short-term variation in catches, driven by climatic conditions such as the presence or absence of El Niño events in the Pacific, some general trends can be observed. Fisheries in the Northeast Atlantic and the Southwest Atlantic have declined markedly in recent years. The Northwest Atlantic and the Mediterranean and Black Seas have stable catches, after years of declining production forcing introduction of stronger management measures. Catches in the Northeast and Northwest Pacific (noted as being 'highly regulated' by FAO) have recorded recent increases in production after years of stable, if not declining, catches having reached their maximum yield two decades ago. Declines in catches in the South East Pacific have been marked, driven in part by the effects of recent El Niño events. The Eastern Indian Ocean and the Western Central Pacific are identified by the FAO as the only regions as having potential increases in total catches (FAO 2007, 5), although greater work on stock assessments is needed. These areas will be targeted by fleets constrained by declining catches elsewhere and increased market demand for fish products in Europe and Asia.

Similar resource pressures exist with respect to forests, which continue to decline in most regions. FAO reports data that indicate a global decline in net forest cover of about 7.3 million ha per annum or –0.18 per cent between 2000 and 2005, only slightly down on the 8.8 million ha per annum (–0.22%) declines of the 1990s (FAO 2006, 20). This overall decline masks high rates of decline in the tropically forested regions of South America, Africa and Asia, which is offset by a rise in 'forest' cover in Europe and East Asia. In Africa, for example, forest area is declining at a rate of –0.62 per cent year on year, while the subregion of South and Southeast Asia is even higher at –0.98 per cent per annum. In South America, the rate of decline remains substantial at –0.50 per cent per annum, with continued massive deforestation occurring in the Amazon despite ongoing commitments by Brazil to exercise greater control.

In addition to the net loss of forest cover through deforestation and the conversion of forests to other land uses, often plantation agriculture, forests are becoming degraded. Degradation occurs when primary natural forests are transformed via logging into modified natural or semi-natural forests. According to the FAO, while the rate at which this is occurring has declined in the past five years over 1990 levels, there is still a significant net conversion of primary forests to other forest categories such as modified natural and semi-natural forest types. Taking both deforestation and forest degradation together, FAO notes that 'About 6 million hectares of primary forest have been lost or modified each year since 1990', that 'there is no indication that the rate of change is slowing down', and that 'this rapid decrease stems not only from deforestation, but also from modification of forests due to selective logging and other human interventions – whereby primary forests move into the class of modified natural forests' (FAO 2006, 26).

The major environmental consequences of these ongoing processes of deforestation and degradation are two-fold. First, biodiversity is on the decline because primary, tropical forests where the majority of the deforestation and forest degradation is occurring are especially rich in species, many of which are understudied. Curiously, the FRA 2005 only focuses on one dimension of this problem – declining species of trees. Forests, however, provide habitat to a huge range of biota and their decline is not only significant for tree species but for the vast range of animals and plants that depend on them. Indeed, it is here where the real importance of forest degradation lies, since the 'forest' can still exist in definitional terms but become so degraded that it no longer provides the required habitat for endangered or threatened species. This is especially the case for species that depend on old-growth forests – tall trees – that may be removed because they are commercially valuable. This removal destroys the habitat of old-growth dependent species such as the much-publicised spotted owl of the Pacific Northwest.

The second major negative environmental consequence of reducing forest cover and degrading forests is related to the role they play with regard to the earth's climate. According to the FAO, 'Forests affect climate globally by reflecting less heat back into the atmosphere than other types of land use that have more bare soil and less green cover. They also play a very significant role in the global carbon cycle that affects global climate change' (FAO 2006, 95). With regard to the global carbon cycle, forests take in carbon dioxide and sequester it in leaves, wood, roots and soil matter and breath out oxygen. FAO estimates that the total stock of carbon in forests in 2005 was 638 gigatonnes (Gt), a figure that is 'more than the amount of carbon in the entire atmosphere' (FAO 2006, 35). In line with the figures on deforestation and forest degradation, FAO reports an overall decline in forest carbon stocks between 1990 and 2005 of almost 17 Gt, with Africa, Asia and South America showing significant declines which were offset by an increase in forest carbon stocks in North and Central America and Europe.

Governance via intergovernmentalism

The governance of global commodity production pits developed and developing countries against each other over the distribution of the revenues arising from trade. The clearest example of this distributive struggle, as outlined in Chapter 1, occurred in the 1970s when countries of the South demanded a 'New International Economic Order' and agreement was reached at UNCTAD to set up the IPC to negotiate international commodity agreements. Despite the general lack of success of the IPC, a few ICAs were negotiated including the atypical International Tropical Timber Agreement (ITTA).[3] The history of the ITTA is interesting because its failure to tackle tropical deforestation and hostility to certification and labelling was one factor convincing environmental civil society organisations (ECSOs) of the futility of intergovernmentalism in the forestry sector.

Colchester (1990), Humphreys (1996 and 2006) and Gale (1998) provide detailed accounts of the establishment, structure and operation of the ITTO. As one of the earliest organisations specifically dedicated to forestry, the ITTO brought together tropical timber 'producers' and 'consumers' in a single organisation for the first time to negotiate issues related to industry development, market access, trade statistics, price transparency and sustainable forest management. Sensing an opportunity to advance the sustainability agenda, environmental organisations, especially the International Institute for Environment and Development, played an important role in encouraging the establishment of the ITTO by mobilising in 1985 to secure ratification of the final negotiating texts (Poore 2003). However, seven years later, most were disillusioned by its failure to take effective and timely action to curb tropical deforestation. Despite some early apparent successes that included the negotiation of a set of guidelines for the sustainable management of natural tropical forests (ITTO 1992), the organisation was beset by compromise in favour of industrial forestry and forest-led development (Gale 1998, 144–57). The final straw came when a commission of inquiry into logging in Sarawak all but ignored indigenous peoples' rights and supported a very high annual allowable cut that appeared to defy the mission's own logic. The Sarawak Mission was a watershed for ECSO attendance at the ITTO and, following a condemnation of the Mission's report, they abandoned the ITTO after 1994 (Gale 1996).[4]

One of the key issues that the ITTO failed to address was certification and labelling. The concept was placed on the agenda in 1989 when a proposal developed by Simon Counsell of FoE-UK and Tim Synnott of the Oxford Forestry Institute was submitted by the UK Government. The proposal, titled 'Labeling Systems for the Promotion of Sustainably Produced Tropical Timber' caused a storm of protest at the organisation's 1989 Yokohama meeting (Gale 1998, 158–77). The Malaysian Government took an especially hard line, supported by Indonesian and Cameroonian delegates. Their opposition to the idea of tropical timber labelling systems resulted in the proposal being completely reformulated to investigate instead the generic issue of 'incentives' for sustainable forest management that included levies, taxes and tariff barriers.

Environmental disillusionment with multilateral forestry negotiations did not just reside with the ITTO. In the early 1990s, negotiations commenced on an international forestry convention in preparation for the 1992 UNCED to be held in Rio de Janeiro, Brazil. The Preparatory Committee (PrepCom) process for these negotiations quickly encountered difficulty over two core issues: sovereignty and funding. Developing countries adamantly opposed the idea, put forward by developed countries, that tropical forests should be considered the 'common heritage of mankind'. They insisted, instead, that a convention should endorse the sovereign right of states to manage their forests according to domestic priorities. Developing countries also demanded new and additional finance to facilitate a transition to 'sustainable forest management', which developed countries were unwilling to countenance. The UNCED PrepCom discussions led to bitter divisions between developing and developed country representatives and resulted in the signing of a weak voluntary agreement at UNCED known as the Forest Principles (Humphreys 1996).[5] The failure of the ITTO to significantly reduce rates of deforestation coupled with the breakdown in intergovernmental negotiations over a Global Forest Convention convinced ECSOs of the potential importance of alternative approaches to global forest governance. Thus, after 1992, they devoted their energy to the development of the Forest Stewardship Council's certification and labelling scheme.

As with forestry, intergovernmental negotiations with respect to fisheries management in the post-war period have generally failed to prevent either conflict or the depletion of stocks. The exploitation of fisheries includes numerous examples of conflict, such as the UK-Iceland 'cod wars', the 1990s Canada-Spain-EU dispute and the more recent problems of 'illegal fishing' in Australia's remote Economic Exclusion Zone (EEZ) off Heard and McDonald Islands (Vince 2007). The idea of developing an adequate international regime to govern fisheries was first mooted in the 1930s and re-emerged in the 1950s with the first United Nations Conventions on the Law of the Sea, held in 1958. These early initiatives, and the (unsuccessful) efforts of the 1960s were succeeded by a major multilateral diplomatic effort – the Third United Nations Conference on the Law of the Sea (UNCLOS III) between 1974 and 1982. UNCLOS III was driven by international concern over the parlous state of the world's seas and oceans (Shotton and Haward 2005).

The final text of the United Nations Law of the Sea Convention (LOSC) was open for signature in December 1982 and finally entered into force over a decade later in November 1994. It now constitutes the cornerstone for the international public regulation of fisheries and has been described as a 'framework convention in relation to the exploitation of marine living resources' (Molenaar 2000, 479). The LOSC is built upon the key principle that the elaboration of rights of states in areas such as fisheries brings related obligations and responsibilities and is anchored in traditional concepts such as flag state responsibility and rights of coastal states. The convention arguably has weaker provisions for governing fisheries outside the EEZ, exhorting states to seek cooperative arrangements to manage these fisheries (Miles and Burke 1989). The convention's reiteration of a 'freedom to fish' on the high seas, albeit with qualifications, has proved highly problematic (Molenaar 2000).

The last decade has seen increasing attention to the problems caused by inadequate controls over fishing. While the problems caused by non-compliance with fisheries regulations are not new, the 1990s brought the problems in managing high seas stocks in particular into sharp relief. This decade also saw unprecedented activity in the development of international fishery regimes linked to broader discussion of sustainable development and ecosystem-based approaches to management embodied in the work of the World Commission on Environment and Development in 1987 and UNCED in 1992.

In addition to extensive development of formal hard law instruments such as the LOSC, the United Nations Fish Stocks Agreement and Compliance Agreement (Haward and Vince 2008), a number of parallel voluntary, hortatory, instruments have been established. The most notable of these is the FAO's Code of Conduct for Responsible Fisheries. The code provides opportunities for the development of subsidiary, specialist, instruments such as International Plans of Action that are key elements of an emerging regime governing high seas fisheries.

The code of conduct's general principles note that states and users should use selective and environmentally safe fishing gear and practices. It is a voluntary instrument that is directly linked to 'relevant rules of international law', including the LOSC. The code contains six thematic areas or chapters for which guidelines should be developed: (1) fishery management practices; (2) fishing operations; (3) aquaculture development; (4) integrating of fisheries into coastal area management; (5) post harvest practices and trade; and (6) fishery research. The Compliance Agreement is an integral part of the code. States should ensure compliance with and enforcement of conservation and management measures. States authorising fishing and fishing support vessels to fly their flag should exercise effective control over those vessels. The code 'provides guidance which may be used where appropriate in the formulation and implementation of international agreements and other legal instruments, both binding and voluntary' (FAO 2010).

One indication of the scope and direction of these legal and policy developments can be seen in the focus given to fisheries at the World Summit on Sustainable Development (WSSD) held in Johannesburg in August–September 2002. WSSD outcomes include encouraging the application of an ecosystem approach to sustainable development of the oceans by 2010; maintaining or restoring depleted fish stocks to levels that can produce the maximum sustainable yield 'on an urgent basis' by 2015; putting into effect FAO international plans of action by the agreed dates; establishing marine protected areas consistent with international law by 2012; establishing a regular process under the United Nations for global reporting and assessment of the state of the marine environment; and eliminating subsidies that contribute the IUU fishing and overcapacity. Regional organisations and states will face significant challenges in the shift to an ecosystem-based approach as identified in the WSSD outcomes. The move away from a focus on management of 'target stocks' will affect decisions regarding total allowable catches and, therefore, allocations.

Governance via certification and labelling

The lacklustre performance of intergovernmental cooperation on forestry and fisheries outlined above led actors in both sectors to consider alternative 'governance' mechanisms. Governance, of course, is a contested term, and there is no single, agreed definition of the phenomenon and, hence, the concept (Pierre and Peters 2000). We employ it here to refer to 'steering and coordinating the affairs of interdependent social actors based on institutionalised rule systems' (Treib et al. 2005, 5). This definition is useful because it does not oppose governance to government as some definitions do, enabling hybrid mixes of 'old' and 'new' governance arrangements to be identified. In referring to an institutionalised system of rules, however, it also clearly demarcates governance arrangements from shorter-term, more spontaneous forms of collective action.

Governance is a complex phenomenon and there has been an unfortunate tendency to simplify analyses by investigating it across a single dimension. Legal scholars have focused on the regulatory instruments used to achieve governance objectives, while public policy analysts have often focused on the institutional arrangements employed (non-governmental actors) and their legitimacy. To better understand an instance of non-state global governance, it is important to investigate all three of its dimensions simultaneously (Treib et al. 2007; Tollefson, Gale and Haley 2008; Tollefson et al. 2010). These dimensions examine the actors that are viewed as legitimate participants in the scheme (the political dimension), the organisational structures and processes that are in place to make decisions (the institutional dimension) and the instruments that are used to achieve objectives (the regulatory dimension).

Political, institutional and regulatory dimensions

Along the political dimension, schemes vary in terms of the number of actors involved in developing them. Building on the Brundland Commission's core idea that sustainability involves economic, social and environmental constituencies, we can distinguish between unipartite, bipartite and tripartite certification schemes. Unipartite schemes are developed by a single constituency, bipartite schemes by two and tripartite schemes by all three constituencies. Analytically, there are seven feasible types of certification schemes (unipartite-business, unipartite-environment, unipartite-social, bipartite-business/environment, bipartite-business/social, bipartite-social/environment and tripartite).

Institutionally, schemes differ in terms of their basic organisational structures. Organisational arrangements can vary from foundations to corporations, associations, federations and quasi-intergovernmental bodies. Each of these governance arrangements has important implications for how decisions are made. Foundations are run by self-appointed boards of directors who are at best indirectly accountable through stakeholder representatives to users of the standard. In contrast, membership organisations with elected boards are directly accountable to their members for the decisions taken. Quasi-intergovernmental bodies like the ISO that are made-up of statutory standards bodies like the Standards Council of Canada, business corporations like Standards Australia and non-governmental organisations like ANSI constitute a third, federated, option. Quite different governance 'qualities' are in evidence as a consequence of these different organisational arrangements (Cadman 2009).

In terms of regulatory arrangements, schemes can differ in the nature of the standard they employ. The literature distinguishes between technology, management and performance standards. Technology standards specify the type of hardware to employ and have been a focus of ISO since its founding in 1946. The aim is to ensure the compatibility of traded products so that differences in standards do not become barriers to trade. In contrast, management standards such as the ISO 9000 and 14000 series are designed to specify the internal processes that a company should undertake to ensure it continually improves its quality (9000) and environmental (14000) performance. These two types of standards can be distinguished from a performance-based approach to standard setting, which aims to establish a set of outputs, leaving it to the individual company to determine what blend of technology and processes it will use to achieve them. For example, a quantitative indicator of performance for a passenger motor vehicle might be 'a fuel efficiency level of at least 30 kilometers per litre of petrol when driving on the highway'. Any combination of technology and processes is acceptable so long as it results in this performance objective being met. In Table 2.6 we map the major schemes discussed in this book using this three-dimensional approach and in Chapter 3 we undertake a more detailed analysis of the FSC and the MSC using this schema.

First-, second- and third-party certification

In contrast to our unipartite, bipartite and tripartite schema, the literature commonly distinguishes between certification and labelling schemes based on the arrangements they use to audit compliance to a standard. Making this distinction between schemes is less relevant today because all major schemes now utilise third-party certification. However, the distinction retains some residual importance by highlighting differences in the contractual distance that can exist between the certifier and the certified. In third-party schemes, firms are audited by a company independent of and at arms length from the company seeking certification. Such schemes are generally viewed as the most credible. In contrast, in first-party schemes the company audits itself, and has a clear incentive to develop weak standards and/or to audit permissively. In second-party schemes, the contractual distance

Table 2.6   Three dimensional analysis of four major certification schemes

	MSC	FSC	ISO 14000	PEFC
Political	Bipartite (mainly business-environment with limited social participation)	Tripartite (all three constituencies equally present)	Unipartite (mainly business with occasional environmental participation)	Unipartite (mainly business with limited environmental and social participation)
Institutional	Foundation (self-elected board)	Association (members elect general assembly and board)	Quasi-intergovernmental organisation (single national organisation, often state agency, represented on International Council)	Federation (single national agency represented on International Council)
Regulatory	Performance-based standard with some management elements	Performance-based standard with some management elements	Management-system standard	Management system standard with some performance elements

between the company and the certifier is somewhat greater than in first-party schemes, although full independence is not achieved. Second-party schemes are typically developed and audited by an industry body.

Type I, Type II and Type III labels

Another common distinction made in the certification and labelling literature is between Type 1, Type II and Type III labels. The distinction comes from ISO documentation and is based on the informational purpose a label is designed to serve. In Type 1 labels, the aim is to signal that the product is superior to other products on one or more criteria. Many nationally developed ecolabels adopt this approach, such as Germany's Blue Angel, Canada's Environmental Choice and Norway's Nordic Swan schemes (Kern et al. 2001). Type I labels, however, differ from each other as some are based on life cycle assessments and thus aim to indicate that the product meets the requirements of sustainability. Others, in contrast, focus only on one specific environmental attribute, such as energy efficiency, and ignore other dimensions.

In contrast to Type I labels, Type II labels are claims made by manufacturers, importers and distributors that have not been independently verified. These labels often accompany first-party certification schemes and, as noted above, result in companies quickly backing away from the claims when challenged for fear of being charged with misleading advertising. Type III labels are rather different, since they do not provide information on the sustainability of a product. Rather, they aim to provide information on the compounds that a product is made from, leaving it to consumers to determine if the product meets requirements. Product lists that appear on packaging are typical of Type III labels. They assist consumers with allergies to avoid consuming allergenic compounds made from peanuts, gluten and so forth. The major schemes we investigate here all utilise Type I ecolabels and are aimed at reassuring consumers that the wood and fish products have been sourced from sustainable sources.

Chain-of-custody certification

When goods are labelled, rigorous certification schemes require a chain-of-custody (CoC) standard and associated certification system to prevent the supply chain from being contaminated with products from unacceptable sources and ensure the claim on the label accurately reflects the product's content (e.g. 100% certified). It is important to note that most certification schemes have systems in place to manage the mixing of certified and uncertified products. Wooden furniture, for example, will almost certainly contain wood from certified and uncertified sources. Both sources require verification through the supply chain. Both the FSC and the MSC specify in detail the CoC standard and certification requirements for their respective schemes.

CoC standards are increasingly important today as concern has grown over illegal logging and IUU fishing. In the case of seafood chain-of-custody standards also address issues of food quality and safety. Although an overly narrow focus on the 'legality' of practices can lead to perverse outcomes – as when powerful, closed policy networks capture the legislative system and use it to promote overuse and/or to exclude others, notably indigenous or small operators, from access – nonetheless a focus on legality provides an important first screen for determining the acceptability or otherwise of products entering the supply chain. As governments in OECD countries become increasingly concerned to halt the importation of illegal timber and fish, they have turned to certification schemes to assist them in providing assurances of the legality of the products entering customs.

Operationalising a CoC system requires that a product entering premises be accompanied by a chain-of-custody certificate which testifies to its origin. The receiving business must also be CoC certified to prove that it meets certain process and technical requirements. These include training staff to distinguish between certified and uncertified material, separating certified products from those not certified and conducting risk assessments of uncertified products to verify they are not sourced from illegal or unacceptable sources. Once the business has completed its own production processes, it can then attach a CoC certificate to the product and forward it to the next firm in the commodity chain.

Scheme design and costs

The diversity of political, institutional and regulatory arrangements underpinning certification and labelling schemes has important cost implications. Politically, schemes can be developed by one or more constituencies giving rise to standards that differ in scope. Narrow standards that focus exclusively on technical aspects of forestry and fisheries and ignore the broader social and environmental dimensions impose fewer costs than standards with a broader scope. Institutionally, schemes can be more or less democratically run, differentially affecting the speed and 'efficiency' of decision making. All things being equal, the more deliberative a scheme is, the more costly it will be to run and the slower will be the decision-making process. Finally, differences in a scheme's regulatory arrangements – notably the balance between management, technology and performance standards – make some schemes notably easier to implement than others. A scheme that is tilted towards management standards and contains vaguely specified performance standards creates a great deal more company and auditor-level discretion than one that is tilted towards highly specific, quantitative performance standards. Given such variations, schemes differ in the costs they impose on companies and, scaled up, on an industry and a country. It is the differential costs imposed by schemes like the FSC's that accounts for the emergence of industry-sponsored competitors. However, while costs are a factor in explaining how states have responded to certification schemes, other factors are also at work. In the next section, we review the literature on what these factors are and then present our own synthetic analytic framework.

State responses to certification and labelling

Two broad approaches to theorising state responses to certification schemes can be distinguished. In the first approach, the state is viewed as a rational actor, developing policy independently from business and civil society in the 'public interest'. The state receives a vast amount of information from line departments, consultants and academics which it synthesises to formulate policy, which is then implemented, monitored and incrementally improved over time. In the second approach, the state is viewed as enmeshed in business and civil society relations with public policy reflecting that enmeshment and the power relations inherent in it. The consequence is that public policy is unlikely to represent an ideal conception of the 'national interest', and most especially when the state is captured by vested interests. Moreover, when policymakers are captured, then needed policy change is unlikely to occur either swiftly or to the degree required to take account of implementation failure, programme evaluations, new science or changing public values.[6]

Both conceptions of the state are evident in the certification and labelling literature. For example, Segura (2004) adopts the first approach, viewing governments as autonomous from business and civil society, and policy as representing the 'will of the people'. In exploring certification schemes such as the FSC's, Segura observes that government responses range from passively ignoring them to actively establishing competing arrangements (2004, 5). Segura highlights how certification schemes may challenge state sovereignty which he views as unimpeachable and calls on the FSC to 'reconsider the important role that these actors [governments] can play in the process' (2004, 14 and 23).

Lister (2005, 2009) also adopts a rational conception of the state in her comprehensive analysis of government responses to certification in Canada, the US and Sweden. According to Lister, governments seek to ensure that certification schemes are 'compatible with policy, laws and international obligations; transparent; open to the full participation of interested parties; non-discriminatory (e.g. to small business operators); not distorting of trade; and of equal and consistent quality' (2005, 11). These criteria are presented as rational, appropriate and achievable based on a conception of government as an independent, neutral arbiter seeking to maintain a liberal, non-discriminatory economic order. Lister goes somewhat beyond Segura in developing a formalised account of the role that governments could play, which range 'from passively observing, to actively facilitating and promoting, to mandating certification' (2005, 12). To analyse government responses to certification schemes, Lister examines the role they have played with respect to three certification components: standard development, implementation and monitoring/evaluation.

A rationalistic, 'normative' approach to the state is evident in Rametsteiner (2002). He employs a new institutional economics framework which views certification and labelling schemes as a solution to informational market failures. Consumers who desire to purchase goods that are 'sustainable' require assurances that they are doing so. However, drawing on Akerloff's (1970) work on the market for 'lemons' in the used-car business, Rametsteiner notes that in the absence of high quality information about a product, fewer transactions will occur than optimal because potential buyers fear being duped by sellers and are unwilling to purchase products at the offered price. He argues that third-party certification and labelling schemes assist in overcoming such informational deficiencies by reassuring customers that the purchased product meets the stated sustainability criteria and that the price premium they are paying, if any, is worth it. Rametsteiner views certification schemes as best run competitively by non-state operators and restricts the role of governments to ensuring that schemes are compatible with 'ongoing activities with laws and international obligations' that deliver 'efficiency and fair play'. However, he also notes that as owners of forests and fisheries, states will also need to consider whether or not to have them certified and, if so, to what standard.

Despite promoting market competition in certification standards, Rametsteiner is aware that his new institutional economics approach implies a rationalist conception of the state. He notes, for example, that 'The only body legitimised through the political constitutional system of voting is the public authority. Governments have, more than any other single stakeholder group, the legitimacy to define sustainable forest management, and thus they are the most legitimate source for any standards that claim "sustainability"' (168). This unqualified conception of state legitimacy is applicable at best only to liberal democratic governments where mainly free and fair elections are held in a context of a separation of powers between the executive, legislature and judiciary. However, even then the conception presumes that liberal democratic governments have the necessary autonomy and capacity to develop policies independently of vested interests. Both of these assumptions are doubtful. The idea that governments always have the capacity to develop policy independently of vested interests is challenged by the large literature on regulatory capture (e.g. Boehm 2007). It is thus to power conceptions of the state that we turn to illuminate the role the state may actually playing with respect to certification and labelling schemes.

In Cashore, Auld and Newsom (2004), elements of a power approach are present in the authors' structural approach to certification and labelling schemes. Government responses to certification are theorised to be the product of three core features of the policy environment. These are 'the place of the country/region in the global economy; the structure of the domestic forest sector; and the history of forestry on the public policy agenda' (2004, 8). The first feature identifies the dependence of the state on import and export markets. Cashore et al. argue that 'a high proportion of forest product *imports* from, or *exports* to, foreign markets appears to have been an important factor in creating an environment conducive to FSC supporters' efforts to achieve forest company and non-industrial forest owner support for the FSC' (2004, 41). The second feature focuses on three structural elements of the domestic political economy: the presence and importance of multinational corporations, the structure of resource ownership (fragmented versus centralised) and coordinated versus diffuse business associational arrangements (Cashore et al. 2004, 45–6). The third feature focuses on the role that environmental groups can play in setting public policy agendas and influencing outcomes. The 'more sustained and extensive environmental group and public dissatisfaction' is with respect to production practices, the more likely producers are 'to be convinced to support' the scheme (Cashore et al. 2004, 47). Conversely, where governments and industry enjoy very close relations, as when 'the subsystem is categorized as "clientelist" or "agency captured"', then support for some certification schemes is less likely since supporting them 'would mean giving up a comfortable policy-making arena in exchange for one in which business ... can no longer dominate' (Cashore et al. 2004, 48).

Despite the apparent determinism of this framework, Cashore et al. argue that the strategies adopted by certification scheme supporters may be inconsistent with the structural situation they confront. Thus analysts 'must carefully explore the interaction between a region or country's structural environment, the specific choices made by the FSC and its supporters, and the path dependencies that are created as they engage in trial and error efforts to gain legitimacy' (2004, 9). Similarly, the roles adopted by states are not predetermined. At the broadest level, Cashore et al. distinguish between two types of states based on their key concept of non-state, market-driven (NSMD) governance. The first type of state does not use its sovereign power to force compliance with certification schemes and can play six roles with respect to certification schemes. These roles are (i) developer of background rules and policies (property rights, contract law, etc); (ii) mediator of traditional interest group behaviour; (iii) adopter of procurement policies; (iv) negotiator as landowner; (v) provider of resources; and (vi) participator in standards development (Cashore et al. 2004, 20–2). The second type of state does use its sovereign power to force compliance. When that occurs, Cashore et al. note, 'our conception of non-state market-driven governance no longer exists, since the government, rather than the market, explains why the certified company or landowner is complying' (2004, 22).

Cashore et al.'s approach evidences a much more nuanced understanding of the state's role with respect to certification and labelling schemes. They aim to strike a balance between key structural features that derive from the political economy of commodity production while recognising that such structures locate actors within a strategic environment in which decisions are taken. Nonetheless, Cashore et al.'s analysis of state responses to certification and labelling schemes is rather thin. This is because they are mainly interested in understanding how supporters of different certification schemes engage in strategic action that is designed to enhance the legitimacy of their preferred scheme in the eyes of industry and state officials. According to Cashore et al., scheme supporters adopt 'converting' or 'conforming' strategies depending on the structural conditions they confront. While aware that states may be captured or clientelistic, the six possible formal roles attributed to them indicate that these authors too are ultimately under the sway of a liberal democratic conception of the state which views it as capable of rational action in the public interest.

An alternative, and much more power-centred approach to understanding state responses to certification appears in Elliott (1999) and Elliott and Schlaepfer (2001). These authors employ Sabatier's Advocacy Coalition Framework to understand state responses to forest certification in Canada, Indonesia and Sweden. The Advocacy Coalition Framework disaggregates the state into 'policy subsystems' which are composed of a variety of concerned public, private and civil society actors. According to Elliott and Schlaepfer, there are four basic premises inherent in the Advocacy Coalition approach. First, a long time frame for analysis (10 years or more) is required to understand policy change. Second, analysts must focus on the policy subsystem rather than governmental institutions, to fully identify the relevant actors. Third, it is important to recognise that policy subsystems involve not just central but also sub-national government actors. Finally, the Advocacy Coalition Framework places a great deal of emphasis on ideas, viewing them as crucial to understanding how advocacy coalitions operate (1999, 645). It also recognises that 'policy-oriented learning' occurs in subsystems and that actor coalitions seek to 'out-learn' each other in order to have their interests represented in policy (Elliott and Schlaepfer 2001, 645). Coalitions adopt a variety of strategies to achieve their objectives (litigation, lobbying, research, advertising, boycotts, etc) and, in contrast to Cashore et al.'s emphasis on structures, the approach stresses the agency of actors in the contest over ideas. Consequently, within the Advocacy Coalition Framework, outcomes are highly contingent and depend not only on the relative power of competing ideas but also on the strategies and tactics of actors in pursuit of policy influence. Thus where Cashore et al. see outcomes as importantly shaped by structural factors with some scope for agency, Elliott and Schlaepfer see them as largely shaped by actors and ideas operating within a specific policy subsystem.

Analytic framework

We situate our approach within the broader literature on power conceptions of the state such as those employed by Cashore et al. and Elliott and

Schlaepfer. Drawing as they do on insights from the policy network literature, we adopt a perspective that sees the state disaggregated into sectoral policy networks with actors inside the network and outside in the broader policy community engaged in a strategic contest to maintain or gain policy leverage in pursuit of their perceived interests. Our basic argument is that the type of policy network in place in a particular region and sector largely explains a network's, and thus a state's, response to certification in that region and sector. However, since policy networks arise from perceived common interests grounded both in the political economy and ecology of a region and refracted through accepted ideas and discourses, we conceptualise them as dynamic configurations with the potential to transform themselves over time in response to shifts in the regional political economy, alterations in power dynamics within the commodity chain, changes in public perceptions concerning a region's ecological significance and the spread of alternative discourses. We develop this basic framework below, commencing with an outline of our approach to policy networks.

Policy networks

There are several alternative policy network typologies and we adopt a modified version of Howlett and Rayner's approach (1995). Howlett and Rayner identify eight policy network types based on the type and number of actors involved and whether the process is state- or society dominated (Table 2.7). The authors identify, at one extreme, a 'bureaucratic' policy network managed by state agencies which excludes societal actors. This network type conforms most closely to the popular image of the autonomous, bureaucratic state that makes policy on its own. At the other extreme are issue networks where 'the principal interactions take place among a large number of societal actors' and the state is relatively powerless (Howlett and Rayner 1995, 386). Between these two extremes lie clientelistic and captured policy networks that are dominated by a single major societal group

Table 2.7 A taxonomy of policy networks

		Number/type of participants		
	State agencies	One major societal group	Two major societal groups	Three or more groups
State-directed Relations	Bureaucratic	Clientelistic	Triadic	Pluralistic
Society-dominated relations	Participatory	Captured	Corporatist	Issue networks

Note: Shaded area includes the four policy network types that are the focus of our analysis.
Source: Howlett and Rayner 1995.

and triadic and corporatist networks where two major societal groups operate in partnership with the state. Pluralistic networks occur when three or more major societal groups are involved in policy under the authority of a facilitating state. The pluralist policy network type conforms most closely to the popular conception of government's role in liberal democracies – as a neutral arbiter of competing interests that brokers compromise in the general interest of all.

Howlett and Rayner's taxonomy highlights the existence of different policy network types, enabling a more nuanced, sectoral analysis of state responses to policymaking in general and certification policy in particular. However, in practice it is difficult to distinguish between all of the framework's policy network types, a criticism equally applicable to an earlier typology of Atkinson and Coleman (1989). Moreover, while it is possible in weak states for societal actors to dominate the policymaking process, our three case studies are of strong states with substantial executive, judicial, bureaucratic and technical capacity. We therefore consider that in our cases the state maintains control of the national and regional policymaking process, but that the configurations it has with major societal actors vary from sector to sector and over time. Focusing therefore on Howlett and Rayner's 'state-directed' policy networks, we identify four different types: bureaucratic, clientelistic, triadic and pluralistic. Bureaucratic and pluralistic networks are as discussed above and represent the opposite ends of our policy network spectrum distinguishing an autonomous state where policy is made by bureaucrats from a responsive state where all major societal actors have input. Between these extremes, policy is dominated by one or two major societal actors. Clientelistic policy networks exist when the state makes policy in conjunction with a single societal group – in modern liberal democracies and depending on the political party in power, this group is either business or labour. A triadic policy network exists when the state makes policy in consultation with two major societal groups; again, in modern liberal democracies, this has tended to be business and labour together.

Our basic argument is that a state's response to a certification scheme depends in large measure on the type of policy network in operation in a sector and the type of certification scheme it is contemplating adopting. With respect to a scheme's political dimension, bureaucratic policy networks will dispute the need to include actors beyond the state and thus of the need to adopt a certification scheme. Instead, they will view existing regulatory arrangements, which they have set in place, as adequate and will resent what will be perceived as a challenge to state authority. Business- or labour-clientelistic policy networks, and triadic networks, will be somewhat more positively disposed to adopting certification and labelling schemes especially if they are being demanded by customers and clients, but will seek to maintain control of the certification agenda and exclude other actors with interests incompatible to accumulation. In contrast, pluralistic policy networks that permit the participation of a larger number of societal groups will be more positively disposed towards certification and labelling in general and towards inclusive schemes that represent interests beyond the narrowly economic.

Across our institutional dimension, bureaucratic policy networks will prefer arrangements that are formal, hierarchical, state-based and that restrict input from non-state actors. These preferences are based on the norms of state sovereignty, bureaucratic management and multilateral bargaining. In contrast, business- and labour-clientelist and triadic policy networks will prefer institutional arrangements that are open to them but closed to interests that they view as incompatible with the promotion of accumulation. All three of these network types will prefer 'efficient' institutional forms – corporations for example – that concentrate decision-making power at the top in a small board that is able to respond quickly to external requirements. Finally, pluralistic policy networks will have a general preference for more open, deliberative institutional forms to enable policy learning to occur across the diversity of interests represented within them.

Finally, with respect to a scheme's regulatory dimension, bureaucratic policy networks will prefer mandated regulations over voluntary standards, although the precise nature of the regulations may vary from state-based codes of conduct to legislated forest management practices. Schemes will validate and endorse the regulatory status quo as that is considered by the regulatory authority as meeting the requirements of sustainability. Business- and labour-clientelist and triadic policy networks will prefer schemes that retain considerable authority within the enterprise to determine the level of effort required to achieve a certificate. This will translate into a preference for management systems approaches that enable the company to establish the level of ambition in the plan, the rigour with which it is implemented and the focus for achieving 'continual improvement'. If performance-based indicators are included in a scheme, a preference will exist for these to be stated at a high level of abstraction, with interpretation delegated to certifying bodies to develop firm-based indicators and verifiers. Pluralist policy networks will prefer performance-based schemes that translate high-level principles and criteria into detailed statements of expected outcomes in the form of indicators and verifiers. This is because environmental, indigenous peoples, community and other groups will be concerned that in the absence of precise statements about what performance is expected, the incentives confronting firms and certifying bodies align in favour of lowering the bar. These relationships between policy network type and certification scheme preference are mapped in Table 2.8.

Policy network formation and transformation

Policy networks do not simply materialise from nowhere, nor do they remain forever locked in the same configuration. Rather such networks emerge and evolve over time as a consequence of changes in the broader political

Table 2.8   Policy network types and certification scheme preferences

Policy network types	Certification scheme preferences		
	Political dimension	Institutional dimension	Regulatory dimension
Bureaucratic	Limit participation to state actors;	Promote multilateral norm of state sovereignty and limited participation by non-state actors;	Promote mandatory over voluntary regulation but regulatory instrument can vary;
Business- or labour clientelist	Limit participation to state and business or labour activists	Promote norm of 'efficiency' with small insulated boards empowered with decision-making authority;	Promote voluntary over regulatory instruments with preference for management systems or vaguely specified performance requirements;
Triadic	Limit participation to state, business and labour activists	Promote norm of 'efficiency' with small insulated boards empowered with decision-making authority;	Promote voluntary over regulatory instruments with preference for management systems or vaguely specified performance requirements;
Pluralist	Open participation up to non-state actors including business, labour, environmentalists, communities and indigenous peoples.	Promote norm of 'deliberation' with larger boards, more open, possibly membership-based decision-making arrangements.	Preference for clearly state, performance standards to secure compliance; however, level at which these are set is indeterminate.

economic, ecological, discursive and political contexts within which they are embedded. To understand why a policy network of a specific type forms and evolves in a specific region and sector, we must examine several mutually conditioning elements. These relate to the political economy of the region, the political economy of the sector, the ecology of the region and the discursive context of the region. We utilise this framework to analyse the policy networks in place in designated countries and sectors in the early 1990s when certification first appeared on the policy agenda. By analysing framework elements, we are able to chart how policy networks responded to specific certification schemes at the outset; and how, over time, changes in one or more of the elements induced shifts in the structure and operation of the policy network leading to a reappraisal of its – and thus the state's – certification response.

Political economy of the region

The political economy of a region shapes the relative power of business in a specific sector at a specific time. In some regions, a single industry may dominate and account for the lion's share of jobs, investment, revenues and exports. When this occurs, industry will wield considerable power over policy since legislation that it perceives as negatively affecting its business can quickly be transmitted into threats of action in the form of job losses and/or relocation to more investment-friendly regions. Under the law of anticipatory reactions, moreover, bureaucrats will be reluctant to bring in such legislation for fear that it might provoke a negative reaction from the region's dominant industry. In economically diversified regions, such sectoral power is less likely to occur, since by definition the region is not dependent on a single industry. Nonetheless policies that affect general industry profitability – export taxes, labour laws, royalties, minimum wages, and so forth – will be contested by business associations which always hold out the threat of an 'investment strike' to persuade governments to maintain the conditions for accumulation.

However, a policy network's dependence on a specific configuration of political economic circumstances makes it at least potentially transformable as changes occur to the political economy of the region and/or sector. Regions that, over time, become more economically diverse become by definition less dependent on the original dominant industry. While the industry may continue to try to exert its policy influence, it may find itself subject to increasing criticism not only from excluded societal groups but from previously supportive elites. As a clientelistic or triadic policy network comes under pressure internally, externally it may try to regain its power by ejecting the dissenting elements from within its ranks or it may begin to open up to previously excluded groups, effectively transforming itself into a pluralistic policy network.

However, the same process of transformation can occur as a consequence of external rather than internal dynamics. If a region is especially dependent on export markets, and if those export markets develop a preference for certified products, then the regional policy network may try to find alternative markets elsewhere to preserve its domestic influence. If that is not feasible, it will then have to shift its position on certification to maintain access to foreign markets. Moreover, in this case its choice of certification scheme will depend less on its own preferences than on the preferences of clients and consumers in foreign markets. However, certification schemes differ in their institutional structures, and the adoption of tripartite, associational,

performance-based schemes can have significant implications for the operation of the regional policy network itself, again opening it up to more societal actors and moving it along the continuum to a more pluralistic type.

Commodity chain

The global political economy of a sector influences producer perceptions
of the need to embrace certification or not. The number of actors in a
sector, their size, the amount of capital required, whether the industry is
producing for domestic or foreign consumption and the relative power
relations between producers, transformers and retailers all matter. One way
of analysing these diverse relationships is via commodity chain analysis
(Gereffi 2001). Analysts employing commodity chain analysis distinguish
between 'buyer-driven' and 'producer-driven' networks. The essential difference between them relates to whether control over access to the network
lies with producers or with retailers. A producer-driven network exists when
industrial capital is concentrated, research and development is intensive
and economies of scale occur. Examples include the automobile, computer
and aircraft industries. In contrast, buyer-driven networks involve products
that utilise commercial capital, employ high levels of design and marketing
expertise and benefit from economies of scope. Typical industries where
these characteristics occur are the apparel, footwear and toy sectors.

Both the forestry and fisheries commodity chains are buyer-driven commodity chains. This means that power lies at the consumer and retail ends of
the chain and that any changes such actors make to product specifications
will have profound consequences for producers and processors further up
the chain. Studies that employ commodity chain analysis trace how actors
producing a specific commodity in a particular jurisdiction access markets
and how vulnerable they can be when wholesalers and retailers further up
the commodity chain alter their specifications and/or pricing arrangements
(e.g. Sun et al. 2008; Smith 2006). In our case studies, we trace these relationships between producers and consumers to better understand the power
relations inherent in the commodity chain. We note how dependence on
different markets – in Europe and Asia for instance – can influence producer
perceptions of the need to embrace certification and of the kind of certification scheme to adopt. We also note that shifts in retail and consumer market
specifications can reverberate down the commodity chain, driving changes
at the transformer and producer level in their perceptions of the need for
certification.

Regional ecology

Another element that influences policy network perceptions of the need
for certification is the real and perceived ecological significance of a region.
Prior to the 1960s, public awareness of ecology, environment and ecosystems was minimal. Instead attention was focused on individual endangered

animals and the immediate habitat required for survival. It was only following Rachel Carson's book *Silent Spring* in 1962 that the public became more aware of the inherent connectivity of physical and natural systems and of how chemicals could accumulate in animal tissues, rendering them toxic to other animals and humans. Since then, there has been an inexorable increase in public awareness of the importance of ecosystems in general and of specific ecosystems in particular, culminating in widespread concern over the potentially devastating effects of global warming and climate change.

Regions differ in terms of their ecological significance. Some regions, like the forests of Amazonia, are especially rich in biodiversity and are globally significant in terms of the role they play in mitigating dangerous climate change. Other regions – like the highly modified landscapes of England and Ireland – are of much less ecological significance and are more in need of restoration. These variations in ecological significance are a product of natural factors related to latitude, altitude and remoteness and social factors related to settlement patterns, population growth and development. Growing ecological literacy has interacted with differences in actual regional ecology to create powerful movements in some jurisdictions for terrestrial and marine parks and protected areas and significantly improved forest and fisheries management practices. Such movements challenge the business model of existing producers who have come to depend on these same ecosystems for profits and jobs.

Changing public perceptions of the ecological significance of a region matters because if the region comes to be viewed as ecologically important, pressure will grow on local producers to limit and/or alter their extractive practices. While environmental groups and some members of the public will start to back certification schemes that set a high bar with regard to ecosystem protection, producers are likely to either reject certification outright in favour of domestic regulation (over which they have influence) or to endorse an alternative scheme that contains less restrictive ecological requirements. In short, we can expect greater conflict in general between environmental and industrial interests in regions of high ecological significance; we can also anticipate that this conflict will play out through certification and labelling schemes, with environmentalists endorsing schemes that set a high ecological standard and industry giving preference to existing regulatory arrangements or to schemes that are less constraining in terms of where harvesting occurs and how it is undertaken.

Management discourse

The coherence of policy networks is achieved through the sharing by actors of a set of hegemonic ideas through which 'reality' is simultaneously created and interpreted. In market capitalist systems, one set of ideas is based on the superiority of the free market, which is linked to such concepts as 'individualism', 'freedom', 'leadership', 'entrepreneurialism', 'progress', 'development', 'efficiency', 'hard work', 'innovation', 'free trade', 'free capital movement', 'free investment' and so forth. These liberal and neoliberal ideas provide a justification for giving preference to markets, criticising regulation and excluding 'socialistic' groups and amateurs who either are not in favour of accumulation and/or who lack the technical expertise to understand the requirements of production.

This mix of liberal and neoliberal ideas has a specific manifestation when it comes to natural resource management, which has historically been based on the concept of 'sustained yield'. Sustained yield involves (a) calculating the existing stock of timber or fish; (b) estimating the annual productivity of that stock; and (c) calculating how much of the annual increment should be harvested and how much should remain to ensure sufficient production the following year. Sustained-yield management deals exclusively in stocks and flows of a species, frequently considered in isolation from the ecosystems on which it depends and which may depend on it. Although dominant in the post-war period, the sustained-yield approach was undermined in the 1980s by biologists and ecologists who investigated how ecosystems operated in practice and how unpredictably they can behave when species are removed. Since then, the sustained-yield approach has been steadily displaced by the ideas and practices of ecosystem-based management.

Ecosystem-based management ideas began permeating the thinking of natural resource managers in the 1980s, although their diffusion has been uneven across sectors and countries. As ecosystem ideas have diffused, however, they have undermined the liberal and neoliberal discourse upon which the sustained-yield approach was based. This is because underpinning the ecosystem-based approach are concepts such as 'system', 'relationships', 'interconnectedness', 'community', 'uncertainty', 'precaution', 'adaptation' and so forth that are fundamentally different from those that underlie liberal and neoliberal conceptions of the world of disconnected things. Once people are convinced of the importance of managing natural resources from within an ecosystem-based framework, it is but a small step to moving towards including excluded groups since the management goals are more diffuse, the management tasks much broader and the knowledge requirements much greater than before (Grumbine 1994).

Conclusion

Analytic frameworks investigating certification have adopted either rationalist or power-centred conceptions of the role of the state. There are two problems to be addressed. First, rationalist approaches overemphasise the capacity of the state to make policy independently of social actors. While such approaches can be defended on normative grounds – that this is how states *should* make policy – most analysts do not appear to employ the approach in this manner. Rather, they adopt a rationalist approach to the state either to defend the right of states to make whatever policy it deems is in the public interest or to provide the basis on which policy recommendations can be made about what they should be, but are not yet, doing. In contrast, we have adopted a power conception of the state that recognises the embeddedness of governments in the wider market-capitalist system and the privileged position of business with respect to policymaking within this context.

The second problem with existing accounts of state responses to certification and labelling schemes is that they treat the state holistically as a single entity. This approach fails to grasp how actors located in different policy sectors are differentially placed with respect to government and society based on their relative power and significance for the national economy and ecology, their location in the commodity chain and the ideas that legitimate their current views on certification and labelling schemes. We thus disaggregate the state by examining responses at the level of the policy network that operates in a sector. We anticipate that differences in the structure and operation of policy networks provide much of the explanation for why the 'same' state responds differently to different certification and labelling schemes and why states alter their position on certification and labelling schemes over time. Policy networks respond to shifts in regional political economic realities, changes in production requirements and new discourses of production. They also respond to public perceptions of heightened regional ecological significance. In short, policy networks become established and endure but also evolve and change over time as a consequence of political economic, sectoral, governmental and discursive pressures. In each of our case studies, therefore, we analyse the structure of the policy networks in place in our selected regions and sectors and chart the way in which the pressures they have been subjected to have altered their structure and operation. In doing so, we aim to explain why they – and the states they are party to – responded as they did at specific times to particular certification initiatives.

3
The Forest Stewardship Council and the Marine Stewardship Council

Certification and labelling emerged on the international policy agenda in the late 1980s in a context of increased public concern over food scares, 'unfair' trade, deforestation, child and sweatshop labour and other ecologically and socially damaging practices. In 1988, following many years by Oxfam and other Alternative Trading Organisation, the first large-scale fair trade certification and labelling scheme was set up in the Netherlands. Taking its name from a nineteenth century Dutch novel critical of the country's colonial practices, the Max Havelaar programme initiated a new era for fair trade by enabling consumers to confidently purchase products that guaranteed a living wage to peasant and artisanal producers. At about the same time as the Max Havelaar scheme was being established the UK-based Friends of the Earth group were promoting a 'Good Wood Guide' and developing a proposal to certify tropical timber certification. And, in advance of negotiations at the UNCED in 1992, business agreed to sponsor the development of an environmental management standard through the ISO leading to the launch of its 14000 series in 1996.

Although certification and labelling was becoming more widespread in the late 1980s, no consensus existed on its focus (fair trade, environment, safety, quality), form (management or performance standards) or institutional set-up (federation, foundation, association). In addition there was no agreement on whether the company, the industry association or an independent third party should undertake the certifying or whether Type I, II or III labels were preferred. Looking back, therefore, we can understand the enormous challenge proponents faced in developing global certification schemes. Not only were they wrestling with numerous practical challenges, they were also operating in a climate of considerable hostility from sceptical state, business and civil society actors.

Notwithstanding the challenges, certification developed apace and is today well established in numerous market sectors including apparel, footwear, coffee, fish, tea, timber and handicrafts. Companies are beginning to view voluntary certification as a requirement of market entry as demand grows from consumers and retailers for assurances concerning the sustainability, fairness, safety, legality and quality of products. While doubt remains as to whether certification schemes can be combined to secure all these outcomes, the remarkable progress made in the past two decades provides grounds for cautious optimism.

Such optimism is partially based on the speed and depth of the institutionalisation of the two certification schemes that are the focus of this book: the FSC and the MSC. Both schemes were established in the 1990s by environmental-business coalitions, both aim to certify products as economically, environmentally and socially 'well managed' and both have experienced rapid growth over the past decade. Despite these similarities, however, considerable differences are apparent. The number of actors involved, the institutional structures adopted and the form of the standard employed differ. Our aim in this chapter is not only to describe in detail the history and operation of each scheme but also to subject them to a three-dimensional analysis based on the framework introduced in Chapter 2. This analysis will demonstrate that despite sharing the 'stewardship' name and heritage, the two schemes differ markedly in their organisational governance arrangements.

Forest Stewardship Council

Historical background

The roots of the FSC can be traced back to several loosely connected initiatives taking place in the late 1980s in Europe and the US. While tropical deforestation in developing countries was a major motivator, there was growing disquiet too over the management of temperate and boreal forests in developed countries. In an effort to improve tropical forest management, the European Parliament passed a resolution in 1988 calling for a ban on tropical timber imports from Sarawak, Malaysia. The ban was aimed at encouraging the Government of Sarawak to resolve an increasingly militant struggle between the Penan, one of the region's indigenous peoples, and loggers. A second EU resolution in May 1989 sponsored by Hemmo Muntingh called on countries to prohibit the importation of tropical timber from countries that lacked forest management and conservation plans (Synnott 2005). These EU resolutions were welcomed by environmental groups in Europe, especially Friends of the Earth International (FoE-International), which was launching a coordinated campaign to raise awareness about the destruction of the world's tropical forests and the role played by the tropical timber trade in facilitating it.

The European Parliament's resolutions did not result in European action because they were not supported by either the European Commission or the Council of Ministers. However, at national and grassroots levels, ECSOs made substantial progress in campaigns to boycott tropical timber imports, a key focus of FOE-UK in the early 1990s (see Chapter 7). The success of boycott campaigns saw large, more conservative environmental groups like WWF-International and IUCN urge caution. These groups argued that boycotts could, perversely, increase tropical deforestation if they depressed the value of forest land and encouraged conversion to more profitable commodities like palm oil and rubber. This proved to be a powerful argument and prompted discussion among ECSOs of how timber from well-managed forests could be distinguished from timber coming from poorly managed or recently converted forests.

From these discussions, the idea of certifying and labelling timber from well-managed forests emerged. In 1987, FoE-UK began working with the Tim Synnott of the Oxford Forestry Institute on a feasibility study to explore the notion. Together they developed a proposal that was shared with the UK Government's Overseas Development Administration, now the Department for International Development (DFID)). Following the preparation of a number of drafts, FoE-UK succeeded in winning ODA's support and the proposal was submitted to the seventh session of the ITTO in Yokohama in 1989 (Gale 1998). Although a modest proposal for a feasibility study into whether certification and labelling of tropical timber could improve forest management practices, it sparked outrage from developing country and timber industry representatives led by Malaysia, Indonesia and Cameroon. Certification and labelling, it was feared, would become a non-tariff barrier to the trade in timber products and a threat to forest-led development. The opposition the proposal encountered forced the UK delegation to reformulate it, much to the disgust of the FoE-UK representative who declined to endorse the new proposal.

The hostile reaction of developing country delegates to the idea of timber certification and labelling meant that no formal discussions could occur at the ITTO, then the dominant forum on tropical timber matters. Momentum continued to build for some kind of certification and labelling system however with a number of proposals canvassed. Of these, Hubert Kwisthout's was the best developed. Kwisthout was the head of a UK firm called the Ecological Trading Company, which imported tropical timber to make bagpipes. Early on Kwisthout had committed himself to sourcing only sustainably produced timber but quickly discovered that no mechanism existed to distinguish such timber from illegal and unsustainable timber. Following discussions with Francis Sullivan of UK's World Wide Fund for Nature (WWF-UK), he developed the idea of an International Forest Monitoring Agency (IFMA), an international body that would verify that imported timber came from well-managed sources (Synnott 2005).

In 1990, Kwisthout outlined his IFMA concept at a meeting of the WARP. Members of WARP were also makers of musical instruments and shared his concern that the purchase of tropical timber was contributing to the destruction of the earth's rainforests. The 1990 meeting, however, included several other individuals that later played a pivotal role in the founding of the FSC. The WARP meeting endorsed Kwisthout's IFMA idea and established a Certification Working Group to explore it further and develop proposals (Synnott 2005).

In a series of meetings held between 1991 and 1993 the Certification Working Group elaborated the IFMA proposal while simultaneously expanding the network of interested groups and gaining support from US philanthropic groups such as the MacArthur Foundation. Over the 28-month period, the notion of establishing a global forest management standard based on a set of international principles and criteria that would be audited by accredited third-party certifying bodies and known as the Forest Stewardship Council took shape. Unlike the original proposal, temperate and boreal forests were included in the scheme in recognition of the conflicts that had emerged in the 1990s in such places as the Pacific Northwest in the US, British Columbia (BC), Canada and Queensland, Australia. Synnott (2005) provides a detailed history of the meetings and extensive consultations that occurred around the world to secure input into, and legitimacy for, the new organisation.

By mid-1993, and following the failure of UNCED to agree a global forest convention, agreement to launch the FSC was reached. The Interim Board of the FSC, as the CWG was now known, organised an October 1993 Founding Assembly. By all accounts, the founding assembly was a fractious affair, at least at the outset. Synnott notes that the German ECSO Rettet den Regenwald released a paper just before the meeting that predicted the FSC would fail because it would be either naïve or 'at worst ... [provide] a framework for the timber industry to achieve a much desired "green veneer" and diffuse pressure to attack the real issues of illegal trade, indigenous peoples rights and over-consumption' (Rettet den Regenwald 1993 quoted in Synnott 2005, 22). At the meeting an initially sizeable coalition, loosely coordinated by Simon Counsell of FoE-UK, objected to the participation of large industry, including Alan Knight of the British do-it-yourself chain B&Q who was a member of the Interim FSC Board. Following several rounds of consultations, however, the number of individuals and groups objecting to the presence of industry dwindled as awareness grew of the importance of their involvement and the institutional arrangements in place to prevent their domination.

The arguments in favour of including industry representation focused on ultimate objectives. If the FSC aimed to make a real difference across the entire forest sector rather than develop a 'boutique' standard, it needed to include a strong voice from the industry in its policymaking arrangements. WWF, which had been working closely with industry in the UK in the establishment of a 'Buyers Group' of companies committed to purchasing certified timber, was a strong proponent of this view. WWF not only contributed funds to set up the FSC but its international forest policy officer, Chris Elliott, chaired the meeting. Synnott notes that opposition to the presence of industry diminished when some indigenous groups distanced themselves from a ban on harvesting undisturbed tropical forests and when social ECSOs recognised a role for dealers and manufacturers (2005, 22).

It is doubtful, however, that the arguments for including industry would have carried the weight they did if the founding assembly had not agreed to two important institutional innovations designed to curb business influence. The first was an agreement to establish the FSC as a membership organisation, not a foundation. This decision was to have profound, but ultimately highly beneficial, consequences for the organisation, enabling it to reach out across the world to encourage diverse parties to join, leveraging the organisation's credibility and legitimacy in the process. The second institutional innovation was the adoption of the 'chamber' approach to interest aggregation which meant that business interests were unable to dominate environmental and social interests in the decision-making process. Without these institutional innovations it is unlikely the FSC would have survived.

FSC governance: The political dimension

As noted in Chapter 2, certification schemes can be distinguished across three dimensions of governance: the political, institutional and regulatory dimensions. The FSC stands out with regard to the political dimension because it seeks to equally represent the diversity of groups with an interest in forest use and protection. Indeed, so unique is the FSC in this regard that it has been termed by one of the authors (Gale) as an instance of global democratic corporatism (Tollefson, Gale and Haley 2008). The 'corporatist' element of this arrangement refers to the FSC's 'chamber'-based system of interest aggregation. When an individual or organisation joins the FSC, they nominate to join one of three chambers depending on whether they represent economic, environmental or social interests. Those assigned to the economic chamber include large industrial forest companies, small woodlot operators and a range of downstream processing, wholesale and retail companies, all of which have a commercial interest in the forests. Members concerned with the ecological values of forests, including especially large and small environmental civil society organisations, are assigned to the environmental chamber along with academics and consultants with professional backgrounds in environmental studies, conservation biology, green political economy and so forth. The third, social chamber encompasses individuals and organisations interested in the developmental, humanitarian, human rights and labour dimensions of forest management. Membership of the social chamber is more diverse than the other two chambers and includes labour unions, indigenous peoples organisations and international development agencies.

The FSC does not aim merely to balance economic, social and environmental interests. It also aims to bridge the North–South divide. Thus a second notable feature of the FSC's political arrangements is that each of the three interest chambers described above is further divided into those representing the global North and those representing the global South. When applying for membership, applicants not only indicate the chamber they are applying for but also whether they come from a developed or developing country. Developed countries are those listed in the World Bank list of High Income Economies and encompass the members of the Organisation for Economic Development and Cooperation (OECD). Developing countries constitute the remainder with most based in Asia, Africa and Latin America. The rationale for this division is that the interests of those based in northern, developed countries differ from the interests of those based in the southern, developing countries and the latter need to be explicitly and equitably included to avoid organisational domination by the North.

An important component of an organisation's political dimension is funding following the general maxim that 'he who pays the piper calls the tune'. Historically, the FSC has been heavily dependent on ECSO and donor funding, receiving early large grants from the Macarthur Foundation, Pew Charitable Trusts and the Ford Foundation (Tollefson, Gale and Haley 2008). While it has consistently sought to grow revenue from its own services, it proved difficult to accomplish this in the early years when the organised was getting off the ground. However, the 2008 financial accounts report a decline in donor dependence which has decreased from two-thirds of total income in 2005 to around one-third in 2007.

FSC governance: The institutional dimension

A unique feature of the FSC is that it is a global association of organisational and individual members. A breakdown of the membership by chamber category and North–South affiliation is provided in Table 3.1. To become an individual member of the FSC, one completes an application form endorsing the organisation's objectives and principles and criteria, secures the support of two other members, and submits the form along with the applicable membership fee. The application is then considered by the FSC and, in almost all cases, accepted. Organisational membership applications follow

Table 3.1   FSC membership by chamber and region, December 2009

Chambers	North	South	Totals	% of total
Economic	182	154	336	41%
Environmental	132	210	342	41%
Social	57	94	151	18%
Totals	371	458	829	
%	45%	55%		

Source: FSC-IC 2009a, 10.

the same basic process although more details are required. The FSC regularly announces new members on its website. For example, in September 2009, five new members were listed including three individuals and two organisations (FSC-IC 2009a).

The FSC is structured as an organisational hierarchy at the apex of which is the General Assembly (see Figure 3.1). The General Assembly meets every three years, with the most recent meeting taking place in Cape Town, South Africa in late 2008. Members can submit resolutions for debate and there is considerable opportunity for grassroots input. A key function of the General Assembly is to debate the structure and operation of the FSC and to scrutinise the performance of its nine-member Board of Directors. A number of critical decisions have been taken at the FSC general assemblies, including a 1996 resolution to approve Principle 10 on plantations and a 2002 resolution entitling government forestry departments to join as members of the Economic Chamber.

The FSC's Board of Directors is elected by members. Elections are conducted electronically: members are notified of an impending election by the Nominations Committee, nominations are sought, ballots distributed, an online forum held and voting undertaken. Elected candidates join

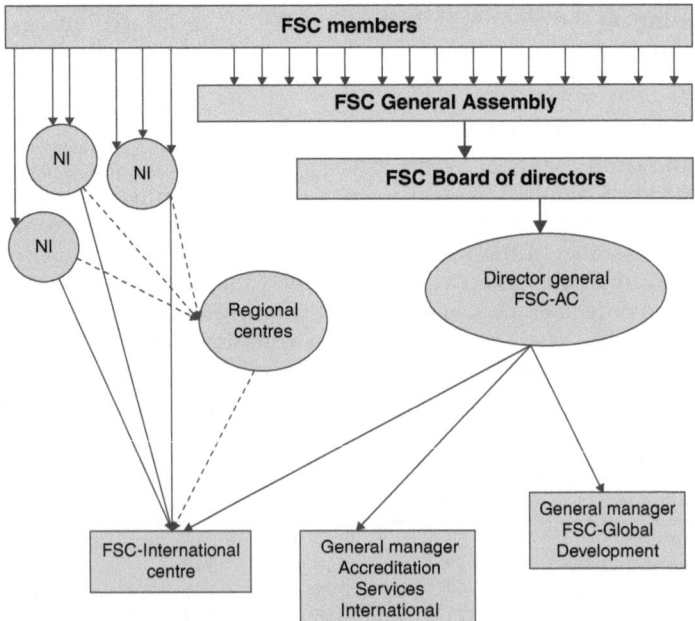

Figure 3.1   Organisational structure of the FSC group
Note: NI stands for National Initiative

a nine-member board equally balanced by chamber and along North–South lines. The board meets virtually and face-to-face during the year to consider the FSC policy and operational matters including those referred to it by its constituent bodies (FSC-IC, FSC-GD and ASI, see below).

Since its inception in 1993, the FSC has grown in size and complexity and a review of its institutional arrangements commenced in 2007 to better integrate its disparate elements. The review resulted in a series of proposals to the 2008 General Assembly to specify more precisely the roles of the board, secretariats, and regional and national initiatives. While the proposal was not accepted in its final form – the resolution requested that further expert advice be obtained along with estimates of cost – it highlighted the structural complexity of the emerging system and illustrates the relationships and division of responsibilities among various bodies.

Between general assembly meetings, members play a key role in voting for members of the FSC board of directors, and in serving on standing and ad hoc committees of the organisation. The board manages the business of the FSC but delegates operational matters to an FSC director general, who presides over three elements of the FSC system: the FSC-International Centre (FSC-IC), the FSC Global Development (FSC-GD) and the Accreditation Services International (ASI).

Collectively, this group of FSC bodies is known as the FSC-AC, which is the formally constituted *Asociación Civil* ('civil association') registered in Mexico. The director general of the FSC-AC is currently Andre de Freitas, who is also managing director of the FSC-IC. The other two bodies have their own managing directors and operate at arm's length from the FSC-AC and the FSC-IC. ASI is responsible for accrediting certifying bodies to carry out certifications to the FSC standard. It processes applications from new certification bodies (CBs), undertakes audits of existing CBs and reviews and proposes amendments to the FSC's accreditation standards and policies. The FSC-GD is responsible for ensuring the appropriate use of the FSC's 'tick-tree' logo and for growing revenue from royalties associated with its use. This involves encouraging companies to become forest management and chain-of-custody certified and to display the FSC logo on their websites and products.

The FSC-AC contains a range of devolved regional and national initiatives that vary in size and significance. Regional centres were set up in 2002 following FSC International's relocation from Oaxaca, Mexico to Bonn, Germany. Designed to enhance the FSC's presence around the world, there are currently four FSC regional centres based in Asia, Africa, Latin America and Europe. Regional offices provide information on the FSC to the region, facilitate a level of coordination among national initiatives in a region and promote national initiatives in countries without them. Regional centres are underspecified within the FSC hierarchy. The FSC Governance Review proposed to formalise their structure and operation to function as Regional Councils composed mainly of relevant national initiatives.

National initiatives are well established within the FSC system. There are two basic types: contact persons and working groups. A contact person's primary role is to provide information about the FSC, field inquiries from potential suppliers and buyers of FSC-certified products and facilitate the establishment of a national working group. The structure and operation of working groups varied significantly in the 1990s resulting in efforts to formalise and harmonise their structure and relationship to the FSC-IC. Synnott highlights the early tensions that existed between national initiatives and the FSC-IC. In his view, matters came to a head in October 1997 at a meeting of representatives of national initiatives in Vermont, where vigorous discussions were held on the need for national initiatives to sign contracts with the FSC-IC. While many reluctantly accepted the need for FSC-IC oversight, members of FSC-US held out. Synnott quotes Jamison Ervin, one of the founders of the FSC as 'increasingly frustrated with the direction and tone of the people in Oaxaca, and their relationship to those of us who are trying to make the FSC more than a small hacienda by the Zocalo in Oaxaca' (Ervin quoted in Synnott 2005, 31).

As of late 2009, there were a total of 57 national initiatives, evenly split between contact persons and working groups (Table 3.2). These were differently distributed across regions, however, with all but one of Africa's national initiatives led by contact persons and all nine of Latin America's working groups. Since working groups signal a more institutionalised FSC presence in a country, the FSC is far better represented in Latin America, Europe and North America than in Africa and Asia. Many African contact persons are recent appointees, and history suggests that it can take many years for a working group to form. In Australia, for example, a contact person was appointed in 2001 but a working group only established in 2006 (see Chapter 5). In PNG, the process took over a decade with the contact person, Yati Bun, appointed in 1996 but a working group only officially accredited in late 2007 (Bun and Bewang 2006; FSC-IC 2010a).

Table 3.2   FSC national initiatives

Region	Contact person	Working group	Totals
Africa	14	1	15
North America	0	2	2
Latin America	0	9	9
Europe and Russia	9	14	23
Asia and Oceania	4	4	8
Totals	27	30	57

Source: Compiled from data available at FSC, 'FSC Worldwide', Online at http://www.fsc.org/worldwide_locations.html, accessed October 2009.

FSC governance: The regulatory dimension

A key role for national initiatives is to develop an FSC standard for their country or, in some cases, for sub-national regions. The process involves establishing a representative national standards committee to develop 'indicators' for the FSC's globally agreed principles and criteria. Technically, national indicators need to be complete, consistent, clear, accurate, measurable, non-duplicative and based on the performance of forest managers. Politically, they need to satisfy the various constituencies involved in their development. Designing a national standard to meet these requirements is challenging although the experience of the past decade indicates it can be done, generating widespread consensus among diverse groups on the meaning of 'sustainable forest management'.

Principles and criteria

At the heart of the FSC's certification system are ten principles of well-managed forests (Figure 3.2). Each principle is further elaborated by criteria, which identify the key elements of the principle to be considered during a forest management audit. To measure whether a criterion is met, national standards elaborate indicators of performance for each criterion. Using a

---

Principle 1: Compliance with all applicable laws and international treaties.

Principle 2: Demonstrated and uncontested, clearly defined, long-term land tenure and use rights.

Principle 3: Recognition and respect of indigenous peoples' rights.

Principle 4: Maintenance or enhancement of long-term social and economic well-being of forest workers and local communities and respect of worker's rights in compliance with International Labour Organisation (ILO) conventions.

Principle 5: Equitable use and sharing of benefits derived from the forest.

Principle 6: Reduction of environmental impact of logging activities and maintenance of the ecological functions and integrity of the forest.

Principle 7: Appropriate and continuously updated management plan.

Principle 8: Appropriate monitoring and assessment activities to assess the condition of the forest, management activities and their social and environmental impacts.

Principle 9: Maintenance of High Conservation Value Forests (HCVFs) defined as environmental and social values that are considered to be of outstanding significance or critical importance.

Principle 10: In addition to compliance with all of the above, plantations must contribute to reduce the pressures on and promote the restoration and conservation of natural forests.

Figure 3.2 Overview of the FSC principles and criteria
Source: FSC 2009e.

national standard, a certifying body measures the performance of a forest manager against each of the indicators to determine if the criteria have been achieved and the principle met. Providing all principles have been met to an adequate degree, the CB will certify the operation and the forest manager permitted to use the FSC logo.

An example helps clarify how the system operates. In the current 2004 version of the FSC-US Rocky Mountain Standard (covering territory in the states of Montana, Idaho, Wyoming, Nevada, Colorado, Utah and South Dakota), FSC's Principle 6 states that 'Forest management shall conserve biological diversity and its associated values, water resources, soils, and unique and fragile ecosystems and landscapes and, by so doing, maintain the ecological functions and the integrity of the forest'. Principle 6 is further elaborated by ten criteria that cover a range of forest management practices that include the requirement to assess the environmental impacts of current management practices (Criteria 6.1), ensure the presence of safeguards to protect rare, threatened and endangered species (Criteria 6.2) and secure the development and adoption of non-chemical methods of pest management (Criteria 6.6). Each of these criteria is elaborated in the form of indicators which, in the FSC-US Rocky Mountain Standard, vary from none (Criteria 6.10) to 16 for Criteria 6.5 covering the preparation of written guidelines on soil erosion, road construction and maintenance and water resource protection.

If a forest management operation performs well on a certification audit but is not fully compliant with the relevant standard, a certifying body has the option of granting certification 'with conditions'. This enables the forest manager to gain certification but obliges the company to undertake the necessary changes to management practices within a set time period and subject to a follow-up audit. In cases where the forest management practices are considered deficient, a certifying body can grant certification 'with pre-conditions'. This obliges the company to undertake changes to management practices before it is eligible to be certified.

Certifying bodies

Certifying bodies are at the heart of the FSC's certification system. A forest manager seeking to certify their operation will approach one of several FSC-accredited certifying bodies and request them, for a fee, to conduct a scoping visit. The purpose of the scoping visit is to identify gaps between current management practices and those required under the relevant FSC national standard. Based on the scoping report, the forest manager will decide whether or not to proceed to a full certification audit. If the decision is to proceed, the company will agree a fee with the certifying body, which will then build an audit team and make arrangements to visit the operation to conduct the audit. Different certifying bodies use different scoring systems to audit an operation. While many use checklists to assess each indicator, the scales used can differ and so can the thresholds for determining an 'acceptable' level of performance. In the early years of the FSC forest certification discrepancies emerged in how certifying bodies rated forest management operations, giving rise to disputes over certification decisions.

The FSC requires that certifying bodies engage in meaningful consultation with a wide range of stakeholders when undertaking a certification audit. The FSC also has an elaborate if rather slow-working dispute resolution system that obliges companies, certifying bodies, the FSC-IC secretariat and, if required, the general assembly, to adjudicate complaints. Parties dissatisfied with a certifying body's decision to certify an operation can bring their complaint before the company and, subsequently, the certifying body. If they are dissatisfied with the response, they can launch a formal objection and have their complaint adjudicated at the national or, if still not satisfied, at the international level. While the complaints process remains somewhat cumbersome and time consuming, it nonetheless provides a systematic process for concerned groups to object to the certification of specific operations (Weber 2005). One example of a dispute was SGS's decision to certify Leroy Gabon's operation in Gabon. It centred on whether an adequate forest management plan was in place and an appropriate level of stakeholder consultation had occurred (Eba'a Atyi 2006). Ultimately it was agreed to withdraw Leroy Gabon's certificate.

National standards

A key role for the FSC working groups is the development and revision of national standards. As of late 2009, several countries had developed national standards, some had revised their original standards following the expiry of the five-year period, and a few were even negotiating a third iteration (Table 3.3). The first ever national standard was negotiated by Sweden in 1997, which contained only seven principles. The structure of the FSC's early standards diverged because the FSC's P&Cs were still being revised and no formal template existed on how they should be written. Thus the 1997 Swedish national standard differed significantly from the two standards that followed: the United Kingdom's Wood Assurance Standard (UKWAS) and the FSC-Maritimes Regional Standard for Canada's eastern provinces. By early 2000, however, national standards were conforming more closely to a template developed, monitored and enforced by the FSC-IC's Accreditation Business Unit. Today, efforts are underway to harmonise FSC standards across regions and countries. The US, for example, is in the process of replacing its nine sub-national regional standards with a single national standard.

Accreditation of certifying bodies

The FSC-IC not only develops and maintains the organisation's standard, but through its ASI arm also accredits certifying bodies enabling them to

Table 3.3 Status of FSC national and regional standards, October 2009

Date of first negotiation	Country	Scope of standard
1998 (May)	Sweden	Standard for Forest Management Certification
1999 (Oct)	UK	Standard for Forest Management Certification
1999 (Dec)	Canada	Standard for Forest Management Certification in the Maritime Forest Region
2001 (Sept)	US, Rocky Mountain region	Regional Forest Management Certification Standard
2001 (Nov)	Germany	Standard for Forest Management Certification
2001 (Oct)	Peru	Forest Management Standards for the Production of Brazil Nuts (*Bertholletia Excelsa*)
2002 (Mar)	Bolivia	Standard for Forest Management Certification of Brazil Nut (*Bertholletia Excelsa*)
2002 (May)	Brazil	Standard for Forest Management Certification On 'Terra Firme' In the Brazilian Amazon
2002 (May)	Peru	Standard for Forest Management Certification for timber products in the Amazonian forests.
2002 (Aug)	US (Lake States Central Hardwoods region)	Regional Forest Management Certification Standard
2002 (Nov)	US, south-eastern US	Regional Forest Management Certification Standard
2002 (Nov)	US, northeast region	Regional Forest Management Standard
2003 (Feb)	Columbia	Standard for Forest Management Certification of Natural Forests.
2003 (Jul)	US, southwest region	Regional Forest Management Certification Standard
2003 (Jul)	US, Pacific Coast region	Regional Forest Management Certification Standard
2003 (Oct)	Columbia	Standard for Forest Management Certification of Plantations
2004 (Jul)	US, Ozark Ouachita region	Regional Forest Management Certification Standard
2004 (Jul)	US, Appalachia region	Regional Forest Management Standard
2004 (Aug)	Canada	National Boreal Standard
2004 (n.m.)	Bolivia	Standard for certification of forest management of timber yielding products in the low lands
2005 (Mar)	Denmark	Standard for FSC Certification

(continued)

Table 3.3 Continued

Date of first negotiation	Country	Scope of standard
2005 (Jun)	Netherlands	National Forest Stewardship Council Standard
2005 (Oct)	Canada	Regional Forest Management Certification Standards for British Columbia
2005 (Dec)	Spain	National Forest Stewardship Standard
2006 (Jun)	Finland	National Forest Stewardship Standard
2006 (Aug)	Czech Republic	National Forest Stewardship Standard
2006 (Nov)	Columbia	National Forest Stewardship Standard for Columbia – Guadua (Bamboo)
2008 (Jun)	Luxembourg	FSC Standard
2008 (Nov)	Russia	National Forest Stewardship Council Standard
2008 (Dec)	Papua New Guinea	National Forest Management Standard

Source: FSC, 'FSC-IC Approved Forest Stewardship Standards,' September 2009. Available at http://www.fsc.org/fileadmin/web-data/public/document_center/national_ FSC_standards/FSC_ Approved_FSS_2009-09-24.pdf, accessed October 2009.

audit forest management operations to the FSC standard. This is particularly important given that national standards exist in only a handful of countries. In the absence of an agreed national standard, the FSC permits certifying bodies to use their own company standards which must be based on the FSC's generic principles and criteria. Since the FSC P&Cs are 'high level' requirements, however, the indicators certifying bodies develop can diverge markedly from each other as can the relative weights attached to them. The FSC's accreditation protocols for certifying bodies provide an important mechanism to prevent this divergence from undermining the programme.

A certifying body seeking accreditation will have its application reviewed by the ASI project manager to ensure the company has the professional expertise necessary to conduct forest certification audits and operates 'within the scope of ASI requirements' and within the scope of the FSC scheme. This review involves ASI assembling an audit team to review the company's documents, visiting the applicant's headquarters for discussions with senior management to ensure the necessary staff, resources and processes are in place and, for re-accreditation audits, undertaking field inspections of certified operations.

Under this system, ASI has accredited 22 organisations (Table 3.4). Thirteen of these are permitted to certify forest management operations, with the remainder accredited to issue controlled wood and chain-of-custody certificates. Companies vary in geographical location and date of accreditation. The 'big four' – Soil Association Woodmark, SmartWood, Scientific Certification Systems (SCS) and SGS Qualifor – were certified early in 1995. Some accreditations issued by ASI can be highly specific: for example, DNV is limited to conducting forest management certification in Sweden

Table 3.4 List of accredited FSC certifying bodies

Date first accredited	Company name and location	Score of accreditation
1995 (July)	Soil Association Woodmark (SA), Bristol, UK	Worldwide for Forest Management, FSC Controlled Wood and Chain of Custody certification
1995 (July)	SmartWood, Rainforest Alliance (SW), Richmond, Vermont, US	Worldwide for Forest Management, FSC Controlled Wood and Chain of Custody certification
1995 (July)	Qualifor, SGS South Africa (SGS), Southdale, South Africa	Worldwide for Forest Management, FSC Controlled Wood and Chain of Custody certification
1995 (July)	Scientific Certification Systems (SCS), Emeryville California, United States	Worldwide for Forest Management, FSC Controlled Wood and Chain of Custody certification
1998 (July)	Institut für Marktökologie (IMO), Switzerland	Worldwide for Forest Management, FSC Controlled Wood and Chain of Custody certification*
2000 (May)	BM TRADA Certification Ltd (TT), High Wycombe, Buckinghamshire, UK	Worldwide for Chain of Custody certification
2000 (June)	GFA Consulting Group GmbH (GFA), Hamburg, Germany	Worldwide for Forest Management and Chain of Custody certification
2001 (Jan)	ICILA Srl. (ICILA), Milano, Italy	Worldwide for Chain of Custody certification
2002 (Dec)	Swiss Association for Quality and Management Systems (SQS), Zollikofen, Switzerland	Worldwide for Forest Management and Chain of Custody certification
2002 (Dec)	KPMG Forest Certification Services Inc. (KF), Vancouver, Canada	Worldwide for Forest Management and Chain of Custody certification
2004 (May)	Technological Institute FCBA (earlier called CTBA), Paris, France	Chain of Custody certification, limited to 32 countries with French as official language
2004 (Oct)	CTBA (please refer to Technological Institute FCBA), Certiquality (CQ), Milano, Italy	Worldwide for Chain of Custody certification
2005 (July)	Bureau Veritas Certification (BV) (former BVQI-Eurocertifor), Paris, France	Worldwide for Forest Management and Chain of Custody certification
2005 (Oct)	Control Union Certifications BV (CU) (former SKAL International), Zwolle, The Netherlands	Worldwide for Forest Management and Chain of Custody certification

(*continued*)

Table 3.4   Continued

Date first accredited	Company name and location	Score of accreditation
2007 (Aug)	Det Norske Veritas Certification AB (DNV), Solna, Sweden	Worldwide for Chain of Custody certification and FSC accreditation for Forest Management limited to Sweden
2007 (Nov)	LGA InterCert GmbH (IC), Nürnberg, Germany	Worldwide for Chain of Custody certification
2008 (Aug)	QMI-SAI Global Assurance Services (QMI), Toronto, Ontario, Canada	Worldwide for Forest Management, FSC Controlled Wood and Chain of Custody certification
2008 (Aug)	HolzCert Austria (HCA), Vienna, Austria	Worldwide for Chain of Custody certification
2008 (Nov)	CTIB-TCHN Belgian Institute for Wood Technology (CTIB), Brussels, Belgium	Chain of Custody certification for Belgium, France, Luxembourg and the Netherlands
2009 (Mar)	TÜV Nord Cert GmbH (TÜV), Essen, Germany	Worldwide for Chain of Custody certification
2009 (Mar)	Forest Certification LLC (FC), Bratsk City, Irkutsk Region, Russia	FSC Forest Management and Chain of Custody certification in the CIS countries excluding the certification of SLIMF operations
2009 (July)	Stichting Keuringsbureau Hout (SKH), Wageningen, The Netherlands	Chain of Custody certification for The Netherlands

Note: * Terminated for FSC Forest Management certification in Chile, as of 16th May 2008
Source: ASI, 'Accredited Certification Bodies for the Forest Stewardship Council (FSC) Program', October 2009. Online at http://www.accreditation-services.com/uploads/media/5.3.1_2009-10-02-ASI_Accredited_CBs.pdf, accessed October 2009.

and Forest Certification LLC is restricted to working in Commonwealth of Independent States' countries of the former Soviet Union such as Armenia, Kazakhstan, Russian Federation and the Ukraine.

Chain-of-custody and controlled wood certification

A key requirement of an effective standard is that the goods be tracked through the product chain to the final purchaser. To reassure retailers and consumers that timber from poorly managed forests has been excluded from the commodity chain, the FSC employs a chain-of-custody standard. Companies that transport, process, stock and sell FSC-certified products obtain CoC certification. The CoC standard consists of several elements covering quality management, labelling and supplementary requirements. Under quality management, the FSC requires companies to identify a responsible manager, ensure staff awareness, develop procedures and instructions, undertake training and keep good records. Under the labelling requirements of the FSC's CoC standard, three types of claim are permitted: a 'pure wood' claim, a 'mixed wood' claim and a 'controlled wood' claim.

The different labelling claims are linked to the type of management system a company uses to process its timber. The simplest arrangement is the 'transfer system' where a company receives 100 per cent FSC-certified wood from suppliers and produces products composed of 100 per cent FSC-certified wood, enabling the product to carry the *FSC Pure* label. However, in many operations, wood is sourced from diverse suppliers: some FSC-certified, some certified under another scheme and some uncertified. Companies in this situation must instead base their claims either on the percentage of FSC timber in the product (percentage system) or by only claiming as certified outputs the same volume of timber it received as certified inputs (credit system).

CoC certification interacts with another important FSC standard – the standard for 'controlled wood'. At a minimum, FSC seeks to exclude wood from controversial sources from its supply chain. The danger of contamination occurs when a company mixes certified wood with wood that is certified to another standard or with uncertified wood. The FSC identifies five controversial sources. These sources are wood that has been obtained illegally, by violating indigenous peoples rights, by damaging high conservation value forests, by extensive conversion of natural forests to plantations or by utilising genetically modified trees. To exclude such wood from the FSC's commodity chain, the organisation requires companies to undertake a risk analysis to ensure that all their non-FSC certified and uncertified wood comes from 'low risk' sources. To facilitate companies in this task, the FSC is building an online risk-assessment tool so that companies can quickly check by country and region what the risks are. National initiatives have also been identified as having an important role to play in determining national and sub-national risk levels and where this has been done it prevails over other risk analyses.

The CoC and controlled wood standards described above are those currently in place. They were updated in early 2008 as a consequence of significant difficulties with the previous arrangements. The catalyst for the revision was the certification of Asia Pulp & Paper (APP) to FSC's controlled wood standard in 2007, enabling other companies to purchase wood fibre from APP's 485,000 ha plantation in Indonesia. SGS certified APP under the old standard which focused only on the direct source of the supply and ignored a company's wider forest management activities. Many environmental groups, including WWF-Indonesia, were outraged that APP could claim to be FSC certified under its controlled wood standard because they viewed APP's forest management practices as highly destructive, verging on the illegal and as endangering a range of endemic species including the Sumatran tiger. Following widespread protests from environmental chamber members, the FSC-IC suspended APP's controlled wood certificate and subsequently revised the chain of custody and controlled wood standards to prevent future abuse. It is notable, however, that although the standards have been tightened up on the environmental side, they remain quite permissive on the social side.

FSC competitor schemes

Governments and industry responded to the FSC's establishment by developing their own schemes. One of the first competitor schemes to be established was the US-based Sustainable Forestry Initiative. Announced in 1993 by the American Forestry and Paper Association (AFPA), SFI commenced operating in 1995 as a 'second-party' scheme utilising a combination of management and performance standards (Gale 2002). Once established, APFA made SFI certification a requirement of membership. The initial standard was very weak. It was based on high-level Montreal Process criteria and indicators that were not further elaborated at the management unit level. It also did not oblige operators to obtain third-party certification leading to wide divergences in practices among SFI-certified operations. Following considerable external and internal criticism, AFPA undertook a range of significant changes to the standard to improve performance. Cashore, Auld and Newsom (2004) note several 'conforming strategies' including the development of a more rigorous standard, the requirement that operations be third-party certified, the institutional separation of the SFI from AFPA and its restructuring to reflect a 'chamber-like' structure. Despite the changes, however, SFI remains industry dominated and has struggled for legitimacy due to a lack of support from mainstream environmental and social civil society organisations.

A second competitor scheme was launched by the Canadian Pulp and Paper Association's (CPPA) in 1993. The CPPA established the Canadian Sustainable Forestry Certification Coalition (CSFCC) (Cashore, Auld and Newsom 2004), raised a million dollars and contracted the CSA to develop a 'sustainable forest management' standard based on the ISO's EMS approach. Industry favoured management standards over technical and performance standards because they retained managerial control over all aspects of a standard's development, implementation, evaluation and improvement within the firm. However, as negotiations continued within CSA's Technical Committee on the CSA standard, and in the face of widespread external criticism from environmental, social and indigenous civil society groups, a hybrid standard was ultimately agreed that combined the environmental management and performance approaches together. The emergence and development of the CSA scheme is described in more detail in Chapter 6.

As the FSC competitor schemes were launched in the US, Canada, Finland, Malaysia, Indonesia and elsewhere, efforts commenced to bring them together under a single, global umbrella organisation. Europe led the way in

1999 by founding the Pan-European Forest Certification (PEFC) scheme to accredit nationally developed schemes to a common international standard enabling wood to carry both the national and PEFC logo. In 2002, and following inquiries from non-European schemes to join, PEFC renamed itself the Programme for the Endorsement of Forest Certification and admitted members from non-European countries. As of early 2010, PEFC had 34 members, 24 of which were based in Europe (including Russia and former members of the Soviet Union). Previously independent schemes such as SFI and CSA became members of PEFC in 2005. Today, therefore, there are basically two major certification schemes available to forest managers. While some still prefer one scheme over another, increasingly managers are certifying to both schemes to meet political and market exigencies. The demand for 'dual certification' has also resulted in certifying bodies seeking accreditation under both schemes and offering dual certification at discounted rates to their customers.

Certified forests in comparative perspective

A recent report by the UN Economic Commission for Europe's Timber Committee (UNECE) provides a comprehensive account of the state of the FSC and PEFC certification (UNECE 2009). The report confirms several previously noted trends in forest certification related to imbalances between schemes, regions and countries. Figure 3.3 provides trend data on the

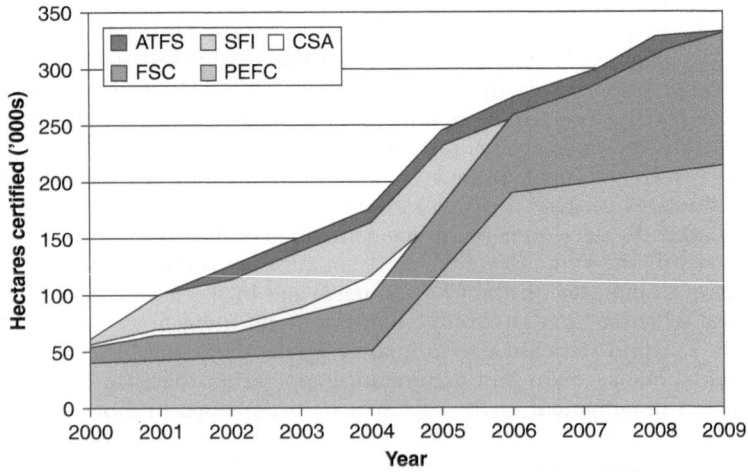

Figure 3.3   Forest area certified by major certification schemes, 2000–9
Source: Adapted from UNECE 2009.

growth of the FSC and PEFC schemes from 2000 to 2009. Several observations are pertinent. First, between 2000 and 2009, previously independent schemes like SFI, CSA and the American Tree Farm System (ATFS) have become members of PEFC and are no longer separately reported. Second, the PEFC scheme grew substantially in 2005 when CSA joined, but then appears to have stagnated at around 200 million ha despite SFI joining in 2006 and ATFS in 2009. In contrast, the FSC certification continued to expand at a steady pace growing from around 85 million ha to 118 million ha between 2006 and 2009, a trend that is made clearer in FSC's own data in Figure 3.4.

It has long been the case that forest certification has been unbalanced across regions, countries and sectors. Figure 3.5 highlights that the vast majority of FSC- and PEFC-certified forests are located in only two regions, Western Europe (approximately 50%) and North America (approximately 38%). The remainder are spread across all other five regions with miniscule amounts of forest area certified in Africa and Asia. The FSC data confirms this broad imbalance in forest certification for its own scheme as outline in Table 3.5.

The regional imbalances in certification are starker at the country level. Total Canadian certified forest area is about 138 million ha, or about 42 per cent of the total certified area of the combined FSC-PEFC total of 325.2 million ha. Total US certified forest area was about 40 million ha (12%), while Finland (20 million ha), Russia (20 million ha) and Sweden (18 million ha) together constituted another 18 per cent of the total. Altogether, these five countries accounted for about 72 per cent of all certified forests,

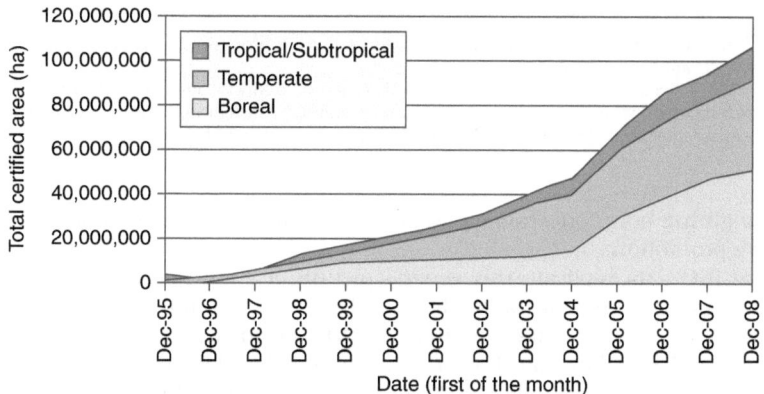

Figure 3.4 FSC-certified forest area growth, 1995–2008
Source: FSC General Assembly, 2008.

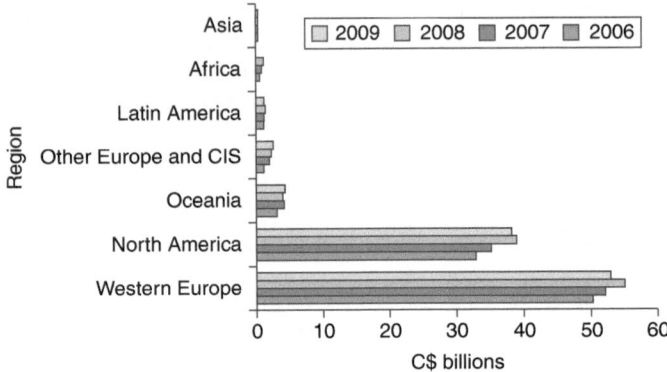

Figure 3.5   Certified forest area as a percentage of total forest area by region, 2006–9
Source: Adapted from UNECE 2009.

Table 3.5   FSC certification by region, September 2009

Region	Forest management certification area (million ha)	Number of CoC certificates
North America	39.903	4,256
Western Europe	28.423	6,921
Russia and CIS	24.814	85
Latin America	10.945	503
Africa	7.095	105
Asia	2.761	2,656
Oceania/Other*	1.659	301
Total	115.6	14,827

Note: * Includes certificates not assigned to a region (6) and countries not in a region (Cyprus (2), Lebanon (1) and Saudi Arabia (1)).
Source: FSC, 'Global FSC Certificates: Type and Distribution', online at http://www.fsc.org/fileadmin/web-data/public/document_center/powerpoints_graphs/facts_figures/09-09-15_Global_FSC_certificates_-_type_and_distribution_-_FINAL.pdf, accessed October 2009.

highlighting how concentrated certification remains after over a decade of active promotion.

The FSC data replicate this general picture of country-level concentration although the rank order differs. The top six countries in terms of hectares of forests certified to the FSC scheme in September 2009 were Canada (27.5 million ha), Russia (21.1 million ha), US (11.6 million ha), Sweden (9.7 million ha), Poland (7 million ha) and Brazil (5.5 million ha). Together these accounted for 50 per cent of all forests certified to the FSC standard highlighting the relative concentration of the FSC certification in developed countries with temperate and boreal forests. Only Brazil is represented among developing, tropically forested countries and major producers of tropical timber such as Indonesia, Malaysia and Papua New Guinea are notably absent. These regional and country-level imbalances are also present in CoC certification. Table 3.5 also identifies significant differences between regions in CoC certification with Western Europe and North America once again dominating. Surprisingly, however, given how little FSC-certified forest exists in the Asian region, is the number of CoC certificates for Asia. This represents a desire of timber importers and manufacturers in Japan, China and Hong Kong to sell products into the environmentally sensitive markets of North America and Europe. Together, these countries and regions account for almost 75 per cent of all Asian CoC certificates.

Although total certified forest area appears to have stabilised in the past three years at approximately 320 million ha, the UNECE report identifies a number of factors that could stimulate future growth. These include increased public concern over climate change, structural changes to the FSC and PEFC that make them more attractive to small operators, government procurement policies and green building initiatives. Concern over global warming has focused government, industry and civil society attention on the potential of forests to sequester and store carbon and mitigate dangerous climate change. Efforts to attract small operators to certification schemes also continue. The FSC has introduced its small- and low-intensity managed forests scheme and group certification. The former relaxes the requirements for a full audit of forest operations, thus reducing costs; the latter enables small operators to band together to obtain certification jointly, also reducing costs.

Governments are also driving certification as a consequence of increased concern to remove illegal timber from the commodity chain. This effort is being led by European countries, with the EU establishing a Forest Law Enforcement Governance and Trade (FLEGT) programme to curb illegal imports. The EU Commission is working with a number of high-risk countries such as Indonesia and Ghana to develop systems to verify the legality of exported timber products. In many countries organisations have been set up to promote 'green building' standards and to reward companies via a points system for their building practices. Schemes include the Green Building Council of Australia, the US Green Building Council's Leadership in Energy and Environmental Development (LEED) programme and the British consultancy BRE's Environmental Assessment Method (BREEAM). As these schemes increasingly enter the mainstream, they are promoting the use of certified timber products in construction design, which is encouraging architects and specifiers to include such requirements in their blueprints.

In summary, then, the FSC has become well established since its foundation in 1993 and continues to grow and adapt to changing market conditions. Over the past 17 years it has evolved into a group of corporate entities – FSC-IC, FSC-GD and ASI – with regional and national initiatives around the world. While that growth is unbalanced, and while the organisation remains dependent on Europe and North America for the majority of its support, it has an established presence in Latin America and is beginning to develop in Africa. It is also beginning to wean itself off donor funding.

Marine Stewardship Council (MSC)

In contrast to the FSC, the MSC was formed later in the decade and built on some of the real and perceived insights of the FSC experience. The MSC celebrated its tenth anniversary as a formal organisation in 2009 and constitutes an initiative to certify the sustainable performance of fisheries on a global scale. It was initiated in 1996 through the joint efforts of WWF-International and Unilever, one of the world's largest consumer product conglomerates that at the time controlled over 20 per cent of the European and US frozen fish markets with global sales of £600 million (Mfodwo 1998). At the outset, Unilever committed to source its fish products from the MSC-certified fisheries, and initially announced that it would achieve this by the target date of 2005 (Fowler 1998), although this commitment was progressively relaxed due to ongoing delays in setting up the MSC and its certification arrangement. At the outset, it was intended that the MSC would 'be an independent, non-profit non-government membership body' modelled on the FSC that would 'create market-led economic incentives for sustainable fishing' (Pellew and Burgmans 1996). The approach built on growing interest in use of ecolabelling in the fishery sector, most notably through the dolphin-safe tuna labels that emerged from campaigns to combat the bycatch of dolphins in the purse seine fishery for yellowfin tuna that was processed into canned tuna (Bergin and Haward 1995; Teisl et al. 2002).

Formation of the MSC

The MSC developed through a shared vision of the long-term viability of fish stocks between a major environmental civil society organisation and a major corporate entity, spurred by concern over the state of the world capture fisheries. The development of the MSC was the result of several factors that coalesced to facilitate its establishment. The 1990s saw increased concern paid to the inexorable rise in wild fish catch in the post-war period and a concomitant concern over the state of the world's fisheries. The increasing number of stocks classified by the FAO as over fished or fully exploited coupled with signs of declines in production gave impetus to a number of initiatives. These included negotiations at the UNCED in 1992 over the United Nations Fish Stocks Agreement (1995), the development of the FAO Code of Conduct for Responsible Fisheries (1995), the Agreement to Promote Compliance with International Conservation and Management Measures by Fishing Vessels on the High Seas (1994, known as the Compliance Agreement) and the extension of the Convention on Biological Diversity to marine areas and species through that convention's 'Jakarta Mandate' in 1995 (Haward and Vince 2008). In addition to these global and regional initiatives, states also increasingly focused on conserving fish stocks within areas of national jurisdiction. The 1970s and 1980s can therefore be characterised as a period of coastal states seeking to extend their sovereign rights over larger areas of ocean jurisdiction through the provisions of the LOSC. In the 1990s, this effort to control fisheries was accompanied by a rise in the rhetoric, if not the reality, of states' 'ecosystem responsibilities' (Haward and Vince 2008; see also Miles 1997).

The 1990s also saw a rise in new governance initiatives and increasing links between business and environmental organisations. These aimed to generate the multiple benefits of a 'triple bottom line' approach to production and to achieve 'sustainable development'. While such partnerships sometimes were more symbolic than substantive, resulting in claims of 'greenwashing', closer links between business, environmental organisations and government provided opportunities for a new political economy of resource management to emerge. Governments, too, continued to shift away from reliance on bureaucratic and regulatory mechanisms, and looked to market-based approaches and economic instruments as management tools. The perception of state-based regulatory failure in fisheries in the 1990s, most notably with the collapse of the northern cod stocks off Newfoundland in Canada but also with declines in fisheries such as the Southern Bluefin Tuna in waters around Australia, contributed to calls for new approaches to fisheries management. The closure of the Atlantic cod fishery resonated around the globe. The cod crisis was caused by a range of factors: overfishing and underreporting of domestic catches, encroachment of large trawlers from European states, a failure to model the 'sustained-yield' and changes in environmental conditions. Environmental organisations were, however, able to focus on the outcome as indicative of the crises in the world's fisheries. Others saw this collapse as a 'crime story' (Harris 1998). Thus the MSC was influenced by a range of factors, including of course the ongoing experiences of the FSC.

The MSC experience has attracted a large and growing literature, with Constance and Bonnano (2000) and Gulbrandsen (2009) providing useful outlines of the MSC's development. WWF and Unilever announced their partnership in August 1996 committing to provide seed funding to establish the MSC for two years. WWF fisheries campaigners had tracked the development of the FSC and its certification model and looked to create a similar model for fisheries (Gulbrandsen 2009, 655). At the same time Unilever was also examining the FSC approach (Gulbrandsen 2009, 655) and considering the merits of third party product certification. The 1996

'Statement of Intent' outlined the form and function for the new organisation (Pellew and Burgmans 1996), noting that the MSC would

> [e]stablish a broad set of principles for sustainable fishing and set standards for individual fisheries. Only fisheries meeting these standards will be eligible for independent, accredited certifying firms. Products from certified fisheries will eventually be marked with an on-pack logo. This will allow consumers to select those fish products which come from a sustainable source.
>
> (Pellew and Burgmans 1996, 16)

The MSC was established after a series of consultations among different groups following a process developed by an international consultancy firm, RESOLVE. RESOLVE operated independently of the WWF and Unilever, an arrangement designed to overcome perceptions from potential participants and stakeholders that the consultation process might be biased if conducted by WWF and/or Unilever (RESOLVE 2007). It was believed that the 'use of a neutral [party] helped ensure that participants felt the workshops were conducted in a manner that provided for the free and open exchange of perspectives' (RESOLVE 2007).

The institutionalisation of the MSC began at a workshop in September 1996 at Bagshot, UK, followed by presentations at fishery-related meetings in Seattle, Seoul, London, Montreal, Honolulu and New Zealand and to the World Bank's environment division in the following months (Constance and Bonnano 2000, 130). After these initial consultations, the next stage was 'a search for a project manager and board chair' (Constance and Bonnano 2000 130). In January 1997 Carl-Christian Schmidt was appointed project manager. Schmidt, of the Organisation of Economic Cooperation and Development (OECD), had extensive experience in fisheries management and administration, including employment in the Danish Ministry of Fisheries prior to joining the OECD (Constance and Bonnano 2000, 130).

Following an announcement by Sainsbury's, a UK food retailer that had heavily backed the development of the FSC, of its support for the MSC in April 1997 the MSC worked to publicise its initiative through 'letters of support' from private sector and civil society organisations. This led to UK retailers Tesco and Safeway following Sainsbury's in supporting the MSC (Constance and Bonnano 2000). In a deliberate decision the MSC departed from the FSC's membership, chamber-based FSC approach and adopted a trust-based organisational arrangement run by an expert-based, self-appointed board. In March 1998, an International Board of Trustees was established. It immediately confronted a funding challenge as in June 1998 full independence occurred when the founding partners' seed funding expired and the MSC was forced to seek funds from a range of organisations, trusts and charities (Gulbrandsen 2009, 655). The MSC's independence from

WWF-Unilever funding served to increase and sharpen external criticisms of the organisation's governance arrangements.

The creation of the MSC met with initial scepticism and concern from fisheries managers, fishing industry associations and environmental organisations other than WWF (Potts and Haward 2001). The concerns of managers and industry associations focused on the degree of consultation over, and scope of, the MSC principles and criteria and the role of an ECSO, WWF International, in evaluating fisheries management systems (Fowler, 1998). Fisher organisations in developing countries articulated similar concerns in an evocative way, describing the collaboration between Unilever and WWF and a 'case of one giant riding on top of another'. The resulting behemoth can either make deep impression [*sic*] on the path it traverses or stumble and crash for lack of balance' (Kurien 1996, 22). This critique highlighted a potential dichotomy between the aspirations of consumers in the first world and the needs of producers in developing countries (Kurien 1996 22).

ECSOs were concerned that Unilever, which committed early on to buy only MSC-certified fish by 2005, could 'water-down' certification standards to maintain supply. Developing country critiques, on the other hand, saw the use of the MSC label as a potential barrier to trade, restricting exports of what were 'important components of easily exportable commodities' (Kurien 1996, 25). This concern over the potentially trade-restrictive effects of ecolabels was enhanced by the introduction by the US and Europe of tougher food safety regulation and the increased use of sanitary and phytosanitary measures in the trade of fisheries products (Bache, Haward and Dovers 2001). Other actors denigrated the role of WWF in setting standards, with Scandinavian countries especially critical of the ECSO's involvement. The Nordic Council of Ministers, for example, developed 'a project group' to 'assess standards for sustainable fish production' in direct opposition to the MSC (Gulbrandsen 2009, 657). Norway's criticisms were also evident at a meeting of the FAO's Committee on Fisheries (FAO-COFI) Sub-Committee of Fish Trade. In June 1998, Constance and Bonnano (2000, 132) reported that 'eco-labeling of fisheries stocks is an issue for fisheries authorities, not NGOs such as WWF and Unilever. All the countries present at the meeting supported the Norwegian initiative'.

While this criticism of the MSC moderated somewhat as fisheries became certified under the scheme after 2000 and as the MSC itself evolved, concerns over the influence of non-state instruments persisted, as shown in regular discussions within FAO forums. The FAO, as the United Nations specialist organisation with responsibility for fisheries, has been a major forum for debate about ecolabelling in fisheries. The Nordic countries supported the FAO's work on ecolabelling, and as a result the bi-annual meetings of the FAO-COFI, the organisation's peak body for fisheries, discussed ecolabelling in 1997 and 1999 (Gulbrandsen 2009, 655). At the 1997 FAO-COFI meeting many states were concerned over the impact of the MSC but there was 'no consensus at that time that FAO should address the subject' (Willman, Cochran and Emerson 2008, 60). Following further work, and discussion at the meeting of the FAO-COFI's Sub-Committee of Fish Trade in June 1998, the committee endorsed Norway's proposal that the FAO 'organise a technical consultation to investigate the feasibility and practicability of developing non-discriminatory, globally applicable, technical guidelines for ecolabelling of fish and fishery products from marine fisheries' (Willman, Cochran and Emerson 2008, 60).

A technical consultation was held in October 1998 but did not reach agreement 'on the practicality and feasibility of FAO drafting technical guidelines'. It did conclude that any agreement eventually reached on these guidelines should be consistent with the FAO's Code of Conduct for Responsible Fisheries. The Code of Conduct for Responsible Fisheries encourages states and fishers to use selective and environmentally safe fishing gear and practices (Haward and Vince 2008, 41). The code is a voluntary, soft law instrument that is directly linked to 'relevant rules of international law', including the LOSC (Bergin and Haward 1995). It provides opportunities for the development of subsidiary, specialist, instruments, such as International Plans of Action, that are key elements of an emerging regime governing high seas fisheries. The code contains six thematic areas or chapters for which guidelines should be developed: (1) fishery management practices; (2) fishing operations; (3) aquaculture development; (4) integrating of fisheries into coastal area management; (5) post harvest practices and trade; and (6) fishery research (Haward and Vince 2008).

The outcome of the technical consultation was discussed at the next meeting of FAO-COFI in February 1999. However, the meeting was divided on future directions, concluding with the recommendation that 'FAO should not get involved in the issues and should leave it to other specialist bodies', a position maintained at the 2001 FAO-COFI meeting (Willman, Cochran and Emerson 2008, 61). By 2001, however, the MSC had elaborated and better publicised its assessment processes reducing some of the concern of fisheries organisations; and the issue of certification and labelling had also been placed on the agenda of the WTO's Committee on Trade and Environment (Willman, Cochran and Emerson 2008). Thus, at the February 2002 meeting of the Sub-Committee on Fish Trade, the issue of ecolabelling was again considered, this time more sympathetically, leading to a change in position at the 2003 meeting of FAO-COFI. At this meeting, the discussion of ecolabelling moved forward and the FAO was invited to host 'expert and technical consultations', subsequently releasing FAO guidelines in 2005 (Gulbrandsen 2009, 655).

MSC governance: The political dimension

Unlike the FSC, the MSC was initially founded with the participation of two major actors – WWF-International and Unilever – as a 'technical' organisation that aimed to depoliticise sustainable fisheries management. Thus, unlike the FSC's sophisticated membership and chamber-based representation system, the MSC established itself as a trust run by a self-appointed board with trustees who initially had broad environmental interests but were highly Euro-, if not UK-, centric. The initial trustees included the Rt Hon John Gummer (former UK Secretary of State for the Environment), Yannis Paleocrassas (former EC Commissioner for Environment and Fisheries), Henrique Cavalcanti (former Brazilian Minister for the Environment and former Chair of the United Nations Commission on Sustainable Development), Sir Martin Laing (former Chair of WWF International) and Shaun Woodward (a member of the UK Parliament). In 1999 the limitations of such a 'narrow' board of trustees was addressed with the creation of the Senior Advisors Group, made up of over 30 members from a range of stakeholder groups. The Senior Advisors Group provided impetus for more formal stakeholder representation within the MSC.

Despite the expression of early general support for the aims and aspirations of the MSC, the lack of formal broad stakeholder representation led to mounting criticism with Tully (2004, 3) noting that the organisation viewed the 'inclusive participation of all interested parties' as not essential. This criticism was matched by critiques of the lack of transparency in the MSC's decision-making arrangements (Gulbrandsen 2009, 655). 'The MSCs governance structure attracted NGO criticism as lacking due credibility, democratic representivity [sic] and effectiveness' (Tully 2004, 3). The Board of Trustees expanded to 14 members in January 2003 with a concomitant increase in the board's expertise. This expanded board also included representatives of the MSC's Technical Board and Stakeholder Council. The board also included Sir Tipene O'Regan from New Zealand and Wicharn Sirichai-Ekawat from Thailand.

Although the broadening of the Board of Trustees addressed some criticisms, the MSC found itself undergoing ongoing reform of its governance structures to enhance participation, accountability and efficiency. Discussions within the Stakeholder Council (MSC 2004b) have focused on improving its role and responsibility in the context of decision making with the MSC Board and Technical Advisory Board. One concern was to improve delegation of authority to the Stakeholder Council and to provide a vehicle that provided adequate representation of the different MSC constituencies. Another area was improving interaction between stakeholders and the MSC board as well as enhancing the links between the board and the broader Stakeholder Council. In response the MSC undertook a review of its governance processes and functions in 2000.

The aim of the review was to move away from inefficiency and increasingly expensive operations. A key goal was to increase transparency and accountability while maintaining financial efficiency. The governance review saw a shift 'to a more transparent governance structure' (Gulbrandsen 2009, 655), detailed later. It is important to note that the MSC has continued to review its performance, both in relation to institutional structures as well as processes. The result of the governance review was the multi-stakeholder approach (Cummins 2004) as the basis for the MSC's operations (MSC 2000).

The MSC's governance was also addressed in the Wildhavens report (*An Independent Assessment of the Marine Stewardship Council* (Wildhavens 2004)), prepared for The Homeland Foundation, the Oak Foundation and the Pew Charitable Trusts. The review 'looked at MSC sustainable fishing standards and how they are applied as well as MSC's performance as a consortium organization of stakeholders' (Wildhavens 2004, i). The review was initiated in response to concerns over the hoki and toothfish certifications, which resulted in environmental groups pressuring the foundations supporting the MSC to reconsider their involvement (*Guardian* 21 February 2004). While attacks on the process of certification identified specific fisheries, criticism was also directed at the management of the MSC.

The Wildhavens Report concluded that 'overall we find that the MSC is not viewed as credible by most of the marine conservation community and conclude that MSC risks failure if it does not seek more balanced support among stakeholders' (Wildhavens 2004 iv). The review made a series of recommendations 'in order to restore its credibility' that included reforms to management processes, improving transparency of decision making in accreditation and rebuilding stakeholder confidences (Wildhavens 2004, iv). In response John Gummer, chair of the Board of Trustees, noted that there were 'difficulties of operating a certification systems acceptable to industry, governments and the environment movement', and that 'some of the criticisms were from a particular American viewpoint which took an absolutist view of what was sustainable' (*Guardian* 21 February 2004).

MSC governance: The institutional dimension

The MSC's institutional arrangements are set out in Figure 3.6. As noted earlier, the MSC's decision making and governance arrangements differ from those of the FSC, and reflect deliberate decisions not to build a 'membership'-based organisation. The Board of Trustees is the executive decision-making body within the MSC. The board has a maximum of 15 members and is the final decision-making authority in terms of the technical, scientific and quasi-judicial organisational functions (FAO 2003). The composition of the board includes the chairman, the chairman of the Technical Advisory Board

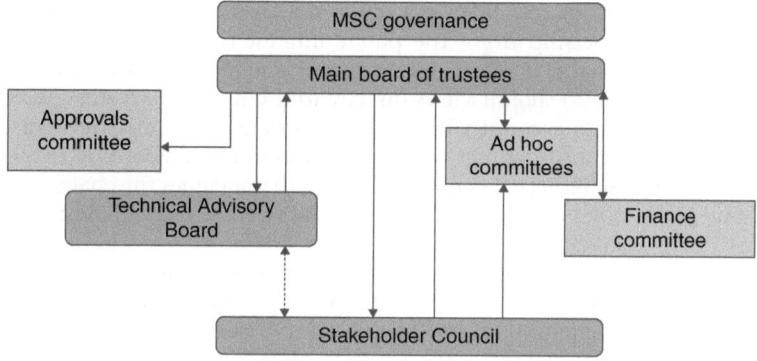

Figure 3.6   Organisational structure of the MSC

and the two joint chairmen of the Stakeholder Council (MSC 2001a). The Board of Trustees is responsible for:

- Ensuring the MSC meets its charitable aims;
- Approving and implementing the strategic direction of the MSC;
- Ensuring that the MSC is financially secure;
- Appointing new board members and key MSC staff;
- Appointing members of the Technical Advisory Board; and
- Publicly accounting for expenditure and income (MSC 2009a).

The Technical Advisory Board (TAB) has 15 members and advises the Board of Trustees on matters that include the setting and review of the MSC standard, logo licenses and CoC certification (MSC 2002). The Board of Trustees is responsible for appointing the TAB while the TAB is responsible for appointing its own chair (MSC 2009a). The TAB is responsible for:

- Advising the MSC board on the MSC standards;
- Maintaining documents relating to the MSC standards;
- Developing methodologies for certification and accreditation; and
- Reviewing the progress of fisheries certifications and providing advice on these (MSC 2009a).

The Stakeholder Council is a body of 30–50 members made up of diverse interests including conservation, industry, academic and developing nations and currently two co-chairs (FAO 2003; MSC 2006). The Stakeholder Council has three categories of members: a public interest category; a commercial and socio-economic category; and a developing world category (MSC 2009a). The council 'meets annually to discuss MSC strategy, activities and other matters' and, drawing its membership from 'the public interest category' (Gulbrandsen 2008, 16) fulfils the role of a participatory forum and representative authority. The council is able to submit views directly to the Board of Trustees, which must take these into account when arriving at decisions. Critics recognised the benefit of the 'broadening' of 'representation on the Board of Trustees from an entirely self-selected committee to one that includes the chairs of the TAB and Stakeholder Council' (Wildhavens 2004 34). As Gulbrandsen noted

> [i]n sum the governance reform resulted in an inclusive multi-stakeholder governance structure, but in order to avoid the inertia and inefficiency sometimes experienced in the membership-based FSC program it left ultimate decision-making authority to the Board of Trustees rather than the Stakeholder Council.
>
> (Gulbrandsen 2009, 655)

Despite the generally positive response to the governance review by governments and industry (Gulbrandsen 2008, 16) the MSC was subject to ongoing criticism over the implementation of its standard, the certification process and the engagement with stakeholders. It is inevitable that conflict and controversy will manifest within the process of certifying fisheries. The University of British Columbia's prestigious Fisheries Centre, initially a supporter of the MSC, removed its support in 2001 over concerns related to a lack of transparency in certification (Pitcher 2001).

Additional governance reform has aimed to improve stakeholder involvement as well as revisions to ensure maintenance of the MSC standard. The MSC admitted to earlier problems with certification that, for example, led to 'blanket approval occurr[ing] with the Alaskan salmon fishery' (Howes 2005). It is a moot point to consider whether a different institutional structure or membership-based model would address these criticisms. In response the MSC initiated internally driven review projects focused on 'monitoring and evaluating the environmental benefits of the MSC certification program' (Agnew et al. 2006), and reviewing the implementation of the MSC standard through the 'Quality and Consistency Project'.

MSC governance: The regulatory dimension

The heart of the MSC process is the certification of 'sustainable fisheries' as defined by meeting a standard set by what are termed the Principles and Criteria for Sustainable Fishing (MSC 1998). The principles and criteria – the MSC standard – and the processes of assessment of fisheries are central elements in the operation and governance of the MSC. Not surprisingly it is these elements that have been at the centre of the MSC development, but also those that have attracted the most criticism. The MSC states that 'it exists to change the way the seas are fished. We seek to reverse the decline of fish stocks and safeguard livelihoods and deliver improvements in marine conservation worldwide' (MSC 2006, 1).

The management of the certification process and the standard are the core functions of the MSC. It is important to note that, as with the FSC, the MSC does not directly perform the certification. To remain independent, the MSC accredits certifying bodies and trains them in the MSC's methodology (MSC 2001a). A list of MSC-accredited certifiers is set out in Table 3.6 and the process adopted to certify a fishery is depicted in Figure 3.7. An MSC certificate applies to the fishery and harvesting operation up until the catch is landed (MSC 2001b).

Principles and criteria

One of the MSC's first tasks was to develop a standard for certifying the sustainability of fisheries. In 1998 the organisation released a set of principles and criteria that followed an extensive 18-month worldwide consultation period and drafting process, concluding with a workshop at Airlie House, Washington DC, US – the 'Airlie House Draft' – 'the MSC Principles and Criteria were handed to the Secretariat of the MSC for them to mould into an assessment standard' (MSC 1998, 2). The principles apply to any fishery with the size, scale, type, location, intensity of the fishery, the resources and ecosystem effects considered in every certification. This includes the

Table 3.6 MSC certifying bodies

Fully accredited certifiers*

- Det Norske Veritas Certification AS
- Food Certification International Ltd (FCI)
- Moody Marine Ltd
- Organización Internacional Agropecuaria (OIA)
- Scientific Certification Systems
- SGS Product and Process Certification
- Tavel Certification Inc.

*These certifiers are accredited by Accreditation Services International GmbH (ASI) to certify fisheries that meet the MSC environmental standard for sustainable fishing.

Certifiers undergoing accreditation*

- Bureau Veritas Certification
- Global Trust Certifications Ltd (Previously I: FQC Ltd)
- Institute for Marketecology (IMO)
- MacAlister Elliott & Partners Ltd
- MRAG Americas

Note: * As of October 2009. These certifiers have applied to ASI for MSC accreditation. 'During, and as part of their accreditation application, they can conduct fishery assessments, which will be evaluated by ASI. Certifiers may not issue certificates until MSC accreditation is achieved. Inclusion on this list implies no endorsement of these organisations, nor of any fisheries operations certified by them' (MSC 2009b).

Principle 1: A fishery must be conducted in a manner that does not lead to over-fishing or depletion of the exploited populations and, for those populations that are depleted the fishery must be conducted in a manner that demonstrably leads to their recovery.

Principle 2: Fishing operations should allow for the maintenance of the structure, productivity, function and diversity of the ecosystem (including habitat and associated dependent and ecologically related species) on which the fishery depends.

Principle 3: The fishery is subject to an effective management system that respects local, national and international laws and standards and incorporates institutional and operational frameworks that require use of the resource to be responsible and sustainable.

Figure 3.7   MSC principles
Source: MSC 1998.

fisheries of developed and developing nations and an approach that does not discriminate towards market access. The MSC's generic principles and criteria cover target and ecological considerations as well as management and governance issues, but unlike the FSC principles, do not have a social focus; that is they do not address indigenous peoples' interests, social well-being more generally, fishery communities or the notion of equity within a fishery. The three principles and 23 criteria form the basis for the elaboration of indicators and completion of a scoring guide.

The MSC has devised a multi-criteria based scoring method that the certifier uses to assess the fishery against the MSC standard using an Analytical Hierarchy Process (AHP) decision tool. This tool aids in weighting, scoring and evaluating performance of a fishery against the standard (MSC 2002b). A means of hierarchical subdivision is employed in translating the principles and criteria into indicators and performance scores. In this process, the certifier moves from the generic principles and criteria to specific scoring benchmarks. For every fishery that is assessed, the hierarchy follows:

- The generic MSC principles;
- The generic MSC criteria;
- Operational sub-criteria: Developed by the certification team as representative of the MSC criteria for the fishery being certified;
- Indicator sub-criteria: The level where scoring is conducted for each fishery;
- Elements: The specific scoring guides for each indicator sub-criterion.

From the generic MSC principles and criteria a series of operational criteria are developed for the fishery based on specific fishery issues. From each operational criterion, several indicator criteria are established. These relate to specific operational issues within the fishery that can be scored by the certification team. Using the AHP pair-wise comparison approach, weighting is conducted at the criteria, sub-criteria and indicator sub-criteria levels according to their importance in the fishery. The weights are developed by the certification team, and justified in their report.

The next step is to establish scores for each indicator criterion on a scale of 0–100. For each indicator, the certifier establishes a set of 'performance guideposts' that must be endorsed by the MSC. Performance on each indicator is then judged against a 100, 80 or 60 guidepost. As defined by the MSC (MSC 2002b):

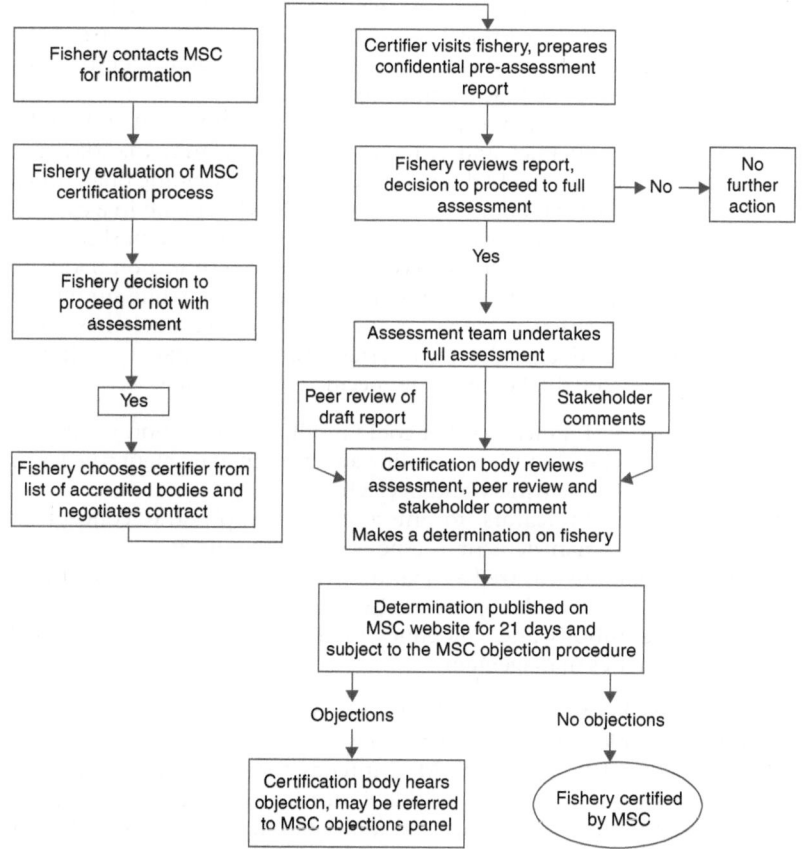

Figure 3.8 MSC fishery assessment process

- A score of 100 denotes the ideal performance;
- A score of 80 defines the minimal requirement for unconditional certification; and
- A score of 60 defines the minimal requirements for conditional certification.

Each indicator is aggregated. Weighted scores are presented for the operational criteria, MSC criteria and each principle. For a fishery to be certified it must obtain a score of 80 for each principle and at least 60 for each MSC criterion (MSC 2001a). Criteria scores of less than 80 can be balanced by scores of greater than 80. Scoring less than 60 on any criterion will fail certification. An advantage of the approach is for the improvement in fisheries management over time. Wherever a score is less than 80 (but >60) in the hierarchy, it indicates performance is deficient and allows specific corrective actions to be set for certification. A conditional certification may be awarded if the fishery adopts the actions (MSC 2001a). Figure 3.8 provides a schematic overview of the MSC certification process.

Another challenging area of the MSC reform is the refinement and implementation of the principles and criteria. Discussion within the MSC has revolved around 'setting the bar' for certification and providing operational guidelines for certification bodies to interpret the MSC standard in a consistent manner, a source of controversy in several assessments. Establishing a set of certification guidelines must preserve the ability of certification bodies to adapt the MSC standard to the specific fishery conditions while simultaneously providing consistency across assessments. Proposed outcomes range from a set of qualitative guidelines to establishing quantitative indicators and metrics (MSC 2004b). Some of the specific issues include:

- Identifying critical indicators that could result in certification failure, for example, the failure to follow scientific advice in management, the levels of IUU fishing in the fishery and bycatch levels;
- Guidelines and indicators to interpret and apply the concept of ecosystem-based management as espoused in Principle 2;
- The use of qualitative as well as quantitative information in the certification process as well as data-poor but information-rich fisheries; and
- Improved definitions and guidelines for interpreting Principle 3 as it relates to fisheries management.

This work has led to considerable effort by the MSC to clarify the intention of the principles and criteria through what was termed the 'Quality and Consistency [Q&C] Project'. 'The MSC intends that the Q&C project will form the basis for a practical "guidebook" for its fishery assessment process, which will improve the clarity and understanding of what MSC fishery certification entails' (MSC 2007a). In addition the MSC has focused on the problems of assessing 'data poor fisheries', those fisheries that lack long time series catch data or lack detailed management arrangements in place. This work focuses on using risk assessment approaches in situations where data is limited and opens up opportunities to extend MSC certification to fisheries in developing states. The fishery assessment and certification process is at the core of the MSC's work. It is also the most controversial, and has been intertwined with calls for reform and change to both structures and processes within the MSC.

Certifying bodies

In the formal assessment, the certifying body's audit team translates the principles and criteria into a set of indicators and scoring guides based on the conditions in the client fishery. As an accreditation body, the MSC must be impartial and independent. The MSC fisheries are certified by an accredited certifier (Table 3.6). It is noteworthy that there are relatively few certifiers accredited, and that these bodies are located in either Europe or North America. As with the FSC, fishing interests seeking to gain MSC certification approach a certifier. Once the certifying body's audit team has completed its draft report, it is sent to the client for review. A Certification Report and Public Certification Summary are prepared in the last stage. If the fishery passes the scoring guideposts and is determined to meet the principles and criteria it can be awarded the MSC label. Official certification takes place when the summary has been received by the MSC for publication on the website. Conflict resolution is an essential part of the certification process and the MSC has developed a revised Objections Procedure to cater for stakeholder input into certification decisions. A key element of the procedure states that a fishery cannot receive the ecolabel until all objections have been fully considered.

Further annual auditing is undertaken by the assessment team to monitor compliance in the fishery with respect to any specific conditions set. The certification team monitors the corrective action requests over time and secures compliance via a process of continual improvement. There are, however, examples of MSC-certified fisheries experiencing compliance difficulties: the bycatch in the Hoki fishery has been an ongoing cause of controversy. However, compliance audits can occur at any time and can occur without any prior notice and certification can be revoked if the conditions set are not met. This approach is in line with the MSC philosophy of continual improvement within fisheries systems.

Fisheries standards

The certification process includes phases of identification and pre-assessment, formal assessment and monitoring. The process is designed to facilitate not only the immediate certification of a fishery but develop ongoing compliance measures and improvements over time. The first phase involves client identification and a feasibility review. The review introduces the principles and criteria and defines the unit of certification. The MSC has determined that the best unit of certification is the operational scale, with the stock as a biologically distinct unit combined with the fishing method and gear (MSC 2001a). Identifying the fishing stock as a discrete unit can be problematic for many fisheries due to a lack of information on life-history characteristics but more fundamentally may reduce consideration of ecosystem impacts, despite the provisions contained in the MSC Principle 2. During certification, the selected scale can influence the assessment due to influences that occur outside the fishery, for example the effects of IUU fishing, economic and trade pressures and environmental influences. These issues have been significant in the certification process for several fisheries, in particular the South Georgian Patagonian Toothfish fishery. The decision to seek certification led to a debate on the effects of IUU fishing affecting the health of the certified stock (ASOC 2001) and how this problem affected the validity of the certification (discussed later). The approach used in the Alaskan salmon fishery that treated the fishery as a single entity failed for Canadian fishers in particular to identify pressure points affecting the fishery.

The reform of the MSC standard and its application to fisheries are critical for the MSC's continued success and expansion. How the standard is applied to a fishery during certification and how fisheries from differing socio-economic and geographical regions can access the programme are ongoing issues. Reform of the principles and criteria must not exclude fisheries from the certification process or provide a technical barrier to trade regardless of size, scale, geography or technological development. On the other hand, the application of the principles and criteria within the certification process must be rigorous and consistent.

Accreditation of certifying bodies and CoC certification

As with the FSC, the MSC-certifying bodies are accredited by Accreditation Services International GmbH, an independent organisation established in Germany. ASI is a limited liability company with the FSC being its sole shareholder. With respect to the MSC, its mandate is to accredit fisheries certifying bodies to undertake assessments. With respect to CoC, and similar to the FSC, any organisation that processes, wholesales or retails the MSC-certified product must be licensed through a 'chain of custody certification'. The purpose of CoC is to ensure that the certified product can be identified throughout its lifecycle (MSC 2004a). The MSC has seen a rapid increase in products with CoC certification, and a ten-fold increase in number of companies certified between 2005 and 2008 (see Figure 3.9). CoC is also important in the catch sector. *The Economist* (23 June 2007, 52) reported a problem with MSC-certified Pollock found on the fish carrier vessel *Polestar* that had been blacklisted for illegal fishing activity by European counties.

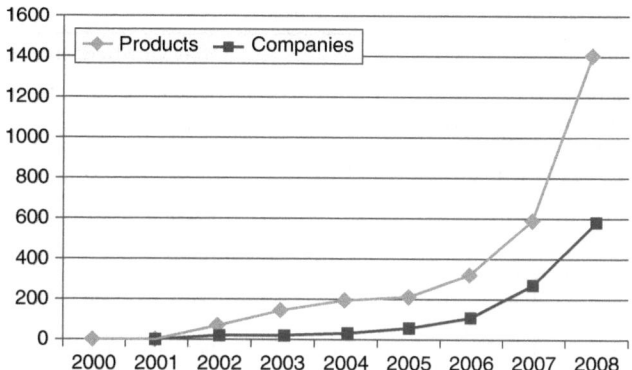

Figure 3.9   MSC products and companies with MSC CoC certification
Source: MSC Annual Reports.

Certified fisheries

The MSC is increasingly attracting a range of fisheries to its programme, with 64 fisheries assessed and 132 fisheries undergoing assessment as of March 2010. There has been a marked increase in the number of fisheries certified in the five-year period (2005–10), but the dominance of North-West Europe fisheries continues. Table 3.7 shows that there are only four fisheries certified in Asia, one in the Iberian Peninsula and none in the Mediterranean. While the vast majority of the certified fisheries are small-scale the MSC programme includes major fisheries such as the Western Australian rock lobster fishery (at the time Australia's most valuable fishery), the New Zealand hoki fishery and the pollock fishery in the Bering Sea and Aleutian Islands and Gulf of Alaska, one of the world's largest fisheries. It is clear that, like the FSC, the MSC is oriented towards developed economies, and has had little practical impact on developing states.

The MSC's certification process and the certification of key fisheries have been criticised. An investigation into the MSC-certified New Zealand hoki, Alaskan salmon, South Georgia toothfish and Aleutian Islands pollock fisheries found that 'none was flawless' (*Guardian* 21 February 2004). The New Zealand hoki fishery has been subject to a 'complaint with the certifier about the certification [on] 25 March 2001 requesting that certification be withdrawn [because the assessment] fails adequately to interpret and comply with MSC Principles and Criteria' (Short 2003, 159). Critics argued that the hoki fishery failed to comply with New Zealand legislation related to avoiding adverse affects on the marine environment (*Guardian* 21 February 2004). The certification of the South Georgia Toothfish fishery

Table 3.7 MSC-certified fisheries, March 2010

Fishery	Date certified/date recertified
Alaska salmon	September 2000/2007
American Albacore Fishing Association Pacific albacore tuna – north	August 2007
American Albacore Fishing Association Pacific albacore tuna – south	August 2007
Astrid Fiske North Sea herring	June 2008
Atlantic deep sea red crab	September 2009
Australia mackerel icefish	March 2006
Bering Sea and Aleutian Islands Alaska (Pacific) cod – freezer longline	February 2006
Bering Sea and Aleutian Islands Pacific cod	January 2010
Bering Sea/Aleutian Islands pollock	February 2005/January 2009
Burry Inlet cockles	April 2001/February 2007
Canada northern prawn	August 2008
Canada Pacific halibut (BC)	September 2009
Danish Pelagic Producers Organisation North East Atlantic mackerel	July 2009
Danish Pelagic Producers Organisation Atlanto Scandian herring	July 2009
Danish Pelagic Producers Organisation North Sea herring	June 2009
Denmark blue shell mussel	January 2010
Domstein Longliner Partners North East Arctic cod	February 2009
Domstein Longliner Partners North East Arctic haddock	February 2009
Dutch Fisheries Organisation (DFO) gill net sole	November 2009
Ekofish Group-North Sea twin rigged otter trawl plaice	June 2009
Faroese Pelagic Organization (FPO) Atlanto-Scandian herring	March 2010
Germany North Sea saithe trawl	October 2008
Gulf of Alaska Pacific cod	January 2010
Gulf of Alaska pollock	April 2005/January 2009
Gulf of St. Lawrence northern shrimp trawl fishery Esquiman Channel	March 2009
Gulf of St. Lawrence northern shrimp	September 2008
Hastings fleet Dover sole (trammel net)	September 2005
Hastings fleet Dover sole trawl and gill-net	July 209
Hastings fleet pelagic herring and mackerel (two fisheries)	August 2008
Irish Pelagic Sustainability Group (IPSG) western mackerel pelagic trawl	August 2009
Iturup Island pink and chum salmon	September 2009
Kyoto Danish Seine Fishery Federation snow crab and flathead flounder (two fisheries)	September 2008

(*continued*)

Table 3.7 Continued

Fishery	Date certified/date recertified
Lake Hjalmaren pikeperch fish-trap	August 2006
Lake Hjalmaren pikeperch gill-net	August 2006
Lakes and Coorong, South Australia	June 2008
Loch Torridon nephrops creel fishery	January 2003/July 2008
New Zealand hoki	March 2001/October 2007
North East Atlantic mackerel pelagic trawl, purse-seine and handline	April 2007
North Eastern Sea Fisheries Committee sea bass	December 2007
Norway North East Arctic saithe	June 2008
Norway North Sea and Skagerrak herring	April 2009
Norway North Sea saithe	June 2008
Norway spring spawning herring	April 2009
Oregon pink shrimp	December 2007
Pacific hake mid-water trawl	October 2009
Patagonian scallop	December 2006
Pelagic Freezer-Trawler Association North East Atlantic mackerel pelagic trawl	July 2009
Pelagic Freezer-Trawler Association North Sea herring	May 2006
Portugal sardine purse seine	January 2010
Scottish Pelagic Sustainability Group Ltd western component of north east Atlantic mackerel	January 2009
Scottish Pelagic Sustainability Group Ltd (SPSG) North Sea herring	July 2008
South Africa hake trawl (two fisheries)	April 2004
South Georgia Patagonian toothfish longline	March 2004/September 2009
South-west handline mackerel	August 2001/February 2007
Stornoway nephrops trawl	April 2009
Tosakatsuo Suisan pole and line skipjack tuna	November 2009
US North Pacific halibut	April 2006
US North Pacific sablefish	May 2006
Vietnam Ben Tre clam hand gathered	November 2009
Western Australia rock lobster	March 2000/December 2006

raised concerns from ECSOs that this process failed to address the problem of IUU toothfish fishing. Critics of this certification argued that 'to certify one small part of the fishery in South Georgia was felt to be an encouragement of the illegal trade' (*Guardian* 21 February 2004). The certification process was also affected by identification of problems in stock assessment and catch data held by the management organisation, the Commission of the Conservation of Antarctic Marine Living Resources. While the Western Australia Rock Lobster fishery was regarded as well managed, the MSC certification in this fishery has been criticised for being 'both inaccurate and misleading' (Sutton 2003: 115). The fishery suffered a major collapse in 2008–9. The virtual elimination of recruitment to the fishery has led to major cutbacks in fishing allocation and a review of the fisheries MSC certification in late 2009.

Fisheries from a variety of jurisdictions and scales are entering the certification process, as are the stakeholders from the entire production chain. This is resulting in new challenges for the MSC related to the consistency of its approach in certification, the management of corrective action requests and site audits in certified fisheries. The proposal to seek certification for a krill fishery is controversial, as this will be the first 'industrial' fishery application for certification under the MSC. This application is likely to be strongly challenged by environmental organisations including WWF-International.

With increasing concern over the sustainability of fishery resources and fishing practices, the MSC initiative is attracting significant interest as a vehicle for ensuring sustainability of these resources. The MSC standard has been remarkably influential in a relatively short time, complementing the work of international and national institutions and industry best practice, yet remains dominated by developed world fisheries. Maintaining the credibility and strength of the 'Principles and Criteria' certification standard – as the basis of a set of indicators and benchmarks against which a fishery is assessed – is critical for the ongoing viability of the MSC, as is maintaining its independence. In a decade the MSC has established a programme of assessment, product certification and promotion of sustainable fisheries, and helped establish ecolabels in fisheries. The original aspiration that through use of overtly market-oriented practices benefits would be secured for fishers, processors, marketers and retailers is not yet, however, fully realised. A key challenge is the creation of suitable markets and promoting ongoing consumer interest to ensure that the incentive remains for commercial involvement. The MSC faces ongoing and recognised challenges in extending the programme to developing states, or in data-poor fisheries. Despite these concerns, the MSC has continued to grow and negotiate these issues while attracting increasing support from industry, governments and NGOs.

The FSC and the MSC: A comparison

The FSC and the MSC differ across each of our three certification scheme dimensions. Politically, the FSC can be characterised as a tripartite certification scheme that formally sought to include a wide range of differently situated economic, social and environmental actors in its chamber-based interest aggregation arrangements. While difficulties at times have been experienced in allocating interests to chambers – and while the Social Chamber remains both more diverse and smaller than the other chambers – the FSC's chamber-based approach to interest representation largely insulated it from claims of illegitimacy from sectoral (economic, environment, social) and geopolitical (North, South) critics. In contrast, the MSC did not structure itself to be inclusive of all actors at the outset and, given it was founded by economic and environmental interests (Unilever and WWF-International), can be characterised as a bipartite certification scheme. Unlike the FSC, the MSC has no constitutional mechanisms to secure the representation of all relevant interests within its Board of Trustees or to ensure that some interests cannot trump others in the making of decisions. Moreover, its response to criticisms of its representational structure has only partly resolved the difficulties.

The FSC and the MSC also differ institutionally in terms of the way in which the organisations are structured. The FSC is a membership-based organisation which enables any individual or group to join on payment of a fee. The FSC's membership-based approach has also contributed to its legitimacy. While the membership-based approach has generated enormous difficulties in terms of practical everyday management – it can be extremely difficult, for example, to get enough members to form a quorum to vote for board vacancies – it has also enhanced the FSC's legitimacy by enabling disaffected groups to bring their concerns to tri-annual general assemblies and to network with others to have policies they disagree with overturned. In contrast, the MSC chose the institutional structure of a foundation which secures some of the benefits of organisational efficiency but at the cost of excluding disaffected members who then have no formal direct avenues to voice their criticisms and who take issue with the 'non-transparent' nature of the decision-making process. Unable to effectively voice their concerns within the MSC, concerned actors have undertaken substantial external criticism of the organisation, obliging it to engage in ongoing governance reform.

Finally, the two organisations differ in their regulatory arrangements although both largely employ a performance-based approach. However, while both have elaborated principles and criteria to secure well-managed forests and fisheries, the FSC's P&Cs are more detailed and more inclusive than the MSC's. Where the FSC has 10 principles and 56 criteria, the MSC has only three principles and 23 criteria; and whereas the FSC's P&Cs include a range of social principles related to indigenous peoples rights, community rights and social benefits, the MSC's include these only at the criterion level, enabling them to be traded off against each other. Moreover, the FSC's P&Cs constitute the basis for the development of national (and sometimes regional standards) via balanced stakeholder consultations, the MSC adopts a more centralised approach that devolves considerable responsibility to certifying bodies to interpret the standards but tries to harmonise these interpretations via the utilisation of its Analytical Hierarchy Process. Given that principles and criteria only become meaningful when they are elaborated in the form of indicators and verifiers, the arrangements adopted by the two organisations authorise rather different actors resulting in potentially different outcomes. The FSC authorises national initiatives to develop national indicators and verifiers based on its tripartite approach to standard setting. In contrast, the MSC authorises certifying bodies to develop fisheries indicators while requiring these to be approved by constituency-based groups.

Conclusion

This chapter has reviewed the history and operation of the FSC and the MSC. The FSC was the outcome of a range of initiatives commencing in the late 1980s that was stimulated in large measure by increasing disillusionment with intergovernmental efforts through the ITTO, the FAO and UNCED to halt deforestation and improve forest management practices. The MSC, in contrast, not only commenced later but built on a set of perceived lessons learnt from the FSC experience. Established in 1996, the consultants advising Unilever and WWF-International considered the FSC's membership and chamber-based arrangements problematic and recommended a more standard foundation and stakeholder model. While this secured greater 'efficiency' in decision making early on, the ongoing criticism the MSC has experienced has been of an altogether different order from the FSC and has necessitated the MSC to undertake a great deal more governance reform than the FSC in the past decade. Moreover, while the FSC remains largely perceived as a legitimate body in the contested area of forest management, it is unclear whether the institutional reforms adopted by the MSC have tackled what may be, in fact, the underlying source of external discontent – the degree of insulation of its trustees from affected interests.

The fact that the FSC and the MSC – despite sharing the same 'stewardship' appellation – are so different helps explain the differential responses they experienced by the policy networks operating in different national jurisdictions. Quite simply, the FSC constituted a far greater threat to the Australian, Canadian and British timber industries than the MSC did to their respective fisheries industries. However, the threat that both schemes posed to accumulation varied from one national jurisdiction and sector to another and also over time. To appreciate exactly how commodity scheme type intersected with policy network type, we turn in the next four chapters to examining state responses to forest and fisheries certification in Australia, Canada and the UK. Before examining certification responses in detail, we compare and contrast the policy context for forestry and fisheries in each country in the following chapter.

4
Forest and Fisheries Management in Comparative Perspective

In Chapter 2 we argued that the structure and operation of policy networks are key to explaining states' responses to certification schemes. Such networks emerge and transform themselves over time as a consequence of interactions between a country's ecology, constitution, commodity-chain structures and resource-management discourses. These natural, political, economic and discursive features not only directly shape the formation of business, environmental and social interests but also influence the institutional framework within which those interests are then mediated. The purpose of this chapter is to compare and contrast these elements of a country's ecological political economy, examining their manifestations in Australia, Canada and the UK at the national and sub-national levels.

Differences across all elements are evident in our comparator countries. The UK inherited a radically transformed ecology as a consequence of millennia of settlement, widespread clearance of native forests for agriculture and more recent heavy industrialisation and urbanisation. In contrast, both Australia and Canada are 'New World' settlements, where land and sea were under indigenous peoples' management until the seventeenth century and where, today, there are still significant areas of wilderness. Our comparator countries also differ constitutionally. While each is a parliamentary democracy operating under the Westminster system, Australia and Canada are formal federations while the UK has been viewed, until recently, as a 'unitary' state. However, the recent devolution of authority in the UK to Scottish and Welsh assemblies means that today all three of our jurisdictions wrestle with issues of 'multilevel governance' and 'regulatory federalism' (Kelemen 2002). Indeed, the UK's engagement is arguably greater than Australia's or Canada's, given its ever-deeper enmeshment in the larger and more institutionalised context of the European Union. To these differences we can also note substantial variation in our comparator countries' commodity chains and discursive contexts.

Our analysis of comparator countries commences with Australia and proceeds to Canada and the UK, in alphabetical order. Within each country, we first examine the ecological inheritance with respect to forests and fisheries before proceeding to outline the constitutional arrangements in place and how these have changed over the past two decades. However, because policy contexts in federal and devolved systems vary widely within a country – and because policy networks operate at both regional and national levels – we also investigate regional arrangements in Australia (Tasmania) and Canada (BC).

Our analysis in this chapter builds on the available national literature on forest and fisheries ecology, comparative politics and comparative political economy. This literature is somewhat uneven across sectors and with respect to different sub-national jurisdictions. Thus, for example, there are many more studies of forestry than fisheries in our three countries; and within each sector, many more studies of forestry and fisheries in BC than in Ontario or Nova Scotia. These differences reflect in part differences in the ecological significance of a region, the economic importance of the industry and the severity of natural resource management conflicts. Where gaps in the academic literature exist, we supplement them by drawing on government, business and civil society reports, on media articles and on occasional interviews with key informants. Since our aim is to provide sufficient context to explain the emergence, operation and transformation of forest and fisheries policy networks in subsequent chapters, only broad overviews are provided here. Interested readers are referred to the extensive literature we cite to obtain the necessary detail.

Forests and forestry in comparative perspective

Comparative forest ecologies

There are significant ecological differences between our comparator case studies in terms of their forest inheritance. The UK lost almost all its original native forests as a consequence of agricultural expansion dating back to the Roman era. The forests that remain are highly modified, resulting in a focus on afforestation, reforestation and forest restoration. In contrast, Canada has vast tracts of intact and unmodified primary forest. These forests provide habitat for a large number of vertebrates and invertebrates including iconic species such as the grizzly bear, spotted owl and marbled murrelet. Australia represents something of a middle case. Unlike Canada, much of the country's forests are woodlands, where trees are dispersed throughout the landscape rather than concentrated in stands. While a portion of the forest is substantially modified and fragmented like the UK's, important exceptions exist at the sub-national level such as Tasmania where extensive primary, old-growth forests remain. Australia's forests provide habitat, moreover, to unique flora and fauna, 80 per cent of which is endemic to the continent.

Australia

Officially, Australia has 149 million hectares of forests, making it the sixth largest forested nation in the world. This surprising statistic derives from the broad definition used by Australia to define a 'forest'. The definition, first used in 1992 to inventory the national forest estate, states that a forest is:

An area, incorporating all living and non-living components, that is dominated by trees having usually a single stem and a mature or potentially mature stand height exceeding two metres and with existing or potential crown cover of overstorey strata about equal to or greater than 20%. This includes Australia's diverse native forests and plantations, regardless of age. It is also sufficiently broad to encompass areas of trees that are sometimes described as woodlands.

(DAFF 2010a)

Two key features of the definition are the height that trees grow to ('exceeding two metres') and the actual or potential crown cover ('greater than 20%'). Taken together, they construct a broader conception of a forest than that used by the FAO, enabling relatively open land where trees exist at considerable distance from each other to be classified as 'forest'. If the FAO definition of a forest were used the amount of 'forest' would be far less, since it would exclude trees that did not grow to a height of more than five metres.[1] Indeed, the Australian government notes that of the 149 million hectares of 'forest', 118 million is 'woodland', 43 million 'open forest' and 5 million 'closed forest'.

Of Australia's 149 million hectares of forest, the vast majority are eucalypts (79%) followed by acacias (7%) (MPIG 2008, 3). Australia's forests are primarily native forests, with plantations constituting only one per cent of the total (currently estimated at about 1.8 million ha). Given the country's geography, forests are concentrated around the seaboard and unevenly distributed within states. From Table 4.1, for example, it can be seen that Tasmania, of all the states in Australia, is the most heavily forested as a percentage of the total land area, followed by New South Wales, Victoria and Queensland.

Forests provide vital habitat for Australia's unique flora and fauna, with old-growth forests being especially important from a biodiversity perspective. Table 4.2 provides data on the distribution of old-growth forests in Australia broken down by state and status. The data are indicative only, however, since the concept of 'old growth' is defined differently in each state and not defined at all by the state of South Australia. Once again, the importance of forests to Tasmania is highlighted, since although its area of old-growth forest is less than either Western Australia's or New South Wales's, old-growth forest constitutes 39 per cent of the total forest area of Tasmania compared to 28 per cent and 22 per cent in the two other states. Tasmania also has the highest level of old-growth forests in reserves although it should be noted that only 635,000 hectares are in 'dedicated reserves' that provide

Table 4.1   Forest as a percentage of land area by state: Australia

	Native forest area	Plantation area	Total land area	Forest as % of jurisdiction
New South Wales	26,208	345	80,064	33
Queensland	52,582	233	173,065	31
South Australia	8,555	172	98,348	9
Tasmania	3,116	248	6,840	49
Victoria	7,838	396	22,742	36
Western Australia	17,664	389	252,988	7

Source: MPIG 2008, 7.

legislative protection against mining and other activities. Of the remainder, 303,000 are in 'informal reserves' that result from administrative decisions which could be overturned.

A 2008 report provides partial data on forest-dwelling vertebrate species broken down by taxon and jurisdiction for five states and one territory. According to the data, New South Wales has the largest number of forest-dwelling vertebrates (760), followed by Southern Australia (574) and Victoria (513). In comparison, Tasmania is relatively impoverished at 137 species of vertebrates, a consequence of its more southerly location and temperate climate. Considerable concern exists for these species, however, given that 1287 forest-dwelling species are listed as vulnerable, endangered or critically endangered. Table 4.3 provides a breakdown of these species by taxon and status.

From Table 4.3, it is evident that a substantial number of species have been rendered extinct since the Australian continent was settled after 1788. One of the most notable has been the Tasmania Tiger or Thylacine (*Thylacinus cynocephalus*) with the last captive animal reported to have died in 1936 (DPIW 2010). It was officially declared extinct in 1986 (DPIW 2010). Another forest-dwelling animal listed as endangered today in Australia is the Swift parrot (*Lathamus discolour*). According to the MPIG (2008, 44), the Swift parrot 'breeds only in Tasmania and migrates to the Australian mainland in autumn to spend winter foraging for lerps and nectar in flowering eucalypts, mainly in Victoria and New South Wales'. Not only does it depend on eucalypts on the mainland for food, but its breeding range 'is mostly restricted to the east coast [of Tasmania] within the range of the Tasmanian blue gum (*Eucalyptus globulus*)'. It is thus doubly dependent on forests for its survival.

Canada

Canada's definition of a forest is even broader than that of Australia's. According to Natural Resources Canada, a forest is:

> An ecosystem characterized by a more or less dense and extensive tree cover, often consisting of stands varying in characteristics such as species

Table 4.2 Area of old-growth forest in areas surveyed for regional forest agreements

	Native forest area in region ('000 ha)	Area of old-growth identified ('000 ha)	Area of old-growth as % of forest in region	Area of old-growth on public land ('000 ha)	Area of old-growth on private land ('000 ha)	Area of old-growth in formal and informal reserves ('000 ha)	% old-growth in reserves
NSW	8,989	2,536	28	1,892	644	1,742	69
Queensland	3,230	270	8	196	71	196	73
Tasmania	3,116	1,228	39	1,118	110	973	79
Victoria	5,774	673	12	673	1	460	68
Western Australia	1,090	5,039	22	4,209	826	3,702	73
Total	23,018	5,039	22	4,209	826	3,702	73

Source: MPIG 2008, 18.

Table 4.3　Status and number of forest-dwelling species

	Extinct	Critically endangered	Endangered	Vulnerable	Total
Mammals	8	1	16	18	43
Birds	4	3	18	13	38
Reptiles			6	14	20
Amphibians	4	1	10	9	24
Fish		1	5	9	15
Invertebrates		3	2	5	10
Flora (higher plants)	34	49	474	630	1,187
Total	50	58	531	698	1,337

Source: MPIG 2008, 37.

composition, structure, age class and associated processes, and commonly including meadows, streams, fish and wildlife.

The reason for Canada's broad definition of a forest lies in the fact that there is, as yet, no common national definition of what a forest is across the country. While BC uses the FAO definition that specifies at least 10 per cent crown cover and single-stem plants that grow more than five metres, Ontario's is closer to the national definition stating: 'A plant community predominantly of trees and other woody vegetation, growing more or less closely together' (OMNR 2006, 744).

Unlike Australia, Canada's statistical information on its forest area, composition and type is difficult to access (Global Forest Watch 2000). This reflects the relative strength of the provinces in resisting moves to a common approach by the Canadian Federal Government on forestry matters. Basic data, however, is reported. According to Natural Resources Canada, the country is home to 402 million hectares of forests and woodlands, making it the fourth largest forested country in the world, with 10 per cent of the world's forests and 30 per cent of its boreal forests (NRCan-CFS 2007). If 'woodlands' are removed, approximately 310 million hectares remain (NRCan 2008a). Boreal forests dominate over other forest categories, accounting for about three-quarters of the total forest and woodland area (NRCan-CSF 2010b). In contrast, montane forests (12.5 million hectares) and coastal forests (8.7 million hectares) constitute only 3 per cent and 2 per cent respectively of the total forest and woodland estate.

Canada's forest, like Australia's, are unevenly distributed across the country. Table 4.4 provides a breakdown by province and forest type. As can be seen from the table, some provinces are heavily dependent on forests including New Brunswick (85%), Nova Scotia (80%), BC (63%) and Ontario (61%). Of these, BC stands out because its total forest land is over five times

Table 4.4 Canada's forest inventory by province and territory

	Forest land	Other wooded land	Total land	Forest land as % of total land
Alberta	27,718	8,670	61,600	45%
British Columbia	57,910	6,337	91,644	63%
Manitoba	18,968	17,386	53,996	35%
New Brunswick	6,091	116	7,129	85%
Newfoundland and Labrador	10,730	9,337	35,378	30%
Northwest Territories	28,352	4,994	111,854	25%
Nova Scotia	4,240	107	5,277	80%
Nunavut	815	125	188,366	0.5%
Ontario	53,758	14,536	88,110	61%
Prince Edward Island	265	8	564	47%
Quebec	73,360	11,215	132,970	55%
Saskatchewan	20,043	4,215	58,816	34%
Yukon	7,884	14,906	46,597	17%

Source: Extracted from NRCAn Statistics Profile, available at http://canadaforests.nrcan.gc.ca/statsprofile/inventory/all?format=print, accessed March 2010.

the size of New Brunswick's and Nova Scotia's combined, while its economy is far less diversified than Ontario's, which is of similar size.

Canada's vast forests vary significantly in terms of tree species and ecology. The country is divided into 20 'ecozones', 15 of which are terrestrial. Forests in the Boreal Plains and Boreal Cordillera ecozones are composed mainly of spruce, fir, larch and pine. In contrast, those of the Montane Cordillera ecozone are more diversified and include Engelmann spruce, lodgepole pine, ponderosa pine, Douglas-fir, western hemlock and western red cedar (NRCan 2008b). The country's forests are havens of biodiversity for a wide range of birds, mammals and insects, including such charismatic megafauna as black bears (*Ursus americanus*), grizzly bears (*Ursus arctos*) and mountain caribou (*Rangifer tarrandus*). Global Forest Watch estimates that forests provide habitat for about two-thirds of Canada's 140,000 species (2000, 31). They report data for forest-dwelling species at risk for 1999 (Table 4.5), noting that most of the species in the at-risk category are in the Coast Forest Region of BC and the Carolinian Forest Region of southwestern Ontario (Global Forest Watch 2000, 32). While most of these are vascular plants, 'the two most widespread forest species at risk are the woodland caribou and wolverine', because 'the woodland caribou requires mature or old-growth coniferous forests, which are disappearing across much of Quebec, Ontario, Saskatchewan, Alberta, and British Columbia' and wolverines are 'like some other species such as grizzly bear ... sensitive to human disturbance' (Global Forest Watch 2000, 32). According to the Committee on the Status of Endangered Wildlife in Canada (COSEWIC), the Eastern population of wolverine remains endangered today, while the Western population is listed as 'Special Concern'. COSEWIC also lists the Northern Mountain Caribou as of special concern.

UK

The definition of a forest differs in the UK from that used in Australia or Canada. Officially, the UK definition of a forest is: 'Land spanning more than 0.5 hectares with trees higher than 5 meters and a canopy cover of more than 10 percent, or trees able to reach these thresholds in situ. It does

Table 4.5   Canadian forest-dwelling species at risk, 1999

Categories	Mammals	Birds	Plants	Reptiles	Total
Endangered	3	4	14	1	22
Threatened	2	4	11	3	20
Vulnerable	13	9	14	6	42
Total	18	17	39	10	84

Source: Global Forest Watch 2000, 43.

not include land that is predominantly under agricultural or urban land use' (Forestry Commission 2010). This definition, which is quite close to FAO's, is complemented by a separate definition of 'Other Wooded Land' where the 10 per cent threshold is not yet met or where the combined cover of shrubs, bushes and trees results in canopy cover above 10 per cent (Forestry Commission 2010).

Using these definitions, the forests of the UK constitute a very small proportion of the total land area, are highly fragmented and largely composed of exotic plantations. The basic data concerning the volume and composition of the UK's forests is set out in Table 4.6. The data indicate that, as of 2008, total forested area in the four territories composing the UK was 2.841 million hectares, the vast majority of which was in England and Scotland and with conifers (mostly plantations) dominating broadleaved forest types. The key factors influencing this UK forest structure have been population growth, agricultural expansion and defence concerns. Each factor has interacted with the others to create fragmented forest ecosystems with no 'wild' places, giving rise to increasing concern over the decline in the UK's 'ancient woodlands' and the loss of the beauty of the countryside.

The factors that created the sparse, fragmented and exotic structure of the UK's forests have also played havoc with the nation's biodiversity. According to a 2008 State of the Natural Environment report, 'there are currently considered to be no more than 12 endemic species of invertebrate in the whole of the UK' (Natural England 2008, 149). Figure 4.1 provides data for wild birds in England between 1976 and 2006 and shows a steady decline in population numbers for woodland and farmland species with, perhaps, stabilisation but not recovery after 2000. For butterflies, the State of the Natural Environment report notes that these are 'good indicators of the environment because they are easily monitored and respond rapidly to changes in their habitats and environment' (Natural England 2008, 159). While populations fluctuate wildly from one year to another, the general trend has been downward, especially for specialist species. The report notes, 'Over the past 30 years, habitat specialists have fared worse and, after a rapid

Table 4.6 Area of woodland in the UK, 2008 (000 ha)

Region	Conifers	Broadleaves	Total
England	366	761	1,127
Scotland	1,045	297	1,342
Wales	157	128	285
N. Ireland	66	21	87
UK	1,635	1207	2,841

Source: UK Forestry Commission, Forestry Facts and Figures 2008, Table 1.

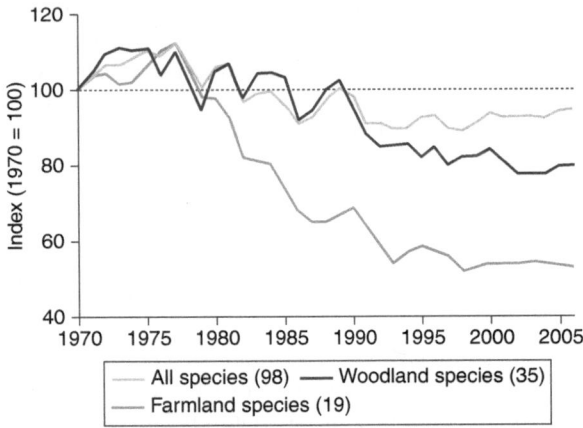

Figure 4.1   Population trends of wild birds in England, 1970–2006
Source: Natural England, State of the Environment 2008, 157.

decline to 1981, have remained low, being at 37% of the 1976 baseline in 2006' (Natural England 2008, 159).

To protect biodiversity, UK governments have developed a large number of habitat designations. One of the most important of these is the Site of Special Scientific Interest (SSSI) designation. SSSIs describe 'the country's very best wildlife and geographical sites' and 'include some of our most spectacular and beautiful habitats' (Natural England 2009a).[2] This includes the New Forest originally notified as an SSSI in 1959, with a larger area placed under that designation in 1971, and further additions made in 1974, 1979 and 1987 (Natural England 2009c). According to Natural England:

> The New Forest embraces the largest area of 'unsown' vegetation in lowland England and includes the representation on a large scale of habitat formations formerly common but now fragmented and rare in lowland western Europe. They include lowland heath, valley and seepage step mire, or fen, and ancient pasture woodland, including riparian and bog woodland. Nowhere else do these habitats occur in combination and on so large a scale. There are about 4600 hectares of pasture woodland and scrub dominated by oak, beech and holly; 11,800 hectares of heathland and associated grassland; 3300 hectares of wet heath and valley mire-fen and also 8400 hectares of plantations dating from various periods since the early 18th century.
>
> (Natural England 2009c)

The area constitutes an important breeding ground for birds including the Dartford warbler (*Sylvia undata*), Kingfisher (*Alcedo atthis*) and nightjar (*Caprimulgus europaeus*). It also 'supports populations of nine rare and twenty-five nationally scarce vascular plants' including Dorset heath (*Erica ciliaris*), and wild gladiolus (*Gladiolus illyricus*) (Natural England 2009c). Finally, the endangered red squirrel (*Sciurus vulgaris*) has died out in The New Forest following the introduction into England of the North American grey squirrel, which out-competes it in its range.

Comparative forest policy contexts

Australia, Canada and the UK differ markedly in terms of their forest policy contexts. These differences are related to the formal federal structures of Australia and Canada and the unitary, if increasingly devolved, structure of the UK. Moreover, although Australia and Canada are formal federations, Canadian provinces have been able to resist federal encroachment over forest policy better than Australian states have, reflecting the relative power of the Australian Commonwealth government compared to the Government of Canada. Finally, while all three countries substantially reworked their forest policies in preparation for the 1992 UNCED, the institutional arrangements adopted to review forest policy were very different and resulted in rather different paths being taken in the three jurisdictions.

Australia

Natural resources and natural resource management are state responsibilities under the Australian Constitution. However, the Commonwealth government has intervened extensively in the 1980s to address state-based natural resource conflicts. In the early 1980s, the Liberal government under Malcolm Fraser used its commercial powers to deny an export licence for sand mined from Fraser Island; and in 1983, the newly elected Labor government under Bob Hawke used its treaty powers to list the area around the Gordon and Franklin rivers in southwest Tasmania as a World Heritage Area (Lane 1999). While such federal 'interference' was deeply resented by state politicians, the Commonwealth's responsibilities consequent on the ratification of international environmental treaties rendered it justifiable. As conflicts multiplied, however, especially in the forest sector, federal politicians found themselves wedged between the demands of environmentalists to halt native forest conversion and logging and state governments and industry who sought its continuation and expansion (Lane 1999; Hollander 2004).

To address the ongoing conflict – and to position Australia for the new era of sustainable development launched by the Brundtland Commission – a flurry of initiatives emerged from Canberra of relevance to forest management and wilderness conservation. These included the Resource Assessment

Commission in 1989,[3] Ecologically Sustainable Development Working Groups in 1990 and the National Forestry Policy Statement in 1992.[4] These initiatives culminated in the launching of the Regional Forest Agreement (RFA) process after 1995. Today, RFAs constitute the basic 'settlement', whereby the Commonwealth government agrees not to interfere in how states manage their forests in exchange for state governments agreeing to reserve sufficient areas and types of forests to preserve biodiversity and heritage values. The criteria used to determine how much and which types of forests to reserve were developed by the Joint Australian and New Zealand National Forest Policy Statement Implementation Sub-Committee (JANIS). JANIS aimed to establish a reserve system that was 'comprehensive, adequate and representative' and negotiations occurred in several Australian states in the late 1990s on which areas were to be set aside to achieve these objectives. These reserves were listed in RFAs, the first of which was signed by the State of Victoria for the East Gippsland region in 1997. The East Gippsland RFA was followed shortly afterwards by the Tasmanian RFA (also signed in 1997) and subsequently several others were signed, the majority with the State of Victoria (Table 4.7).

Although RFAs were designed to end forest conflict in Australia, there is little evidence that this has occurred. In Tasmania, federal and state politicians continued to be dragged into debates over forest management. One reason for the ongoing conflict was the speed with which RFAs were negotiated. Kirkpatrick (1998, 35) observes that 'the timeframe for reaching regional agreements have been such that data collection has largely consisted of the assembling and cleaning up of accessible existing information'. Unfortunately not only was the data deficient but even when it did exist 'apparent failures to use available data occurred in both the East Gippsland and Tasmanian processes' (Kirkpatrick 1998, 35). Another reason relates to the fact that

Table 4.7   Australian regional forest agreements

State	RFA region	Date signed
Victoria	East Gippsland	3 February 1997
Tasmania	Tasmania	8 November 1997
Victoria	Central Highlands	27 March 1998
Western Australia	South West	4 May 1999
Victoria	North East	23 August 1999
NSW	Eden	26 August 1999
NSW	North East	31 March 2000
Victoria	Gippsland	31 March 2000
Victoria	West	31 March 2001
NSW	Southern	24 April 2001

Source: DAFF 2010.

RFAs are designed to provide long-term resource security to industry in a dynamic context characterised by evolving public opinion. Thus, although the Tasmanian RFA committed both levels of government to a 20-year settlement, it was substantially amended in 2005 to reserve additional areas of forest as well as to undertake a commitment to phase out the conversion of native forests by 2010. The Western Australia RFA experienced a much more dramatic fate, and was effectively cancelled six months after it was signed as a consequence of public outrage (Brueckner and Horwitz 2005).

The institutional arrangements for managing forest policy at the intergovernmental level also underwent changes in the 1990s. In 1994, the Australian Forestry Council merged with the Australian and New Zealand Fisheries and Aquaculture Council to form the Ministerial Council on Forestry, Fisheries and Aquaculture (MCFFA). MCFFA subsequently constituted an important intergovernmental forum for the discussion on a large range of forestry matters throughout the 1990s including certification and labelling. In 1999–2000, the Australian government reorganised its internal arrangements to reflect the growing importance of natural resource management issues. The MCFFA was disbanded (it held its last meeting in 2000) and its functions were taken over by the Natural Resources Management Ministerial Council (NRMMC) and the Primary Industries Ministerial Council (PIMC). Since then, it is the PIMC that has played the major role as a forum for considering forest certification and took the lead in developing a competitor standard to the FSC, the AFS.

Canada

Ownership of forest land in Canada resides with the provinces, creating a serious constitutional barrier to the formation of a unified forest policy. Under Section 109 of the British North America Act, provincial governments are granted 'ownership of lands within their borders' (Cairns 1992). Those crafting the Canadian constitutional settlement aimed to establish a clear division of powers between 'local' and 'national' matters and avoid the 'unfortunate' experience of the US, where states rights were enshrined in a complex arrangement that delegated 'residual powers' to the individual states. The aim of the Act was to create a strong central government that would rule over all matters deemed 'national' in character. Ironically, however, and despite this constitutional effort to empower the federal government, the evolution of Canada's division of powers has tended in the opposite direction. In contrast to Australia, Canadian provinces have gained considerable powers over the years, primarily as a consequence of the early rulings of the Judicial Committee of the Privy Council in Great Britain (Cullen 1992). In several rulings on key issues in the late nineteenth and early twentieth centuries, the JCPC ruled in favour of the provinces and against extending the powers of the national government in the direction originally foreseen.

The development of Canadian forest policy has occurred in the shadow of longstanding conflicts over forestry. The most renowned of these occurred in BC, where environmentalists and First Nations groups battled then timber giant Macmillan Bloedel over the logging of forests in Clayoquot Sound on the west coast of Vancouver Island.[5] Forest conflict also occurred in the central Canadian province of Ontario in the late 1980s when a conflict erupted at Temagami over government and industry plans to log the area (Hodgins 1992). In another similarity to Clayoquot Sound, the conflict eventually escalated to such an extent that environmental and First Nations protesters engaging in civil disobedience were arrested and jailed.[6] In the Maritime Provinces of eastern Canada, forest conflict took a different form. In part, this was because the tenure arrangements there differed from other provinces, with the majority of forested land in private rather than public hands. Forest conflict in the Maritimes focused on the technical aspects of forest management including clear-cut logging, riparian buffers, aboriginal rights and the use of 'biocides' (pesticides, herbicides and fungicides). The use of such chemicals was a key issue for many of those negotiating the FSC-Maritimes regional standard (Cashore and Lawson 2003; Gale 2004).

Lacking direct power, Ottawa has sought to generate compliance with the growing number of international treaties related to forestry via the judicious use of incentives and through inter-provincial dialogue. An important fiscal programme of the 1980s and 1990s was the negotiation of federal-provincial Forest Resource Development Agreements (FRDA).[7] FRDA provided the provinces with the means of funding reforestation and forest research projects in return for vague and non-enforceable commitments to embrace more sustainable forestry practices. For example, in 1984, the governments of Canada and Ontario signed a five-year FRDA for $150 million, the vast majority of which (90%) was spent on reforestation (Ross 1997, 35). In another example, the British Columbia-Canada FRDA I agreement was signed in 1985 and provided $300 million, mainly for reforestation of 'not-satisfactorily restocked' (NSR) lands (Ross 1997, 45; Drushka 1999, 23). This agreement was renewed in 1991 to provide $200 million for another five-year period (Ross 1997, 45).

Ottawa's efforts to promote more sustainable forestry practices within the provinces improved in the 1980s following the establishment of the Canadian Council of Forest Ministers (CCFM). Composed of the 14 elected ministers responsible for forestry from the federal, provincial and territorial governments of Canada, the objectives of the CCFM are to promote cooperation between governments on forestry matters, develop and maintain scientific forestry information, show leadership on SFM, promote Canadian forest management, liaise with other ministerial councils, share information and create a framework whereby forest agreements can be signed (CCFM 2008). A notable CCFM achievement was the 1992 National Forest Sector Strategy titled *Sustainable Forests: A Canadian Commitment* (CCFM 1992).

Several subsequent forest policy developments can be traced back to the 1992 document. These include efforts to establish a national ecological classification system for forests lands (Para 1.1) and to set up Canada's Model Forest Network (Para 2.17). Another important initiative of great relevance to forest certification was the commitment to develop, by 1993, 'a system of national indicators to measure and report regularly on progress in achieving sustainable forest management' (CCFM 1992, 25). This set Canada on the path to developing a set of 'criteria and indicators' (C&Is) that could be used to both define what SFM meant and determine the degree to which it was being practised in the country. Finally, the 1992 forest strategy was important because it foreshadowed that governments and industry were seeking mechanisms to promote 'means of identifying and promoting Canadian forest products' that would reflect Canada's commitment to SFM (Para 4.13). Exactly how this was to be done is not spelt out in the 1992 strategy, but there is a commitment that 'by 1994, forest-based industry associations will adopt self-regulating codes of environmental practice' (Para 4.17).

UK

For almost the entire twentieth century, Britain was a unitary state, with its component territories governed from Westminster, in London. The exception, of course, was Northern Ireland, which had an independent House of Assembly at Stormont, until it was suspended as a consequence of the 'troubles' in the 1970s. As the century ended, however, the 'New' Labour government of Tony Blair was swept into office and was committed to devolving power to Scotland and Wales, a promise fulfilled with the passage of the 1998 Scotland Act and Government of Wales Act. Today, therefore, the UK (formally consisting of the territories of England, Scotland, Wales and Northern Ireland) is a quasi-federal state while simultaneously being deeply enmeshed in a larger supranational political unit, the EU. Devolution has resulted in diverging strategies for forest management in the different jurisdictions within the UK.

Modern forest management in the UK has been the prerogative of the Forestry Commission. The origins of the commission lie in concerns over Britain's 'forest security' during World War I when the country was forced to rely on domestic resources. In 1916, the Reconstruction Committee issued a report by Lord Acland recommending the establishment of a forestry authority to undertake the task of post-war reforestation and afforestation to build up a three-year timber reserve. In 1919, Prime Minister Lloyd George implemented the Acland recommendations and founded the Forestry Commission. The commission reported directly to parliament until 1946, when it was restructured to require commissioners to work under the direction of the Minister of Agriculture and Fisheries and the Secretary of State for Scotland (James 1981, 237). The Forestry Commission survives to this day despite recommendations that it be abolished in the 1920s and privatised in the 1990s.

In the 1970 and 1980s, the commission was criticised both internally and externally for its forestry practices and their impact on the UK's 'ancient woodlands'. These are defined as:

> Woodland that has existed since at least 1600 AD and possibly much longer. Two broad types of ancient woodland can be identified, Ancient Semi-Natural Woodland, that is composed of native trees and shrub species which have not obviously been planted, or Plantation on Ancient Woodland Sites, that which has been continuously wooded since 1600 AD but where the former tree cover has been replaced with planted trees (often conifers).
>
> (Natural England 2009d)

An inventory of Ancient Woodlands conducted by the Nature Conservancy Council (NCC) revealed that there was remarkably little remaining in the UK: 'only 2.5 per cent of the land surface of England and around 2.7 per cent of the land surface of Wales' (Tsouvalis 2000, 102). Moreover, a great deal of semi-natural ancient woodland had been lost between 1933 and 1983 as a direct consequence of aggressive changes in land use resulting from agriculture expansion and monocultural exotic timber planting. In the face of this criticism, the government introduced a new Broadleaves Policy in 1985, which established that timber production was no longer the primary purpose of the Forestry Commission. From then on it was permitted to give grants for other purposes including protecting ancient woodland and enhancing natural heritage.

The 1980s saw intensified conflict over exotic monocultures which culminated in a dispute between the NCC and the forest industry over afforestation in 'the Flow Country', the counties of Caithness and Southerland in far-north Scotland. These counties were covered by peatlands and bogs and were viewed by the Scottish forest industry as wasteland ripe for afforestation. Claiming that the area constituted 'one of the largest blanket bogs (moorland and peat bogs) in the world' (Tsouvalis 2000, 82–3), the NCC sought to protect the region by publicising its value and by listing many of its areas as protected SSSIs.

In the 1990s, responsibility for UK forest policy fragmented. Increasingly, Europe began to influence national forest policy developments through the Ministerial Conference for the Protection of Forests in Europe (MCPFE). While the EU lacks a Common Forestry Policy unlike the fisheries sector, there are some signs that forestry is moving up the EU agenda as concern mounts over climate change, biodiversity loss, illegal timber imports and amenity. While the MCPFE has limited itself to a coordinating and information role in the past, in 2008 it hosted a first meeting to explore options for a Legally Binding Agreement on Forests, with a number of options canvassed and with the relationship to other EU treaties and arrangements discussed (MCPFE 2008). It is as yet too early to tell whether this initiative will bear fruit and lead to a Europe-wide approach to forest policy.

At the same time as the EU is playing a more influential role in national forest policy the formal devolution of power from Westminster to the Scottish Parliament and the Welsh Assembly that followed the 1998 acts establishing the same have reduced the power of the Forestry Commission. While a level of devolution already existed in the forestry sector following the 1945 Forestry Act's restructuring of the commission's responsibilities for forestry matters (James 1981, 237), formal devolution saw much greater power shift from Westminster to Edinburgh and Cardiff. As a result the UK's forest strategy actually consists of four forest strategies: one each for England, Wales, Scotland and Northern Ireland that were 'brought together (quite literally without adjustment) in *Sustainable Forestry in the UK*: the UK's national forest programme' (Weldon 2004, 6). Weldon also notes that 'the four devolved strategies are very different in their emphasis on specific contexts (Scotland's, for instance, still has a strong emphasis on commercial wood production' (Weldon 2004, 6), whereas in England, there is much more emphasis on the protection of ancient woodlands, on the establishment and maintenance of Areas of Natural Beauty and on the expansion and maintenance of community forests in peri-urban regions for amenity, recreation and well-being.

Comparative forestry discourses

Discourses on forests and forest management have evolved over time in our three comparator countries. In each country, a broad shift from a 'sustained-yield' forestry paradigm to a 'sustainable forest management' paradigm is evident. However, the precise meaning of SFM varies from one jurisdiction to another. In Australia, SFM has been conceptualised in terms of a shift from natural forest to plantation management, a move promoted by the states, the Commonwealth government and environmentalists. In Canada, where growing conditions do not favour plantation establishment and where vast areas of natural forest remain, the meaning of SFM focuses on how to manage natural forests 'sustainably' with an emphasis on 'ecosystem-based forestry'. In the UK, a timber-security discourse based on reforestation and afforestation dominated public and private forestry for most of the twentieth century. It was displaced by a 'multiple-use' discourse in the 1980s which still underlies the meaning of SFM in the UK today.

Ecological discourses have also differed, varying from those that emphasise the 'wilderness' value of forests to others grounded in ecoforestry and amenity. The wilderness discourse, prominent in Australia, views native forests as the world's remaining 'wild' areas which require protection from logging to preserve such values. This discourse differs considerably from that promulgated in the UK, where the emphasis has been on forests' amenity values, which is tied to the desire for forest-based recreational pursuits and the preservation of the character of the countryside. In Canada, a discourse of ecoforestry and ecosystem-based forestry has emerged alongside a wilderness discourse. Once sufficient wilderness areas have been reserved, ecoforestry mandates a range of environmentally sensitive logging practices on the remaining landbase.

Australia

Following World War II, Australian states embraced the discourse of sustained-yield forest management, but encountered difficulty in implementing it (Dargavel 1995, 76). To meet the growing demand for fibre, the Australian Forestry Council, set up in 1964 to coordinate state activity on forestry, advocated the establishment of softwood plantations. Dargavel notes that the AFC recommended that the targets set for plantation establishment by states be increased from 16,000 to 28,000 hectares a year to achieve the objective of 1.2 million ha of plantations by 2000. To encourage plantation establishment, the Commonwealth government provided loans to states under Softwood Forestry Agreements (Dargavel 1995, 77), a programme that led to a steady expansion in softwood plantations through the 1970s and 1980s. As public environmental awareness grew in the 1980s, the forest industry found it increasingly difficult to justify the conversion of native hardwood eucalypt forests to exotic, softwood, pine plantations. The plantation issue became framed in an interesting way, however. Rather than being condemned outright, the environmental movement argued in favour of plantations replacing native forest logging, provided they occurred on marginal agricultural land.

In the 1990s, plantations were strongly linked to the new discourse of sustainable forest management and a new phase of plantation establishment occurred. They were aggressively promoted by the Liberal-National Commonwealth Government led by John Howard, which in 1997 established its Plantations 2020 programme (DAFF 2010a). The new plantation programme promoted industrial plantation establishment via Managed Investment Schemes (MIS). High-earning individuals were able to obtain tax breaks on income earned by investing in such schemes, a system that saw several large companies emerge in the plantation sector in the 2000s, including Timbercorp and Great Southern Plantations. These companies aggressively purchased agricultural land, which were then planted with fast-growing hardwood saplings (especially *Eucalyptus globulus* and *Eucalyptus nitens*). Extensive criticism of MIS schemes ensued, especially from rural communities. The criticisms focused on the economics underlying the schemes and on such issues as increasing agricultural land prices, rural depopulation, pesticide use and water table depletion (Schirmer 2005). In response, the Commonwealth government introduced substantial changes to the tax regulations for MIS schemes in 2007 which made them generally less attractive to investors. These changes interacted with the substantial downturn in the price of forest products during the global financial crisis and the increase in borrowing costs, causing highly leveraged companies such as Timbercorp and Great Southern Plantations to go into receivership (ABC 2010a).

Another forest discourse strongly advocated in Australia is 'wilderness'. The strength of Australia's wilderness movement is linked to the unique and 'alien' nature of the flora and fauna, to the outdoor lifestyle Australians lead and to the human desire to directly experience and form a connection with 'untransformed' nature. The discourse has also been developed in contrast to modernism's instrumental view of nature that values it for its utility in meeting human needs. Modern Australian environmentalism has its foundation in a conflict between a nascent wilderness movement and developers over the flooding of Lake Pedder in Tasmania. Although wilderness enthusiasts lost that battle, they went on to form The Wilderness Society (TWS), today one of Australia's pre-eminent conservation groups.

The wilderness discourse has seen most Australian states and territories introduce legislation to define and safeguard the values that define it. Herath notes, for example, that TWS's definition of wilderness as 'a large tract of land remote at its core from access and settlement and substantially unmodified by modern technological society and of sufficient size to make practicable the long term protection of natural systems' was sufficiently influential to become embedded in state legislation. The influence of this definition can be seen in New South Wales' 1987 legislation, where declarations by the minister must consider the following three elements: whether '(a) the area is ... in a state that has not been substantially modified by humans ... (b) the area is of a sufficient size to make its maintenance in such a state feasible, and (c) the area is capable of providing opportunities for solitude and appropriate self-reliant recreation' (Government of New South Wales 1987).

Canada

Howlett and Rayner identify four historical forest *regimes* in Canada: unregulated exploitation, revenue enhancement, conservation and timber management. These regimes were based on different forest management discourses. The regime of unregulated exploitation was, as its name implies, largely associated with pre-federation Canada when forests were viewed as an impediment to productive use of the land underneath. Once colonial governments became consolidated, they turned their attention to earning revenue from deforestation via rent, stumpage and export duties (Howlett and Rayner 2001, 26). It was only in the early years of the twentieth century that a fully fledged body of ideas emerged on the management of forests based on Gifford Pinchot's approach to conservation or 'wise use' forest management (Beyers and Sandberg 1998). Economic development lay at the base of Pinchot's notion of conservation. The conservation regime put in place in Canada in the early twentieth century aimed to achieve the wise use of forests by eliminating natural and human wastage (Beyers and Sandberg 1998; Howlett and Rayner 2001). Natural wastage occurred mainly as a consequence of forest fires and dying trees. In the post-war period, this conservation regime was replaced by a more explicit focus on industrial production for the booming pulp and paper industry based on timber management.

According to Ross (1997), although most provinces based forest practices on these four forestry discourses they did so at different times. In Ontario, for example, the conservation regime commenced in 1898, with the passing of the Forest Reserve Act, which implemented the recommendations of the 1897 Royal Commission on Forest Protection (Ross 1997, 34). Over a decade later, and in part influenced by the Canada's first-ever forestry congress held in Ottawa in 1906, BC launched a Royal Commission of Inquiry into Forestry, with its recommendations later embedded in BC's first Forest Act of 1912 (Drushka 1999, 39; Ross 1997, 44). In contrast, Nova Scotia's entrance into the conservation regime was delayed until 1926 when the passing of The Land and Forests Act gave the minister 'responsibility for conservation and preservation of forest and timber land' (Ross 1997, 24).

A notable feature of the conservation and timber management regimes in place in Canada was the degree to which provincial governments vested ownership of forested land in the Crown. The idea of public ownership was central to Pinchot's conception of conservation and justified the vast network of federal parks established in the US in the early years of the twentieth century. Wilson (1998) notes that Pinchot's ideas 'reflected a limited but still significant notion of the public interest' which 'held that in the name of efficiency and conservation, government had a right to control overly expedient development practices' (Wilson 1998, 13). Where forest land is under public control, the basic tenure arrangement for logging wood is the forest licence. While the mechanism underpinning forest licences remains common across the country – governments granting permits to cut timber up a certain volume on condition that royalties are paid – there are a bewildering array of alternative licence arrangements.

The timber management regime is in the process of being transformed. Writing in 2001, Howlett and Rayer do not name the new regime although in hindsight we can see it is based around the idea of forest *ecosystems*. The origins of the new regime lie in the 1992 Canada Forest Accord which was signed by industry, First Nations and environmental groups and which set out the beliefs, vision, goals and actions to be taken. Two subsequent accords were signed in 1998–2002 and 2003–8. The later accords evidence greater concern for the environmental importance of forests, with the 2003–8 accord committing signatories to the management of Canada's natural forests 'through an ecosystem-based approach' and to 'accommodating Aboriginal and treaty rights in the sustainable use of the forest' (NFCS 2003a).

The ecosystem-based approach adopted in the most recent Canada Forest Accord has seen broader-based participation by affected constituencies. Thus, while there were only 15 signatories to the original 1992–7 accord, the 2003–8 accord was signed by 60 groups and individuals. Moreover, while the majority of the signatories to the first accord were from the forest sector, the most recent accord includes, for the first time, the Sierra Club of Canada.[8]

The ecosystem-based approach has been defined in the Canadian context as 'a strategy for the integrated management of land, water and living resources that promotes conservation and sustainable use in an equitable way' (McAfee and Malouin 2008, 7). The ecosystem-based approach is a fundamentally different way of conceptualising forest management and McAfee and Malouin (2008, 15) note several ways it diverges from the current, dominant 'management' regime (Table 4.8). The management regime approach is associated with a tendency to focus on the natural environment (e.g. trees), adopt a small spatial scale (e.g. stand level), promote a single sector (e.g. timber production) and to intervene to achieve short-term objectives. In contrast, ecosystem-based forestry aims to integrate the natural and social aspects of the environment, which are conceptualised to operate over multiple scales and encompass many sectors, with adaptive management prescribed to secure the long-term sustainability of economic, social and cultural values.

UK

From 1919 until 1957, the Forestry Commission was dominated by a timber security rationale and traditional sustained-yield forest management ideas. Because so much deforestation had occurred prior to and during World War I, after the war the commission launched an aggressive planting campaign, especially in Scotland. The afforestation programme built on a long legacy of forest planting, that James (1981, 169) traces back to the sixteenth century to the Inglewood Forest in Cumberland, when almost two million trees were planted. Later, trials of exotic species such as Sitka spruce and Douglas fir

Table 4.8 From environmental management to ecosystem-based forestry

From environmental management that	To ecosystem-based management that
Views the environment as a single issue	Also takes into account economic and social aspects
Involves small spatial scales	Involves multiple scales (ecosystem)
Focuses on one sector	Encompasses many sectors, activities and users
Has a short-term perspective	Has long-term sustainable goals
Intervenes for specific species or outcomes	Addresses the whole ecosystem

Source: McAfee and Malouin 2008, 15.

were found to be very successful and played a key role in twentieth-century British forestry (James 1981, 184). The focus was on growing trees quickly for timber production, with little concern for other forest values.

It was only in 1957 following the *Enquiry into Forestry, Agriculture and Marginal Lands* (the Zuckerman Report) that the original national defence argument for public support of forestry was challenged. The Zuckerman Committee concluded that nuclear war had obviated the need for a strategic timber reserve and that 'the rate of return on capital invested in forestry should be the crux of future forest policy' (Oosthoek 2003, 3). By the mid-1980s, forestry discourse in the UK had been transformed from one based on exotic plantations for timber security to one that emphasised multiple-purpose forestry. Despite the challenges, efforts were also being made to translate the new discourse into practices on the ground. The Forestry Commission thus considered that it was already practising 'sustainable forest management' when the discourse arose following the publication of the Brundtland Commission's report, *Our Common Future*, in 1987. Indeed, the reforms undertaken in the 1980s in response to pressures from the environmental and amenity movement had significantly closed the gap between public demands concerning forestry and Forestry Commission practices. The discourse of multiple-use forestry, reinterpreted as SFM, thus constituted a basis for compromise in the UK that did not exist in either Australia or Canada.

Comparative forest product commodity chains

Countries differ in the structure of their forest commodity chains. These differences concern the products produced, imported and exported, and the environmental sensitivity of domestic and foreign markets. Our comparator countries differ in the structure and operation of their forest product commodity chains. The UK is a major importer of forest products including timber, paper and furnishings. Not only are UK consumers environmentally discriminating but they have access to a range of certified timber products from around the world, including Sweden, Finland and Canada. In contrast, the Australian commodity chain is 'extroverted': it depends heavily on environmentally less discriminating markets in Asia. While Australia also imports considerable quantities of forest products, it does so from China (furniture) and New Zealand (paper and lumber). Canada is the paradigm case of a forest product exporting country. Its forest industry is heavily dependent on the US market, which is environmentally discriminating and manipulable by transnational environmental advocacy groups.

Australia

Australia has a diversified forest industry that produces most categories of wood products including woodchips, sawnwood, panels and pulp and paper. However, the commodity composition of its forest products industry coupled with high levels of demand within Australia means that the country is a net importer of high value-added forest products. Figure 4.2 provides data on Australia's forest exports by value which highlights the dominant role of woodchips in the export mix. In volume terms, the sum represented about 5.2 million bone dried tonnes of hardwood woodchips in 2007–8, of which one-third came from Tasmania. These were shipped by Gunns Limited, the island's major producer of hardwood and softwood woodchips. The vast majority of Australia's woodchips are shipped to the Japanese market.

In trend terms, Australian imports of forest products have increased by about 36 per cent between 1997 and 2008 (Table 4.9). While the highest rate of increase has occurred in wood-based panels, this is from a low base. More significant is the slight decline in imports of sawnwood and the steady growth in paper and paperboard imports, the latter large in value terms. On the export side, sawnwood has grown significantly, almost tripling in value from $43 million in 1997 to $125 million in 2008. While paper and paperboard and woodchips have grown more modestly in comparison, increases of 92 per cent and 70 per cent are still significant given the very high base of exports in 1997. Indeed, in value terms, woodchips still account for over 40 per cent of Australia's total forest products exports, illustrating the comparatively low value-added nature of its forest export industry.

Australian companies source their timber from a large number of locations, but the dominant countries are New Zealand, Indonesia, the US, Finland and Germany in that order. New Zealand is by far the largest source country

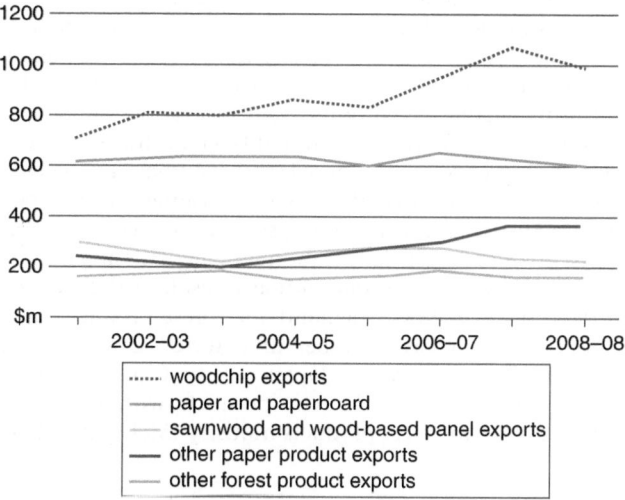

Figure 4.2   Forest sector exports, Australia
Source: ABARE 2009a, 2.

Table 4.9   Australian forest products: Imports and exports ($million)

Imports	1997–8	2002–3	2007–8	% Change (1997–2007)
Sawnwood	417	502	405	–2.8
Wood-based panels	143	190	271	+89.5
Paper and Paperboard	1756	2137	2277	+29.7
Total	3,264	3,995	4,459	+36.6
Exports				
Sawnwood	43	74	125	+190.7
Wood-based panels	108	151	101	–6.5
Paper and Paperboard	314	635	605	+92.7
Woodchips	586	794	997	+70.1
Total	1,347	2,040	2,342	73.9

Note: Total includes other forest products, not included in table.
Source: ABARE 2009, 7.

by volume and by value, accounting for almost one-fifth of total imports by value in 2008–9. The five countries combined accounted for almost 50 per cent of Australia's total timber imports in the same period. On the export side, Australia is very heavily dependent on Japan, which accounted for 41 per cent of exports in 2007–8, the vast majority (98%) of which were wood-chips (ABARE 2008, 44). In the same year, the next most important destina-tion for Australian forest products exports was New Zealand, closely followed by China, which accounted for 16 per cent and 15 per cent respectively. However, exports to the Chinese region were greater than to New Zealand if wood products to Chinese Taipei and Hong Kong, China are included.

The Australian timber industry's dependence on exports to Japan and China, historically relatively environmentally insensitive markets, has meant that feedback through the timber commodity chain for forest certification has been muted. Moreover, when signals did eventuate in the early 2000s, Australian producers calculated that Asian consumers would be satisfied with a domestic Australian certification system accredited to the Programme for the Endorsement of Forest Certification scheme. This apparently rational calculation has been called into question as a result of the global financial crisis. The downturn in global demand for forest products has resulted in falling prices and increased timber inventories. Under pressure from environmentalists for many years, Japanese and Chinese purchasers are beginning to prefer FSC-certified forest products with important conse-quence for the Australian forest products industry (McLaren 2010).

Canada

Canadian producers manufacture the full range of forest products including lumber, panels, pulp, newsprint and paper. While each of these markets has its own distinctive structure, all production depends heavily on exports to the US. The fortunes of Canadian manufacturers are closely tied to the health of the US economy, and experienced a serious decline as a consequence of the global financial crisis. Critical factors included a slump in US housing starts, a decline in advertising and the appreciation of the Canadian dollar.

The dependence of Canadian producers on US consumers is evident from an examination of individual forest product sectors. For example, Canada is the world's largest producer of softwood lumber, which 'accounts for ¼ of Canada's forest product exports' (NRCan 2008e, 13). Figure 4.3 illustrates the country's dependence on exports to the US in this product. Apart from the US, the only other country that Canadian producers export to in relatively high volumes is Japan. Canadian producers also depend on exports to the US market for structural panels. These are very important in the housing industry and the overwhelming majority are shipped to the US. From Figure 4.4, it is evident that only a tiny fraction of Canada's panel production is exported to other countries; however, following the downturn in the US housing in 2006 and 2007, there is evidence of some diversification to other markets.

Although somewhat less dependent on the US to purchase its newsprint and pulp, Canada nonetheless exports large quantities of both products to its southern neighbour. In 2003 approximately 80 per cent of Canada's newsprint was sent to the US. Four years later, and following a significant reduction in total output (from C$5.5 billion in 2003 to about C$4

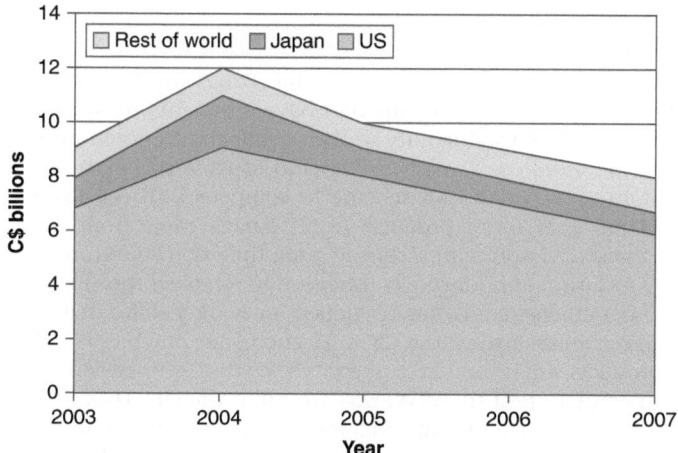

Figure 4.3 Canadian softwood lumber exports, 2003–7
Source: Adapted from NRCan 2008.

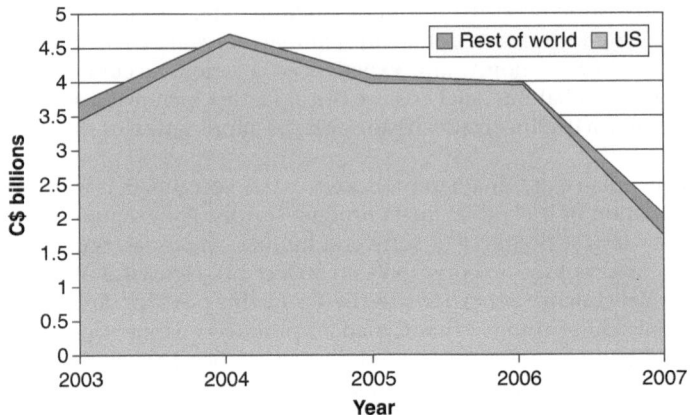

Figure 4.4   Canadian structural panel exports to the US and rest of the world, 2003–7
Source: Adapted from NRCan 2008.

billion in 2007), total exports to the US had declined to 63 per cent of the total, with the rest of the world accounting for the balance. Of its several categories of wood products, Canada is most diversified in pulp exports. From 2003 to 2007, it averaged around C$7 billion worth of market pulp production annually and shipped only about 40 per cent to the US. China is becoming an increasingly important export market for Canadian pulp, doubling from about one billion dollars in 2003 to about C$2 billion in 2007 (NRCan 2008e, 15).

Consumer awareness of environmental issues in the US is high. Consequently, foreign forest companies have been obliged to substantiate claims regarding the sustainability of their products. As noted in the previous chapter, in 1993 the Canadian Pulp and Paper Association announced it was initiating Canada's CSA scheme to compete with the FSC's and to reassure clients that forest products from Canada came from sustainably managed forests. At approximately the same time the American Paper and Forestry Association announced its intention to establish the SFI scheme. As these schemes emerged, a battle developed for market share that pitted the CSA and SFI schemes against the FSC's. To encourage purchasers to buy FSC-certified products, environmentalists established the Certified Forest Products Council in emulation of the WWF 95+ Group in the UK. They also worked with the LEED green building programme and the Environmental Paper Network to gain preference for FSC-certified products. These efforts produced some important 'victories' in the certification wars, one of the most notable being Home Depot's 1999 adoption of a responsible purchasing policy that gave preference to forest products from well-managed forests (Fisher et al. 2005). The importance in the 1990s of offering certified forest products when exporting to the US market stands in stark contrast to the situation then prevailing in Asia and highlights how different the commodity chain was for Australian and Canadian producers during forest certification's formative years.

UK

The UK timber industry produces largely for the domestic market, where it experiences ongoing competition from low-cost imports. The essential problem confronting UK producers emerges from a consideration of Table 4.10, which compares total British production of forest products to imports from elsewhere. By inspection, it can be seen that UK manufacturers produce only a fraction of the total timber consumed within the UK and experience significant competition from foreign sawnwood and paper and paperboard producers. However, the high volume of imported wood-based panels, which is almost equivalent to those produced domestically, and the absence of any especial commitment to buy locally produced wood products means that even domestic producers of panels must meet the average import price to remain competitive.

Mason highlights the economic difficulty confronting UK timber producers, therefore, especially those in the more remote regions.

In 2003, imports of coniferous industrial roundwood from the three areas [Scandinavia, Russia and the Baltic States] were priced at about £35 m3 (FAO, 2005), thus setting a benchmark for Scottish grown timber. Average harvesting costs (i.e., fell and extract to roadside) are about £10 m3 (I. Murgatroyd, Forest Research, pers. Comm, 25 July 2005), which leaves a comparatively small price margin for the grower.

(Mason 2007, 48)

Table 4.10 Domestic production and imports of forest products in the UK

Year	Sawnwood		Wood-based panels		Paper and paperboard	
	Domestic 000 m3	Imported 000 m3	Domestic 000 m3	Imported 000 m3	Domestic 000 tonnes	Imported 000 tonnes
2003	2750	8714	3361	3492	6226	9112
2004	2783	8653	3533	3813	6240	9251
2005	2783	8223	3398	3552	6039	9434
2006	2709	7963	3498	3685	5588	9332
2007	3146	8402	3549	3891	5228	9396

Source: Adapted from the Forestry Commission 2008.

Mason's comments are echoed by Lawson and Hemery (2008) who observe that 'A crucial issue ... is the collapse of standing timber and sawlog prices over the past 20 years. ... In real terms coniferous sawlogs were worth around 30% of their value in 1998, and standing sales of conifers only 20%' (Lawson and Hemery 2008, 27).

It is in this context then that UK producers have considered certification. On the one hand, with small profit margins they have been highly resistant to any initiatives that place additional costs on the firm. On the other hand, increasing volumes of certified timber imports have put pressure on them to obtain certification. Moreover, the UK is also environmentally sensitive and a large and active environmental movement ran highly effective publicity campaigns on the dangers of purchasing uncertified wood products. WWF formed a buyers group in 1991 in the UK with the participation of important do-it-yourself stores like B&Q. It and other large companies committed to purchasing only FSC-certified timber by 1995 (WWF 95+ Group), placing substantial pressure on domestic producers to become certified or lose market share.

A recent investigation conducted by the Timber Trade Federation for the DFID highlights the pressure domestic producers in the UK remain under to maintain their certification status. According to the report, the certified proportion of all timber and panel products imported into the UK rose from just under 56 per cent in 2005 to over 81 per cent in 2008. The increase in certified forest products imports occurred at the expense of uncertified forest products. While the total import market declined from 11.48 to 9.15 million m³, certified forest products increased from 6.41 to 7.45 million m³, while uncertified forest products declined from 5.07 to 1.7 million m³.

Fish and fisheries in comparative perspective

Fish are important commodities in each country, but with significant differences in the industry's size and scope. Australia has a much smaller fishery that either Canada or the UK. Wild or capture fisheries are fully developed in each country, and in some cases stocks are over-exploited. Each country has seen a rapid increase in aquaculture operations that has changed the traditional form of the industry and commodity chain and a reorientation to higher-value fisheries. Fish trade is also an important feature in each country, with each exporting and importing fish, although the trade balances differ, with Canada importing less fish relative to either Australia or the UK.

In each country fisheries management has evolved under similar pressures within increasing concerns over the state of fish stocks and the impacts of fishing. While all countries are characterised by multilevel governance frameworks, the context for fisheries policy is also shaped by the ecology of the respective resource base. Shifts to ecosystem approaches to fisheries management have been promoted as part of a broader sustainability agenda, and while some progress has been made, and the approach included in contemporary policy documents, this management paradigm has yet to be fully implemented in any of our case studies. At the same time fisheries commodity marketing has changed markedly, with similar trends observed in each of our selected countries. Fish products are now primarily marketed through supermarkets rather than traditional fishmongers, and in processed or semi-processed or prepared forms, rather than whole fish. The fish sold in these supermarkets is part of a global fish trade that provides important supports for certification, initially focusing on food safety and quality but also addressing sustainability of supply.

Comparative fisheries ecologies

Wild or capture fisheries targeting fish, crustaceans and molluscs are fully developed and in some cases over-exploited in Australia, Canada and the UK. The resource base is very different with Australia having a much smaller fishery than either of the other two countries. Ground fisheries are also more important in Canada and the UK than in Australia while crustaceans (rock lobster and prawns) have been major fisheries. Noticeable declines in recruitment to the rock lobster fishery off Western Australia have had a major impact on the scale of this fishery, once Australia's most valuable. The ground fishery off Atlantic Canada suffered a collapse in the 1990s with major consequences to the fishery and to the economic and social well-being of the region. In response fisheries in this region have moved to targeting crustaceans. While levels and values of aquaculture production have risen markedly in each country, there have been a number of impacts. Concerns in Canada by indigenous peoples over the farming of wild species have been matched by concerns by industry over impacts of aquaculture operations on wild fisheries through increasing the possibility of, and vectors for, disease. In Australia, as will be noted below, the largest fishery by production is the small pelagic fishery targeting sardines, an industrial fishery with the majority of product used for aquaculture feed.

Australia

Australia has a significant maritime domain, the third largest maritime jurisdiction in the world, with an EEZ of 16 million square kilometres (twice the area of the continental land mass) extending from tropical to Antarctic waters (Haward 2003). While fish and fisheries have been and remain important to the Australian economy, the industry ranks around 50th in the world for production. The industry has undergone significant development and change over the past two decades illustrated by the fact that in 2007–8 salmonids (aquaculture species) were the largest catch by production and second largest by value to rock lobster.

In relation to wild-caught stocks, most of Australia's wild fisheries are fully, if not over, exploited. Of 98 Commonwealth-managed species (referred to officially as stocks), 13 are classified as 'overfished' and 44 as 'not overfished'. The status of the 41 remaining species is uncertain, with some of these possibly subject to overfishing. Wilson et al. (2009: 8–12) note that 'the high proportion of stocks that remain classified as uncertain ... is a continuing cause for concern. The extent of uncertainty highlights the importance for AFMA [Australian Fisheries Management Authority] of applying the precautionary approach in fisheries management' (Wilson et al. 2009: 12).

While uncertainty is a factor that generally affects fisheries management because the physical stock of biomass cannot be easily measured or counted, it can be especially high with regard to low-value fisheries where there is reduced funding available for research (McLoughlin 2006). Uncertainty can also increase when fish species are caught incidentally as bycatch or as by-product species to high-value target species.

Australia has a diverse fishery base, with operations targeting molluscs, crustaceans and finfish (Table 4.11). Most fisheries are in waters close to the coast, although the past 15 years have seen operations targeting pelagic (shallow or mid-water) species such as tuna in the Indian and Western Pacific Oceans and demersal (bottom-dwelling fisheries) such as Patagonian Toothfish in the Southern Ocean thousands of kilometres from the Australian coast. The emphasis on high value, relatively low production fisheries, is also shown in this table. Sardine production targeted as industrial fishing for aquaculture feed, pet food and recreational fishing bait 'rose by more than 700 per cent between 1999–2000 and 2004–05', a harvest that accounted for 14 per cent of the total Australia catch in 2007–8 (ABARE 2009b, 2). In contrast the abalone fishery (a marine mollusc) had a relatively small catch totalling 5300 tonnes. It is based on small fishing grounds around Tasmania, South Australia and Western Australia and was worth an estimated A\$188.5 million (ABARE 2009b). Concerns over the state of Australian fish stocks have led to actions limiting catches, restructuring fisheries and developing new approaches to management such as harvest strategies that focus on the impact of fishing on stock biomass. The development of aquaculture operations for finfish such as salmonids and

Table 4.11 Australian fisheries production by volume and value, 2007–8 (top five fisheries)

Fishery	Tonnes	Value (A\$ millions)
Australian sardines	33,600	NA
Salmonids	25,500	299
Prawns	22,400	268
Tuna	14,700	210
Rock lobster	13,800	407
Abalone	5,300	189

Source: ABARE 2009b.

southern bluefin tuna has led to an increased harvest of small pelagic stocks such as sardines for aquaculture feed.

Canada

Canada's marine and freshwater fisheries landed 1.9 million tonnes of product in 2007. The fisheries included important salmon fisheries, demersal ground fisheries (halibut, cod and haddock), pelagics (herring, tuna and mackerel), crustaceans (lobster, crab and shrimp) and molluscs (scallops and clams). Commercial fishing operations employ approximately 100,000 people, and fish and seafood exports were worth C$3.9 billion in 2007 (DFO 2007). The value of marine capture fisheries was C$1.9 billion in 2007, a decline from C$2.26 billion in 2004. Aquaculture and fresh water fisheries production had a value of C$64 million in 2004, a reduction of 25 per cent over 2002. The decline is attributed to a reduction in the volume of aquaculture salmon production of almost 30,000 tonnes, which was down by 23 per cent on previous years (DFO 2007, 1) (see Table 4.12). Fisheries contribute less than one per cent of Canada's gross domestic product (DFO 2007), but they provide important contributions in the coastal communities across the country.

Canada's fisheries have been marked by collapses of major fisheries on each of Canada's coasts in the last decades of the twentieth century. In addition to shaping debates over Canadian fisheries management these events had significant impact on broader Canadian politics. The collapse of the Atlantic ground fishery for cod off the province of Newfoundland in the early 1990s provided a shattering blow to coastal communities whose community and economic well-being was intrinsically bound to the cod fishery. The collapse also had significant impacts on the Atlantic region's economy, both in relation to loss of jobs and income and also in relation to the increasing dependence on federal government welfare payments in a region already characterised as suffering from dependency (Harris 1998; Mathews 1983; Charles 1995). Lobster and crab fisheries have replaced finfish fisheries as major export fisheries in Atlantic Canada, where the 'cod crisis' also served as a catalyst for a reassessment of management policies and tools used to govern Canada's fisheries (DFO 1993). 'In the early 1990's,

Table 4.12 Canadian fisheries production by volume and value, 2006

Fishery	Tonnes	Value (C$ millions)
Salmon	24,000	60
Ground Fish	258,000	329
Pelagic Fish	296,000	132
Crustaceans	338,000	1165
Molluscs	117,000	176

Source: DFO 2008, pp. 20–1.

groundfish played a major role in the fish harvesting and processing sectors, but over time the dominance of groundfish decreased. In 2006 groundfish as a whole represented less than 17% of the total landed value of marine commercial fishing in Canada' (DFO 2008, 5). In a marked change in the Atlantic Canada region 'crustaceans replaced groundfish as the main species harvested' and as a result of their higher value than ground fish stocks 'Canada's total landed value has increased despite declining overall landings' (DFO 2008, 5).

Concern over salmon stocks on the West Coast in the 1990s, too, led to considerable federal–provincial conflict. As in the Atlantic Canada conflict, 'the salmon wars' (Brown 2005) between the federal government and other stakeholders coloured much of the 1990s in the Pacific Salmon fisheries. This conflict clearly shaped responses to initiatives such as MSC certification.

UK

Fishing has long been important in the UK, and was a major activity around the coastline and in the North Sea and the Atlantic. Cities such as Aberdeen, Grimsby and Hull developed as ports and market towns on the base of 'the fishing' (Raban 1986). Fleets targeted fishing grounds in the North Sea, English Channel, Irish Sea and Atlantic, moving further offshore as developments in technology allowed larger, powered vessels greater range and opportunities to market products to major population centres. The development of fishing grounds in the North Atlantic led inevitably to conflicts with vessels from other countries also keen to exploit stocks of fish such as cod and assertions of jurisdiction over such stocks. This resulted in the long-running conflict with Iceland from the 1950s to the 1970s termed the 'cod wars'.

The decline in the levels of production and value of catch taken from offshore or distant water fishing industry from the mid-1970s has meant that in- or near-shore fisheries and aquaculture operations have increased in importance. The first years of the twenty-first century saw fishing policy and management again at the top of the policy agenda, after a period of relative neglect, paralleling the decline in the industry. A key driver of this new phase in UK fisheries was concern with sustainability, both in terms of stocks and the marine environment and also the industry. This has also led to increased attention to fisheries by environmental civil society organisations (see Gray, Gray and Hague 1999). As a result the UK has a greater level of overt support towards external (i.e. non-governmental) eco-certification of fishers in general and the MSC in particular than either Australia or Canada. This is driven by market pressures and influenced by a sophisticated environmental lobby actively engaged with and concerned over the state of European fish stocks.

The fishery is dominated by demersal (bottom dwelling or groundfish) species such as cod, haddock and whiting (see Table 4.13). Pelagic species

Table 4.13   UK fisheries production by volume and value, 2008

Fishery	Tonnes	Value (pounds millions)
Demersal Fish	82,000	52.0
Pelagic Fish	45,000	23.4
Shellfish	2,200	3.0

Source: DEFRA 2009.

such as herring and mackerel are important fisheries, while shellfish production from scallops, nephrops and crab fisheries provide smaller catches
but relatively high-value fisheries. There have been marked changes in the
make-up of the UK catch in the last decade. Cod catches have declined a
third from 15,000 tonnes in 2005 to 10,000 tonnes in 2008. The haddock
catch provided almost a third of all demersal landing in 2008. Of pelagic
catches herring, a mainstay of the UK fishery for centuries, has remained
important but mackerel provided approximately half the volume of production (DEFRA 2009).

The UK fishery is responsible for about 12 per cent of the European catch
(UKRCEP 2004, 39–40). In 2002 the fishing industry in the UK had 7590 registered fishing vessels (UKRCEP 2004, 39–40), most under 10 metres in length.

> Scottish registered vessels accounted for 65 percent of the weight and 60
> percent of the total value of total landings by the UK fleet. English and
> Welsh vessels accounted for 30 percent of the quantity and 34 percent
> of the value of the total, whilst Northern Ireland vessels represented
> 5 percent by quantity and 5 percent by value.
>
> (FAO 2008)

The sector supports almost 12,000 direct jobs (UKRCEP 2004, 40) although
this level of employment is significantly less than the mid-1990s. Full-time
employment declined by one-third and part-time by 40 per cent in the
period 1995–2004 (URCEP, 2004, 40). 'While fishermen account for a small
percentage of the national workforce (0.2% in Scotland and 0.1% in England
and Wales), they make a significant contribution to some local economies as
they tend to be regionally and locally concentrated. Around 20% of UK fishermen are located in the south west of England and 13% in Aberdeenshire,
Scotland' (Prime Minister's Strategy Unit 2004).

The FAO has noted that

> 686,000 tonnes of sea fish were landed into the UK and abroad by the
> UK fleet with a total value of £546 million. Around 75% of UK demand is
> met through imports, with some £344 million of fish imported in 2002.
> Around 50% of UK catch (by value) is exported.
>
> (FAO 2008)

The loss of access to Icelandic fishing grounds as a result of the diplomatic settlement of the cod wars was one factor leading to the decline of the UK's offshore fleet, with direct impact on employment in the industry. Many of these vessels were sold to operators targeting new fisheries in the Southern hemisphere, in particular the orange roughy fisheries off New Zealand and Namibia (Bergin and Haward 2000).

Comparative fisheries policy contexts

Fisheries are important primary industries in each of our selected countries. Fisheries policy in each country is shaped by the nature of the resource base but also by the governance frameworks within which fisheries management is conducted. While all our countries are characterised by multilevel governance frameworks, each is distinctive with resultant impacts on the context for fisheries policy. Fisheries management in Australia is shared between the national and sub-national government while in Canada the federal government has major responsibility over fisheries. In the UK traditional unitary governance has given way to first the influence of supra-national policy frameworks through the EU and more recently internal devolution. The policy context is shaped by internal factors such as the interactions between governments and between governments and non-governmental interests and stakeholders. External influences include the growing international environmental agenda of the past 35 years. The outcomes of these processes, such as concepts of (ecologically) sustainable development and ecosystem-based approaches to fisheries management, have helped shape the policy context. 'Management' – actions undertaken by the state to develop and/or implement measures to regulate fishing – is itself an important element, overlaying the 'conservation' of stocks, 'community' interest, and the 'economic' performance of industry.

Australia

Management of Australian fisheries is shared between the Commonwealth and state and Northern Territory governments. As a result intergovernmental relations derived from the federal division of powers and responsibilities provides a further dimension to the politics of fisheries management and adds another element to the policy context (Haward 1995). The question of jurisdiction is one of a number of factors that influence the management of Australia's fisheries. The interaction between these factors contributes to a particularly complex, yet relatively under-examined, policy environment. Commonwealth power is based on section 51 (x) of the constitution: 'Fisheries in Australian waters beyond territorial limits', and state legislation under provisions of their constitutions. Jurisdiction is divided by a boundary three nautical miles offshore from low water mark (LWM) defining 'state waters'. This boundary, introduced by the passage of the Coastal Waters (State Powers) Act 1980 and the Coastal Waters (State

Titles) Act 1980, provided the basis of management of fisheries under the Offshore Constitutional Settlement (OCS) (Haward 1989). A key element of the application of the OCS to fisheries was the classification of fisheries (Haward 1989). Fisheries can be managed by either:

- the Commonwealth;
- a state, given responsibility to the edge of the AFZ;
- retain the status-quo with State and Commonwealth legislation; or
- a 'joint authority' to manage fisheries which cross jurisdictional boundaries.

While the OCS encouraged the use of joint authorities, there has been little progress towards the establishment of these bodies (SSCISTTCI 1993). A number of factors contributed to the limited use of such arrangements, including tensions between governments. Initial criticism has been overcome by a more flexible, cooperative approach. Despite an early view that OCS agreements were 'an essential precondition for good management' (SSCISTTCI 1993, 8), such agreements are conspicuously absent in New South Wales fisheries and have been notoriously difficult to achieve in some other fisheries (Haward 1995, 2003).

A major turning point in the administration and management of Australian (and particularly Commonwealth) fisheries occurred in early 1985. The Australian Fisheries Conference led to the establishment of a new peak industry body and advisory bodies. The 1980s ended with the release of a major Commonwealth policy statement, *New Directions for Commonwealth Fisheries Management in the 1990s* (Government of Australia 1989), which provided the basis of major legislative and administrative changes implemented in 1991. The development of institutional arrangements governing Australian fisheries in the 1990s clearly illustrates the interdependence of these paradigms. These arrangements, contained within the Fisheries Administration Act 1991 and the Fisheries Management Act 1991, established the AFMA in early February 1992.

AFMA has continued to evolve, being re-chartered as a commission in 2008, as opposed to its initial form which saw it administered by a board. One interesting development has been the establishment of Management Advisory Committees (MACs) within Commonwealth fisheries. The MACs serve several purposes: they provide a means for 'co-management' of the fishery, increase transparency of decision making and increase the efficiency and effectiveness of industry–government relations. The MAC structure is replicated (sometimes in slightly different forms) in fisheries under state jurisdiction. These arrangements are necessarily diverse, ranging from liaison committees to fishing industry councils. Developments in Queensland augmented the MAC concept with Zonal Advisory Committees (Tanzer 1995). A House of Representatives inquiry into AFMA, which reported in June 1997, reiterated the importance of industry involvement in management through the broadening of the role and membership of MACs. Contemporary bodies are more representative of broader stakeholder interests.

One cannot understate the influence of these legislative and administrative reforms on the fishing industry. One response was the change in name of the peak industry body from the National Fishing Industry Council to the Australian Seafood Industry Council (ASIC). ASIC, however, had a chequered history, being wound up in 2005 after experiencing financial problems. Industry also responded to increasing responsibilities provided by these reforms that facilitated their involvement in management.

Canada

Canada's national government is responsible, broadly speaking, for management of marine fisheries, through the Fisheries Act. This responsibility is exercised primarily through the Department of Fisheries and Oceans (DFO).[9] The provinces manage aquaculture operations and are responsible for land-based activities associated with fishing activity, such as fish-processing facilities. Provincial governments are also responsible for habitat and, through administrative agreements with the federal government, for management of most fresh-water fisheries (Haward et al. 2003; Pross and McCorquodale 1987). In addition, local government and First Nations, as well as a wide range of non-governmental stakeholders have significant interests in Canada's marine fisheries. This complex overlap of interests and responsibilities provides a complex framework for governance. VanderZwaag, for example, described the Canadian fisheries management system as a ghost ship: 'everyone knows the system exists but it often lies veiled under a mysterious mist of flexibility and informality' (VanderZwaag 1983, 172).

Canada's Fisheries Act originated in 1868 and enables government to regulate 'for the proper management and control of seacoast and inland fisheries' as well as controlling pollution of the marine environment. This legislation has been progressively amended, with developments in the 1990s recognising the influence of new international and regional agreements. In 1996 the federal government introduced legislation to 'streamline and modernize' the act after considerable opposition from commercial fishers and First Nations (Haward et al. 2003). One impetus for this action was the conclusion of the United Nations Fish Stocks Agreement in which Canada had taken a strong and leading role as part of its campaign to address the problem of 'straddling stocks', referring to the lack of coastal states' control over stocks that straddled the EEZ and the High Seas, as occurred on the 'Nose' and 'Tail' of the Grand Banks off Newfoundland. Foreign fishing in these areas assumed the status of high politics in Canada, and particularly in the Atlantic region, as the ground fishery stocks collapsed (Haward et al. 2003).

The relations between the federal and provincial levels of government have a longstanding and ongoing significance in fisheries. Considerable efforts to reduce intergovernmental conflict have occurred, most notably with the development of an intergovernmental Agreement on Interjurisdictional Cooperation with Respect to Fisheries and Aquaculture in June 1999 (Canada 1999; see also Haward et al. 2003). Among other things, this agreement created a continuing Canadian Council of Fisheries and Aquaculture Ministers intended to meet annually to promote cooperation and coordination. The Pacific Council of Fisheries Ministers was created in July 2003, paralleling the previously established regional council, the Atlantic Council of Fisheries and Aquaculture Ministers.

These events occurred at a time of increased activism by First Nations over access to fishery resources and later concern over development in aquaculture operations (see VanderZwaag et al. 2006). Salmon aquaculture, in particular, has been the focus of concern from 'wild' salmon fishers over animal health and disease and possible vectors affecting wild stocks. As in other places, developments in aquaculture provide opportunities but also highlight a range of environmental issues. Coastal communities too, concerned over declines in fisheries that were the economic and social bulwarks of these communities, generated substantial pressures for restructuring and reorientation of fisheries management.

First Nations interests in fisheries have been reinforced by key Supreme Court cases, such as the Sparrow case and (particularly with respect to Atlantic Canada) the Marshall case. The Sparrow case was decided in 1990, when the Supreme Court of Canada issued a landmark ruling in a case involving a native fisher from British Columbia. In the Sparrow decision, the Supreme Court upheld the rights of aboriginal people to fish for food, social and ceremonial purposes. The ruling noted that under the Constitution Act 1982, aboriginal peoples' rights to the food, social and ceremonial fishery have priority over other uses of the fishery, including commercial fishing, but are subject to overriding considerations such as conservation. The court also said that it was necessary for the Government of Canada to consult with aboriginal groups when their rights might be affected (Haward et al. 2003). The Marshall case arose when Donald Marshall, a Mi'kmaq, was charged with commercial fishing without a licence. His defence was that the Mi'kmaq have a treaty right to fish for commercial purposes. In September 1999, the Supreme Court of Canada ruled in favour of the defendant, thereby recognising Mi'kmaq treaty rights to the commercial fishery. In the west coast the Haida Nation have taken court action that includes a claim to the sea and seabed surrounding Haida Gwaii (Queen Charlotte Islands) on Canada's Pacific Coast (Haward et al. 2003).

Canadian fisheries policy is also shaped by its engagement at international and bilateral levels. As noted earlier, arrangements with the US have been significant in the management of Pacific Salmon stocks, and boundary agreements in the East coast with the US have also been driven to a certain extent by fisheries issues. Canada has been 'adventurous' in its pursuit of its interests in relation to fish stocks within its maritime jurisdiction. A sharply drawn example is Canada's actions over straddling stocks in the 1990s. Although Canada did not ratify the United Nations Convention on the Law of the Sea until November 2003, it had actively asserted its interests in its offshore waters in the 1970s. Canada, along with many other coastal states extended its maritime jurisdiction as part of customary law arising from deliberations at the United Nations Third Conference on the Law of the Sea (UNCLOS III) held between 1974 and 1982 (see Haward and Vince 2008). Canada declared a 200-mile EEZ in 1977 and saw this as a significant element in increasing Canada's share of the Atlantic cod fishery. The declaration of the EEZ was a means to counter the problems of distant water fishing pressure on the Grand Banks off Newfoundland (VanderZwaag 1992).

Canada had also been active in arguing for a broader application of sustainable development to fisheries in forums such as UNCED and the federal government argued that its reform to the Fisheries Act in the mid-1990s indicated its commitment to post-UNCED principles (Haward and VanderZwaag 1995). A DFO Backgrounder declared at the end of 1995 that 'The Canadian government continues to demonstrate leadership by being the first country to amend its domestic legislation to implement the precautionary approach to fisheries management following the signing of the UN *Convention on Straddling Fish Stocks and Highly Migratory Fish Stocks*' (DFO 1995 in Haward et al. 2003).

UK

Fisheries policy and management in the UK is shaped primarily by its engagement with the European Union's Common Fisheries Policy (CFP).[10] The UK's entry into the EU in general and the EU's Common Fisheries Policy in particular had marked effects on UK fisheries. Traditional heartlands of UK fisheries declined and operators left the industry or shifted focus, while at the same time the remaining fisheries looked to a range of approaches to add value to their production. These actions took place as fisheries (globally, as well as within the UK) were subject to increasing scrutiny by civil society groups. The UK policy environment differs markedly from Australia and Canada, therefore, as the latter countries are not subject to a supra-national policy framework that affects national fisheries management decisions. While the CFP is the single most important influence over UK fisheries policy and management, the process of multilevel governance that it engenders (Kelemen 2002; Haward and Vince 2008) have similarities to practices within Australian and Canadian federations, where responsibility is shared between central and sub-national jurisdictions, and where policy is shaped by processes and practices of inter-governmental relations.

The frameworks established by the EU in areas such as the CFP have led to a 'transfer of authority for fisheries policy making from national government to European institutions' (Symes and Phillipson 2009, 1). The CFP 'was created to manage a common resource and to meet the obligation set in the original Community Treaties'. It was developed in accordance with the understanding that a common policy creates 'common rules adopted at Community level and implemented in all Member States' (Europa – Common Fisheries Policy 2006). The CFP addresses four areas: (a) conservation of fish stocks (b) structures (such as vessels, port facilities and fish-processing plants) (c) the common organisation of the market, and (d) external fisheries policy which includes fishing agreements with non-Community members and negotiations in international organisations. The UK (and its constituent governments) does, however, retain important responsibilities in addition to implementing and enforcing the CFP. These responsibilities include

- the ability to take non-discriminatory measures for the conservation of stocks and to minimise the impact of fishing on marine ecosystems up to 12 nm offshore, provided they are in line with the CFP, and to restrict fishing by non-UK vessels in waters up to that limit to vessels that traditionally fish in those waters;
- the right to take temporary emergency measures where there is evidence of a serious and unforseen threat to the conservation of living aquatic resources or to the marine ecosystem resulting from fishing activities; and
- the ability to take conservation and management measures in all waters provided they are applicable solely to UK flagged vessels.

(UKRCEP 2004)

The CFP uses two types of instruments to conserve fish stocks. The first involves setting total allowable catches, which are upper limits for total amount of fish that can be landed from particular areas and gear restrictions, closures and size limits. In addition, the CFP also includes measures that attempt to control the capacity of the EU fleets (Daw and Gray 2005, 189). Despite its sophistication and its focus on developing the EU's fisheries, the CFP has been criticised for failing to achieve its major objectives (e.g. Daw and Gray 2005; Gray and Hatchard 2003). Attempts to restructure fleets and reduce capacity have had limited impact in reducing catches. Overfishing of major stocks in areas like the North Sea and the Mediterranean has been a cause of growing concern over the last decade. Most commercial species in the North Sea are heavily exploited, some to the point of collapse (UKRCEP 2004).

A key part of the CFP was the provision of reviews of the policy. These reviews have taken place in 1992 and 2002. The first review of the CFP

> showed that if there are too many vessels for the available resources, technical measures and control alone cannot prevent overfishing. The amount of fishing has to be regulated too. In order to make the common fisheries policy more effective the link between its component parts was reinforced. Control measures were also developed to ensure that rules are respected throughout the industry. New technologies are being used to transmit data to the authorities and to monitor larger vessels through satellite tracking systems
>
> (Europa – Common Fisheries Policy 2006)

The 2002 review highlighted that 'more effective conservation and management of fisheries resources is a clear priority' (European Commission 2002, 7). The review process included extensive consultation including publication of a 'Roadmap' document that provided a detailed analysis of the issues facing the CFP, noting significant changes in both the internal and external policy environment in which the CFP operates. The 2002 reforms strengthened mechanisms to reduce over-exploitation, greater emphasis on reduction of the fishing fleet and decommissioning of vessels, a focus on environmental protection, addressing illegal fishing and discards and improving sustainability of aquaculture (UKRCEP 2004, 71). This included a clear attempt to improve communication with industry and formalise a devolved institutionalised arrangement for management advice. A 'significant strand of the reform is the move to regionalise the management of CFP through establishment of seven Regional Advisory Councils' (UKRCEP 2004, 71).

In 2003, following the introduction of significant limitations on fishing time as a result of declines in cod stocks, and concomitant impacts on the UK fishing fleet including decommissioning of vessels (DEFRA 2007), the UK prime minister, Tony Blair, commissioned his Departmental Strategy Unit to investigate and report on the UK fishing industry. The strategy unit completed its work in 2004 and its report, *Net Benefits: A Sustainable and Profitable Future for UK Fishing* (Prime Minister's Strategy Unit 2004), provided an important impetus for discussion and policy development. The standing Royal Commission on Environmental Pollution's 25th report, *Turning the Tide: Addressing the Impact of Fisheries on the Marine Environment* (UKRCEP 2004), also released in 2004, provided further reinforcement to concerns over the state of the marine environment and the impacts of fishing. The Royal Commission noted that in relation to the marine environment 'collateral damage ... to the environment can be severe. Fishing can be likened to unsustainable tropical logging. ... We need to care as much for the seas around us as we do for the rainforest' (UKRCEP 2004, 3).

Net Benefits 'proposed a possible long-term strategy for improving the fortunes of the fishing industry and the communities that depend upon it'

(DEFRA 2008). The report 'recommended that industry should work to achieve Marine Stewardship Council (MSC) certification (or equivalent) for key stocks by 2015' (DEFRA 2005, 46). Working groups of stakeholders and officials met between autumn 2004 and summer 2005 to develop the policy framework in *Securing the Benefits*, the response to *Net Benefits* (DEFRA 2008). *Securing the Benefits* was a joint response from the Department of Environment, Food and Rural Affairs (DEFRA) and the Department of Agriculture and Rural Development and the Scottish Assembly and the Welsh Assembly Government. The Scottish Executive, no doubt given the relative importance of Scotland's contribution to the UK fishing industry, decided to develop its own implementation plan, part of a broader strategy related to marine resources and environmental management.

In 2007 DEFRA released *Fisheries 2027*, subtitled 'A Long-term Vision for Sustainable Fisheries' (DEFRA 2007). This statement recognised that 'a sustainable fisheries sector is essential for delivering the Government's vision of *clean, healthy, safe, productive and biologically diverse oceans and seas*' (DEFRA 2007, 5, original emphasis). The report's 'vision statement are ambitious ... all stakeholders will need to work together' to achieve sustainable fisheries (DEFRA 2007, 10). *Fisheries 2027* noted that responsibilities of 'certification bodies and standard setters' included:

- Making schemes as accessible as possible;
- Considering the impact of schemes on developing countries and small producer organisations;
- Being open about the benefits and limitations of schemes; and
- Ensuring consistency with eco-labelling guidelines of the Food and Agriculture Organisation

(DEFRA 2007, 12)

Implementation of the *Fisheries 2027* vision is a major challenge. As stated in this report 'delivering sustainable fisheries will require big changes in the way we think about, manage and exploit fisheries in England, the UK, the European Union and internationally' (DEFRA 2007, 17).

Comparative fisheries management discourses

The late twentieth century saw significant reforms to approaches to fisheries management. Concern regarding the limitations of traditional methods of management such as regulating the number of fishers or boats or type of gear (tools known as input controls) saw the adoption of new approaches that focused on the volume caught such as quotas or catch limits. These latter approaches (known as output controls) attempted to restrict 'effort creep' – ever-increasing catches as new technologies and techniques overcame the constraints set by input controls. At the same time increased attention was focused on the impacts of fishing, both in relation to the target stock, but also in terms of bycatch (simply any species of fish or non-fish that is caught while targeting a stock). Shifts to ecosystem-based approaches to fisheries management have been promoted as part of broader sustainability agendas, and while some developments have been made this management paradigm is yet to be fully embraced.

Australia

Australia, in general, conforms to a pattern of broadening and deepening of discourses on fisheries management, from early regulatory efforts in the colonial period to developing fisheries management systems that have increasingly addressed concerns at impacts of fishing on both target stocks and the broader ecosystem. Contemporary management, focusing on ecosystem impacts and utilising harvest strategy, adopts an ecosystem-based management approach. ESBM aims to overcome the problems associated with past techniques that attempt to see fisheries reach maximum yield, and constitutes a significant improvement on earlier approaches. In turn ESBM utilises sophisticated modelling and stock assessment techniques and improved monitoring and surveillance approaches.

European settlement from the late-1700s saw fisheries develop, with early government interest and action directed at expanding fisheries. At the same time, the first regulations to 'manage' fisheries were enacted in the 1850s.

> Dependence on fisheries increased with European settlement, and as pressure on fish stocks increased so too did early attempts to regulate use. Fisheries management rested largely with the colonies and, after 1901, with states until the passage of the *Fisheries Act* 1952 (Cth). In the ... years since this act was proclaimed in 1955, the Commonwealth has increased its influence to the point where it is a major actor in management of Australia's fisheries
>
> (Haward 1995)

The discourse for managing fisheries in Australia has shifted from input controls (limits to gear, number of boats or fishing days) to output controls (limiting the level of catch through imposition of quotas), marking a significant shift in fisheries management and the activity of fishers. Allocation of 'quota' has altered the legal and economic basis of managing fisheries. Access privileges (the traditional fishing 'licences') have no property rights attached to them, but may accrue market value in high-value fisheries governed by limited entry criteria. The introduction of Individual Transferable Quotas (ITQs) in Commonwealth and in some state-controlled fisheries has provided a form of property right in these fisheries. The Fisheries Management Act 1991 (Cwlth) required AFMA to establish a system of statutory fishing rights in Commonwealth-controlled fisheries. Questions of statutory rights in wild or capture fisheries (and the level of resource rents accruing from these rights) have been raised in several Federal

Court cases in the 1990s. Economic instruments have increased in salience through policy to recover costs of management from participants in fisheries through licence fees and levies. Cost recovery was the subject of a major study by the Industry Commission in 1992. The 1993–4 Commonwealth budget emphasised a commitment that users should pay for management services in proportion to benefits received, and from 1994–5 the industry has been responsible for payments of 100 per cent of all attributable costs for management of Commonwealth fisheries.

Australia's indigenous peoples have utilised fisheries resources for thousands of years, and in some case developed complex tenure and management arrangements to govern 'sea country' (Haward 2003). Traditional fishing activities by Australia's indigenous peoples raise important management issues such as access to resources and sea claims (see Bergin 1991, 1993; Exel 1994), and can result in direct conflicts with commercial fishing and recreational interests. Recreational fishing is also significant in Australia and has salience in discussions over environmental performance and assessment of Australian fisheries. Recreational fishing is now more regulated but some concern has been expressed at the size of catch in specific areas and fisheries. Surveys indicate that 4.5 million Australians undertake at least one fishing trip a year with over 800,000 people being regarded as 'serious' fishers (Industry Commission 1992, 203). Recreational fishing introduces another social dimension to fisheries management, with a range of tools being used to manage this sector. Impacts of recreational fishing may be significant in specific fisheries with ongoing consequences for management. Fishing-related tourism is an important activity in coastal towns in regional Australia (see Haward 2009).

Emphasis on improving the sustainability of fisheries is mandated by legislative requirements that focus on principles of ecologically sustainable development. This increased focus on ecosystem approaches to management has emerged as a driver in fisheries policy. Developments in broader government policies, primarily through the implementation of Australia's Oceans Policy, reinforce the requirement that fisheries policy and management needs to reorient towards an ecosystem-based approach. Ecologically sustainable development (ESD) has become an important driver of Australian fisheries policy and management. Key operational principles underpinning the use of sustainable development as a driver for fisheries management at international and domestic arenas developed from UNCED in 1992. These parameters include integrated management, the precautionary approach, intergenerational equity, polluter pays, public participation and indigenous rights. Agenda 21's Chapter 17, embraced by Australia, deals with oceans and coasts (including EEZs) and provides a specific action plan which emphasises the requirement for 'new approaches to marine and coastal area management at the national, subregional, regional, and global levels, approaches which are integrated in content and are precautionary and anticipatory in ambit.'

Canada

The evolution of discourses regarding fisheries management in Canada has strong parallels to Australia's in the early years of development. However, the fisheries crises in Canada in the 1990s had important impacts on discourses over fisheries management. As Charles (1995) has noted, one result was that fishers no longer accepted government-imposed regulations, particularly as the fishers saw 'government' as part of the problem in relation to the fishery collapse in Atlantic Canada. Alternatives, including development of community-based approaches to fisheries management, emerged in Canada, as in other countries. These approaches, in general, attempt to shift from a 'top-down' model of government control and towards greater 'empowerment' of local organisations and participatory structures. In Canada this saw the move to a consultative model, rather similar to Australian fishery Management Advisory Committees, in which the government discussed management measures with the industry prior to implementation but did not delegate decision-making power. More participatory approaches, engaging actively with the fishing industry have, however, been developed (Pinkerton 1999). Canada, at both federal and provincial levels, has attempted to provide a vehicle for co-management with commercial and First Nations fishers as well as other stakeholders. These arrangements use a range of tools including 'territorial management', 'community-based management' and 'community quotas' (Haward et al. 2003).

Territorial management has a relatively long history in Canada originating from communal and kinship linkages of fishers as well as the practices of the First Nations peoples prior to European settlement and the location of their harvesting grounds. These arrangements link the strong community base of the fishery to practices at sea (Haward et al. 2003). The contemporary development of this arrangement is seen in the concept of territorial use of rights in fishing (TURFs) (Haward et al. 2003). TURFs are rights assigned to individuals and groups or both to fish in certain locations, generally, although not necessarily, based on long-standing traditional use and harvest practices. It is well established that under suitable circumstances TURFs can serve as relatively stable and socially supported components of a fishery-management system (Haward et al. 2003).[11]

Community quotas defined on a geographical basis tend to bring people together in a common purpose, with fishers in a given community managing themselves, perhaps also with the involvement of their community. In the Scotia-Fundy region, the implementation of community quotas has strengthened the management role of regionally based fishermen's associations, including those located along the eastern part of the Scotian Shelf (Haward et al. 2003). The concept of community-based management has become increasingly popular across Canada in a wide variety of natural resource sectors, including fisheries and forestry. In the fishery sector, community-based management may be found on all three of Canada's coasts, involving both aboriginal and non-aboriginal participants. On the Pacific coast, a number of community-based fishery and coastal management initiatives have been developed. A particularly innovative example is that of the West Coast Vancouver Island Aquatic Management Board which evolved from a grass-roots local initiative into a regional management body involving federal, provincial, municipal and First Nation representatives as well as local citizen members (Foster and Haward 2003).

UK

Concern over the management of UK fisheries is not new. The impact of fishing on the marine environment including the impact of gear on the seafloor and on bycatch species has long been recognised in the UK. In 1376 a petition was sent to King Edward III 'requesting that he ban a new and destructive type of fishing gear [causing] great damage of the commons of the realm and the destruction of the fisheries' (Roberts 2007, 136–7). These concerns over the impacts of bottom trawling have re-emerged in the 1990s and 2000s as part of broader debates over the state of the marine environment and the impacts of industrial fishing (Roberts 2007). A second and related impetus is derived from consumer and governmental concern over quality and safety of seafood products.

Fisheries management in the UK has developed from input controls and gear restrictions to more complex quota allocations within the context of the EU's Common Fisheries Policy. While technically a 'rights-based' system of management, this quota allocation system is not as sophisticated as the transferable quotas used in Australian or Canadian fisheries nor is it linked to formal harvest strategies that have been developed in Australia and to a lesser extent in Canada (Beddington, Agnew and Clark 2007). While lagging behind Australia and Canada in the introduction of harvest strategy approaches to management the UK is constrained by the policy context set by the Common Fisheries Policy. The current review and reform of the Common Fisheries Policy is central to the UK's future fisheries management, with the government committing itself to the WSSD objectives that, inter alia, encourage an ecosystem-based approach to fisheries management.

A range of international forums are engaging with and responding to environmental pressures for responsible and sustainable fisheries (see Haward and Vince 2008). This agenda includes promotion of ecosystem-based approaches to fisheries management and the integration of environmental concerns into the more traditional 'developmental' or extractive concerns of fisheries management. The second source is the pressure from consumers, governments and civil society groups to maintain food safety and quality with seafood products. Clearly these two sources are related and increasingly linked through certification and ecolabelling arrangements. Fisheries which are managed sustainably and which produce food of high quality will be in better shape to cope with increased environmental scrutiny.

Comparative fisheries commodity chains

Fish are important commodities, both in terms of domestic consumption and export products in each country. Imports of fish are important in each country but Canada stands alone as having a positive fish 'trade balance', exporting vastly more fish than it imports. It is the fish trade that drove certification with an initial focus on food safety and quality but also addressing sustainability of supply. Fisheries commodity marketing has changed markedly over the past five decades, with similar trends in each of our selected countries. Fish products are now primarily marketed through supermarkets rather than the traditional fishmongers, and in processed or semi-processed or prepared forms rather than whole fish. Dependence of local fleets and stocks has given way to products shipped from significant distances, and moves away from 'fresh' fish to frozen product. This has increased the salience of consumer buying power and enhanced opportunities for ecolabels to be used in consumer education. While each country has experienced similar broad trends, outcomes are different, with greater take-up of certification and labelling and use of CoC certification in product lines in the UK than in either Australia or Canada.

Australia

Australia has a distinctive fishery resource base, which influences the structure of the commodity chain through a focus on the export of high value fish species such as rock lobster, abalone, tuna and salmon (the last two produced from aquaculture operations). The country also imports large quantities of lower-grade fish. Australian wild fisheries experienced a period of impressive growth in the 1990s, in relation to the production and value of catch, but with production and value declining in the following decade.

> In real terms (2005–06 dollars) the value of Australian fisheries production has been declining since 1999–2000. In 2005–06 the gross value of fisheries production was 2.12 billion, 25 per cent less that the real value in 1999–2000.
>
> (ABARE 2007, 1)

The value of seafood exports has also declined by 36 per cent in 2005–6 compared to 1999–2000 while the quantity of imports of edible fisheries products increased by 35 per cent (ABARE 2007). As demand for seafood has increased, so too has the level of imports. 'Australia imports vast quantities of seafood from Asia, New Zealand, North America, South America, Europe and elsewhere' and the 'appreciation of the Australian dollar since

2002–03 has simultaneously made exports less competitive and imports more attractive to consumers' (Yearsley, Last and Ward 1999, 3). Declines in production and value of wild fisheries have been offset by an increase in value of aquaculture production (Haward 2006). Between 1996–7 and 2005–6 aquaculture's share of total value of Australian fisheries production grew from 25 to 35 per cent (ABARE 2007). The value of aquaculture has grown in value from \$237 million in 1990 to \$748 million in 2005–6.

Australia's balance of seafood trade in 1997 indicated that the value of exports was almost double that of imports. This trade balance was achieved by offsetting high value exports for high levels of imports of lower value but significant amounts of fish for domestic consumption. Exports of high-value fisheries including rock lobster and abalone to Asian markets have been a mainstay. By 2005–6 this gap had closed. Seafood supply is clearly a significant element in future planning. While Australian fisheries exports continue to grow, limited production for the domestic market means that increasing levels of lower-value imports will be needed to maintain expected growth in demand over the next 20 years. Aquaculture of new species may provide further opportunities for development, although such opportunities are limited by a number of environmental factors and available species and areas (Haward 2006).

While the Australian fishing industry is generally small relative to the major fishing powers, the industry is diverse and includes small owner-operator 'bay and inlet' fishers and offshore and 'distant water' fishing operations, the latter operating in the Southern and Indian Oceans (Haward 2009). Australian wild capture fisheries together with associated processing activities are vital rural and regional industries, sustaining a number of communities around the Australian coastline. Australia also has significant interests in aquaculture with a range of farmed species contributing to the growth in total value of the industry (Haward 2006). Aquaculture production counted for 40 per cent of the gross value of Australian fisheries in 2007–8 (ABARE 2009b). Ongoing development of aquaculture operations has led to increased concerns over sustainability of operations and impacts. For example Australian sardines, mainly caught for aquaculture feed, were the largest catch of any Australian fishery in 2007–8 at 33,600 tonnes (ABARE 2009b). The fishery for sardines has contributed to total Australian fisheries production growing by one per cent by volume over the decade to 2007–8. Over the same period, however, the value of production has declined by 22 per cent, driven by reduced catches in high-value fisheries such as rock lobster as well as an appreciating Australian dollar, which seriously affected export earnings. Comparative data for the value of different Australian fisheries is given in Table 4.14.

Australia is a significant importer of seafood (Table 4.15). In the five years since 2003–4 'the real value of Australian fisheries imports has risen by \$153 million (12 per cent), being mostly driven by greater imports of fresh,

Table 4.14 Australia's top four seafood exports by value, 2007–8

Fishery	Value (A$ million)
Rock lobster	401
Abalone	217
Tuna (whole)	202
Prawns	69

Source: ABARE 2009b, 17.

Table 4.15 Australia: Top four seafood imports by value, 2007–8

Product	Value (A$ million)
Canned fish	257
Frozen fish fillets	228
Fresh, chilled or frozen prawns	167
Canned crustaceans and molluscs	128

Source: ABARE 2009b, 18.

chilled and frozen prawns and frozen fish fillets ... the share of edible fishery imports from China and Viet Nam has risen, although Thailand and New Zealand remain Australia's main source of edible imports' (ABARE 2009b, 18). In summary, therefore, Australia's wild-capture fisheries are fully developed. Aquaculture is a major contributor to the industry that is generally focused on exporting high-value products, while at the same time importing significant quantities of seafood.

Canada

Canada ranks 20th in the world in terms of volume of marine capture fisheries (DFO 2008), but is ranked sixth in relation to value of seafood trade. This was a significant decline from Canada's position as the world's second largest exporter of fish products in terms of value of catch in 1990. This decline was caused by the collapse and closure of the cod fishery in the 1990s (DFO 2007, 30). Salmon was the major export, with Sockeye, Pink and Chum salmon mainly exported to Europe, while the aquaculture salmon, Chinook, Coho and sockeye salmon were exported to the US (DFO 2008, 11). Canada exports two-thirds of the world's smoked herring and over half the world's lobster exports (DFO 2008, 3) (see Table 4.16).

The US is the principal export market for Canadian product with over 60 per cent of Canadian exports by value. The US was the major market for Canadian lobster, while 'Japan was the main destination for herring, sea urchin, albacore tuna, Pacific Dungeness crab, Greenland turbot and sablefish' (DFO 2008, 11). The value of imports increased slightly over the

Table 4.16 Canadian fishery exports, 2008

Product	Value (C\$ million)
Groundfish	442
Pelagic fish	992
Shellfish	2278

Source: DFO 2008, 10.

preceding year: in 2006 imports were valued at C\$2.12 billion. The main imported species were shrimp, lobster, tuna, salmon, cod and haddock. Together, these species represented almost half the total value of Canadian fish imports in 2006 (DFO 2008, 12), but 'Canada imports far less fish and seafood than it exports' (DFO 2008, 4). The Canadian seafood export commodity chain, particularly to Europe and the US, is significant in leveraging concerns over fisheries' sustainability and certification in these countries back to Canadian industry, government and civil society organisations.

UK

Fisheries in the UK have been subject to considerable stress over the past three decades. The loss of, or reduced access to, 'traditional' fishing grounds in the 1970s following the 'cod wars', and the negotiation of the LOSC with the concept of 200-nautical mile EEZ becoming customary law and practice, therefore limiting access of UK vessels, had a major impact on UK fisheries. While greatly reducing the scale of the industry it has in turn affected the commodity chain.

> In 2008, landings by UK vessels into the UK (based on landed weight) fell by 27 thousand tonnes. Imports rose by 33 thousand tonnes and exports fell by 51 thousand tonnes. The net effect is an increase of 56 thousand tonnes in the amount of fish available for domestic use.
>
> (DEFRA 2009, 59)

UK vessels landed 397,000 tonnes of fish in the UK in 2008, worth £518 million, while importing 781,000 tonnes of fish worth £2,207 million (DEFRA 2008, 59).

The increased dependence on imports is interesting and provides impetus for ECSO campaigns on sustainability of stocks (see Table 4.17). A range of states provide product to the UK market: 'the largest exporters to the UK in 2008 were Iceland (103 thousand tonnes), Germany (71 thousand tonnes) and China (55 thousand tonnes)' (DEFRA 2009, 64). In turn the UK also exported considerable amounts of product, with the Netherlands (75 thousand tonnes), France (75 thousand tonnes) and Russia (44 thousand tonnes) respectively the largest export markets (DEFRA 2009, 64).

Table 4.17 Fisheries imports to the UK

Product	Tonnes	Value (pounds millions)
Demersal and pelagic fish	671 000	1,770.6
Shellfish (crustaceans and molluscs)	110 000	436.0

Source: DEFRA 2009, 61.

Enhanced industry interest in value adding product together with increased significance of supermarket chains as major points of sale and consumer pressure on food safety and quality have provided important elements in enhancing opportunities for fisheries certification. The UK government initiated the compulsory registration of buyers and sellers of fish in 2005, an initiative that was seen to have ended an earlier 'fairly permissive regime' (Symes and Phillipson 2009, 4).

Conclusion

The management of forestry and fisheries has occurred in markedly different historical, ecological, constitutional and discursive contexts in Australia, Canada and the UK. In the UK, the experience of timber shortages during World War I had a marked impact, creating a forest security discourse that justified an aggressive strategy of afforestation by an entrenched Forestry Commission that lasted most of the twentieth century. In Canada, the traumatic events of the collapse of the cod fishery in the 1990s exercised a similarly profound effect on fisheries management discourse, but in this case undermined the dominant regulator, the DFO, and created momentum for industry and environmental groups to push for more community-based approaches. While Australia has not experienced events of similar magnitude, the exotic nature of its endemic flora and fauna has seen the articulation of an especially vibrant wilderness discourse that contrasts markedly with the 'ancient woodlands' and 'amenity' discourse in operation in the UK forest sector. Numerous other cross-country differences can be noted too. Australia's export dependence on Asian markets for forest products contrasts sharply with the UK's import dependence on Scandinavia; Canada's centralised federal administration for managing fisheries differs surprisingly from Australia's complex multilevel governance arrangements managed through AFMA; aquaculture is relatively more important to Australia than for either Canada or the UK; and the UK's massively modified ecology shares little in comparison with Canada's vast tracts of old-growth, native primary forests.

Despite these significant differences between sectors and countries, there are also some notable similarities in trajectories. In each country, the general historical pattern of resource management that moves from 'mining' to 'sustained-yield' to 'sustainable natural resource management' has occurred. While the pattern occurred relatively early in some jurisdictions and sectors and significantly later in others, the 1992 United Nations Conference on Sustainable Development gave strong impetus to convergence towards 'sustainable' forestry and fisheries management programmes. While the precise content of these programmes varies, there is also an emerging consensus that 'sustainable development' has failed to deliver on its promise resulting in an increasing focus on ecosystem-based approaches to achieving natural resource sustainability. Canada appears to have moved furthest along this path in the forestry sector, while Australia appears to be further advanced in fisheries.

Finally, a notable feature of our comparisons is the dynamic nature of the interactions between discourses, constitutions, ecologies and commodity chains. Each has mutually influenced the other with no single element 'causally' responsible for determining how events unfold. This simply highlights the fact that the FSC and the MSC have been enmeshed in a wider set of ecological, constitutional, economic, social and discursive relations that need to be understood if we are to explain the differential responses of states to certification's emergence, establishment and development. It is to the responses by states to the FSC and the MSC that we now turn in the following chapters.

5
Forest and Fisheries Certification in Australia

The 'resourcist' and 'developmentalist' orientation of Australia's economy (Walker 1999) creates opportunities for producers to engage in regulatory capture via clientelistic and triadic policy networks. In the forest arena, triadic policy networks exist and are especially effective at the state level. While fisheries policy networks are generally less organised, conflicts over fisheries policy (as with water, agriculture and mining) are often resolved in favour of producer coalitions over the objections of environment and community groups. Environmental and community coalitions can, however, utilise Australia's federal system to stimulate national debates over what were previously considered parochial or sectoral concerns. For example, forest issues have been at the centre of Australia's resource wars since the 1960s, pitting powerful state-based triadic policy networks against state and national environmental and community groups. While the profile of fisheries was much lower during this period, the sector is now subject to greater external scrutiny. In this chapter we examine the development of forest and fisheries certification in Australia, commencing with the struggle over voluntary certification in the forest sector, which was deeply coloured by Australia's unique and unforgiving forest policy context.

Australian forest politics

In order to comprehend the Australian state's visceral reaction to the FSC and its high-level involvement in developing a competitor scheme in the shape of the AFS, it is necessary to examine the actor networks involved. This description of actor networks provides the necessary background to grasp the opportunities and threats presented by forest certification when it first appeared on the national agenda in the early 1990s. Once on the national policy agenda, each constituency tried to bend certification to fit its predetermined goals and objectives. In the 'no holds barred' Australian context, this meant that forest certification was enlisted as a weapon by each constituency to further the fight and gain an advantage. Its potential to act as a forum to bring warring parties together to debate the meaning of ecologically sustainable forest management has not been really explored until recently.

Environmental coalition

The Australian environmental movement consists of a large number of state environmental groups linked to nationally based ECSOs. These organisations span the spectrum of the environmental social movement ranging from 'radical' groups like TWS and Friends of the Earth Australia (FoE-Australia) on the one hand to 'reformist groups' like the World Wide Fund for Nature-Australia (WWF-Australia)[1] and The Nature Conservancy on the other (Beder 1991; Hall and Taplin 2007), sometimes united in campaigns under the national umbrella group, the Australian Conservation Foundation (ACF). These organisations fashion complex, non-linear relationships around specific campaigns and issues, while jealously guarding their individual autonomy. They adopt different modes of action that vary from 'insider' strategies related to political lobbying, research and policy development to outsider strategies associated with media 'stunts', demonstrations and public blockades. With respect to forest politics in Australia, TWS, FoE-Australia and WWF-Australia have played the dominant role at both state and federal level, often in collaboration with the ACF.

National ECSOs cooperate with state-level counterparts, which have their own state-based conservation councils. Doyle lists the membership of the Conservation Council of South Australia, noting it is composed of over 50 groups including Adelaide Bushwalkers, Bicycle Institute of South Australia, Field Naturalists Society of SA, People for Public Transport, South Australian Ornithological Association and TWS (SA Branch) (Doyle 2000, 98–101). These broader-based conservation councils often encounter difficulty in agreeing strong policy positions, unlike TWS, FoE-Australia, WWF-Australia and the ACF. Significant tensions can also arise between state-based conservation councils when some members collaborate on campaigns with one or more of the big four ECSOs over the strong objections of other members.

Of the big four, WWF-Australia has a reputation for operating mostly as an 'insider' ECSO, preferring to utilise existing political institutions to influence policy development. Doyle observes that 'WWF is the most business- and government-friendly environmental NGO', and that 'There is some debate as to whether WWF is still actually part of the movement, or whether it has simply become a high-level, wise-use-style front group for industry' (Doyle 2000, 80).[2] WWF-Australia was founded in 1978 in Sydney and claims to have 80,000 supporters within the Asia-Pacific region, where it runs projects, lobbies governments and carries out research and policy advocacy work. The organisation is run by 57 'governors' who are 'appointed because of their commitment to WWF's mission, their standing in the community and their ability to contribute to our success' (WWF-Australia 2008a). Because it is a company limited by guarantee, only the governors are allowed to attend annual general meetings and vote. Existing governors nominate new governors. New governors are formally appointed by the existing board and their membership ratified at a subsequent annual general meeting.

Like WWF-Australia, FoE-Australia is a 'branch' of a larger body, FoE-International. However, the FOE system operates as a loose network of affiliated but independent groups and the institutional connections between FoE-Australia and FoE-International occur mainly through International Liaison Officers operating out of Melbourne. This loose network arrangement operates within FoE-Australia, which has affiliated local groups based in Victoria, Queensland, New South Wales and Western Australia. While there are regional contacts in Tasmania and the Northern Territories, there is as yet no formal FoE-Australia Local Group in these latter locations. Structurally, FoE-Australia is a 'grassroots' or 'bottom-up' organisation in contrast to WWF-Australia's top-down, elitist structure. Members join local groups (for example the FoE-Adelaide Local Group) and in practice the organisation appears to encourage a two-tier participatory arrangement, with 'supporters' contributing money and time at the outset and then later making the transition to membership when they have proved themselves (FoE-Australia 2006).

Whereas WWF-Australia emphasises the natural environment, conservation and wise use, FoE-Australia stresses the importance of its 'environmental justice' approach to environmental issues. Specifically its members believe that 'pursuing environmental protection is inseparable from broader social concerns' and that a 'strong social perspective means FoE is well placed to develop alliances at the grassroots level with other organisations and communities' (FoE-Australia 2008a). Its grassroots approach, focus on environmental justice and modest financial resources mean that FoE-Australia engages in a wide range of strategies that reflect its outsider position with respect to relevant policy networks. Key campaigns run by FoE-Australia include opposition to uranium mining and nuclear energy, demands for climate change justice (within Australia and between the Global North and South), protests against unsustainable plantation expansion (especially in Victoria) and calls for the regulation of nanotechnology and clean and renewable energy. Its environmental justice values and support for indigenous peoples has led FoE-Australia members to object to the concept of 'wilderness'. FoE-Australia argues that the concept of wilderness 'effectively removes Indigenous people from ecological history' and that the 'environment movement has been complicit in the dispossession of Indigenous people on a number of significant occasions in Australia'. Hence FoE-Australia 'does not endorse the use of the term wilderness', believing the 'natural landscapes of the continent are effectively cultural landscapes that have been formed over thousands of years through the interplay between natural processes and human management regimes' (FoE-Australia 2008b). This position puts it in direct conflict with the third of our big-four Australian environmental civil society organisations, TWS.

Originally known as the Tasmanian Wilderness Society, TWS was founded in 1976 following the failed efforts of Bob Brown and others to halt the flooding of Lake Pedder by Tasmania's Hydro-Electric Commission (HEC) (TWS 2008a). TWS came to national prominence during the Franklin Dam campaign. Following the commencement of construction, TWS organised a blockade that eventually saw over 1400 protestors arrested, including Bob Brown who spent three weeks in jail. TWS backed the Australian Labor Party (ALP) in 1983 national elections, contributing to the party's electoral victory under Bob Hawke. On coming to power, Hawke delivered on his pre-election promise of halting the construction of the Franklin river dam, precipitating a constitutional struggle over the relative powers of the Commonwealth and Tasmanian governments over land rights. The majority ruling of the full bench of the High Court adjudging the Commonwealth to have the power to declare state lands as a world heritage protected area catapulted TWS onto the national stage. The organisation changed its name to The Wilderness Society in 1986 in recognition of the fact that it had already established campaign bases across the country in most state capitals (TWS 2008b).

TWS is an incorporated association under the Tasmanian Associations Incorporation Act 1964. Like WWF-Australia and in contrast to FoE-Australia, TWS has a multimillion dollar budget that amounted to $12.152 million in 2006–7, up from $11.952 million the previous year. The major campaigns that TWS is involved in today revolve around its central concept of WildCountry. Branded as a 'completely new approach to nature protection in Australia', TWS defines WildCountry as 'protecting the best of what is left of Australia's natural environment and restoring important areas' (TWS 2005a, 11). TWS' WildCountry approach constitutes an overarching master plan for Australia, where 'environment plans or conservation strategies will be developed for every part of Australia' forming a 'continent-spanning, scientifically-planned network of WildCountry landscapes, with wilderness (the strongholds of nature) at its core, surrounded and linked by a combination of different protected and restored landscapes' (TWS 2005, 11).

The last of the big four ECSO's involved in Australian forest politics is the Australian Conservation Foundation. Following a suggestion by the Duke of Edinburgh in 1963 that an Australian branch of WWF be established, a distinguished entomologist, Francis Radcliffe, convened a meeting in 1964 in Canberra to discuss the idea, and two years later the ACF was born under the presidency of Sir Garfield Barwick, then chief justice of the High Court (ACF 2008a). Under this conservative and politically well-connected elite, ACF chose as its first campaign the Cape Barren Goose on the basis that it would be better to proceed cautiously and 'choose issues for which there was a lot of scientific information' (Alec Costin quoted in ACF 2008b).

Doyle notes, however, that despite some early successes, ACF's conservative approach to environmental issues was increasingly criticised from inside by some of its more radical members. Matters came to a head following the flooding of Lake Pedder, where ACF was perceived by some to have been 'a "collaborator" in the final decision' (Doyle 2000, 89).[3]

Warhurst identifies three 'stages' of ACF development. Although he does not label them, they could be considered 'conservative', 'radical' and 'moderate'. The first, conservative, stage ended in 1973 as a result of a 'palace coup', with members voting in a more radical council, appointing Dr Geoff Mosley as the new director, and adopting a motion 'to become a vigorous champion of conservation using every feasible means' (Warhurst 1994, 77). ACF's second, 'radical' stage lasted until 1986, during which it took on a more complete range of environmental issues including nuclear power and aboriginal land rights. It also became politically aligned to the ALP, with 'support for the coalition parties ... never a serious option because these parties rejected a strong national role for the Commonwealth government in environmental policy and considered the ACF too radical' (Warhurst 1994, 79). In 1986 the organisation moved more to the centre and in 1989 it elected former Midnight Oil lead singer Peter Garrett as president of the ACF. Strong ties with the ALP remained and Garret himself joined the ALP in 2004 and took over as Minister of the Environment in 2007 when the ALP won government under the leadership of Kevin Rudd.

The environmental movement led by the above four organisations played a key role in Australia's forest certification politics. However, the evolution of certification in the country was affected by tensions between the four organisations over the environment–people nexus, insider versus outsider tactics and organisational cultures. These ongoing tensions within the Australian environmental movement help explain the relatively cold reception the FSC received when it was launched in 1993 and are further discussed later.

Forest industry coalition

The forest industry coalition encompasses a diverse array of state and commonwealth organisations representing business, workers and communities. For much of the period under consideration, these groups were united under the industry's peak body, the National Association of Forest Industries (NAFI). While NAFI's authority has diminished somewhat recently with the establishment of the Australian Plantation and Paper Products Industry Council which aims to more directly present the interests of the plantation and paper sectors, the organisation nonetheless remains a powerful voice representing forest interests in general and native forest interests in particular. NAFI's power derives not only from the fact that it represents state-based forestry associations and large companies but also because it represents workers and forest-dependent communities. Workers are united within the Forestry and Furnishing Products Division (FFPD) of the Construction, Forestry, Mining and Energy Union (CFMEU), which is a NAFI member. Similarly, forest-dependent communities have become organised by Timber Communities Australia (TCA), also a NAFI member. NAFI's capacity to articulate the combined interests of business, workers and communities is enhanced by the professionalism of its staff, which mastered the media to pressure state and federal politicians to respond to its demands.

In Australia, forest companies are organised into state-based producer associations such as the Forest Industries Association of Tasmania (FIAT), the Victorian Association of Forest Industries (VAFI) and the Forest Industries Federation of Western Australia. For example, until recently FIAT's membership included two of Tasmania's largest forest products producers: Gunns and the now defunct Forest Enterprises Australia. Formed in 1983, FIAT's purpose is to 'represent the interests of processors of Tasmanian forest products' including woodchips, panel products and hardwood and softwood timber. It does this by lobbying governments, mainly at the state level, concerning forest policy. FIAT is well connected to the Tasmanian state, both directly and via other institutions such as the Forests and Forest Industry Council of Tasmania (FFIC). FFIC brings together not only processing interests but all associations 'with an interest in forest and land use issues' (FFIC 2010). Created in 1989, FFIC is now both a producer association and an industry council with an executive committee consisting of private and public interests. Private interests include the CFMEU's FFPD, Tasmanian Country Sawmillers Association, Tasmanian Farmers and Graziers Association and Tasmanian Forest Contractors Association. Public interests include Forestry Tasmania, Private Forests Tasmania and the Department of Infrastructure Energy and Resources.

As noted, state associations are integrated at the Commonwealth level within NAFI, which also has a number of high-profile corporate members including Gunns, Gunnersens, ITC, Great Southern Plantations, McCormack Timbers and Timbercorp. Since being founded in 1986 'on the initiative of senior forest industry executives, State Forest Services, State forest associations, unions and CSIRO' (NAFI 2008), NAFI has sought to represent the interests of its members vigorously to governments and the public. To do this, it monitors all aspects of the forest industry and regularly issues media releases commenting on developments. It also produces briefings for its members and the wider public on such topics as pulp and paper, woodchips, renewable energy, softwood plantations, forests and the economy, native forests, old growth forests and many others. Finally, NAFI's major mode of communication has been its newsletter, which it publishes every month and which contains a wealth of information on developments in the global, regional, national and local forest industry.

The major union in Australia with responsibility for forestry and related workers is the CFMEU. The CFMEU represents 120,000 workers, with 20,000 organised into the FFPD (CFMEU 2008). The CFMEU views itself as a 'progressive, militant and class struggle based union' that seeks to educate its members about 'the broader political forces that impact them at work and affect other aspects of their lives' (CFMEU 2008). The FFPD, under the leadership of Michael O'Connor since 1992, coordinates activities via state secretaries. As with other modern unions, the CFMEU not only aims to provide an effective voice for its members in state and federal policy-making but also provides a range of services to members in the workplace including information, industrial relations representation, education and training.

The FFPD has been actively engaged in disputes over conservation for two decades, continuously siding with producer interests and against environmentalists and the range of 'new economy' interests including tourism, sustainable agriculture and viticulture. In Tasmania, unions strongly promoted forest-led development, endorsing the Wesley Vale pulp mill in 1987, the Tasmanian Regional Forest Agreement in 1997 and the Tamar Valley Pulp Mill in 2004. The FFPD has also taken on ALP governments on forest policy issues. In 1995, the CFMEU in conjunction with NAFI and the Forest Protection Society organised a week-long blockade of federal parliament in Canberra to protest the 'locking up' of more native forest from logging (Ajani 2007, 6–17). A decade later, it actively campaigned against the federal ALP's generous A$800 million transition package for the Tasmanian forest industry, supporting instead the incumbent government's campaign promise to guarantee jobs in the industry.

The major group representing forest-dependent communities in Australia today is TCA. TCA describes its mission as 'to secure long term access to natural resources to generate employment and a future for regional communities' while ensuring that 'our unique Australian forests are scientifically evaluated and sustainable managed for the benefit of future generations and genetic diversity' (TCA 2008). TCA adopts a 'wise-use' approach to environmentalism, arguing in favour of 'productive conservation' and aiming to counter 'the misinformation promoted by those who seek to impose unrealistic, unfair and unnecessary levels of forest preservation and deny economic and community growth opportunities' (TCA 2008).

TCA was founded in 1987 as the Forest Protection Society and at that time was heavily supported by producer interests, notably NAFI. While environmentalists viewed its name as ironic, TCA argues that its aim was simply to protect all forest interests which included the 'protection of forests, protection of communities, protection of history and heritage, protection of livelihoods, families, individuals and jobs' (TCA 2008). Its early close association with producer interests tainted it in the eyes of environmentalists and Beder

(2004) argues, along with many others in the environmental movement, that it was and remains an industry 'front group' that promotes producer interests under the guise of a grassroots community group. This argument is shared by Bob Burton of TWS who argues that FPS/TCA is an Australian example of the North American phenomenon of environmental public relations employed by large producer groups to defend their interests (Burton 1997).

Today, TCA claims to have over 13,000 members organised into about 76 regional branches across Australia. It focuses on educating primary school children about sustainable native forest and plantation management (TCA 2008) as well as on public awareness campaigns, forest and industry tours and community relations. At the state level, TCA coordinators are heavily involved in forest politics, siding with producer interests against environmental, new economy and wider social interests. For example, as with the CFMEU's FFPD, TCA members in Tasmania strongly supported the Wesley Vale Pulp Mill, the RFA and the Gunns pulp mill.

In summary, industry interests are organised via a hierarchy of state and industry associations. Companies belong to state-wide industry associations such as FIAT and VAFI, which are in turn members of state forest industry councils (e.g. FFIC). State producer associations are members of NAFI, which also has direct membership from the same companies. Hence, larger forest interests are doubly represented within NAFI – by themselves and by their own domestic associations, which they dominate. In short, forest industry interests were extremely well placed in the 1990s to strategise around the issue of forest certification and, as we will see later, there is clear evidence they took advantage of the triadic policy network it was able to form with government to block movement towards certification initially and subsequently to endorse a competitor standard to the FSC's.

Forest certification in Australia

The emergence of forest certification in North America and Europe presented business, environmental civil society and state actors in Australia with a set of strategic challenges to which they were obliged to respond. At the outset, each constituency took a cautious approach. Environmentalists were suspicious, fearing certification was a Trojan horse to legitimise native, especially old-growth, forest logging. Business likewise considered certification unnecessary because Asian clients were not demanding it. At the governmental level, Commonwealth officials from the Department of Agriculture, Fisheries and Forests (DAFF) were prepared to adopt a wait-and-see approach and monitor developments in North America and Europe. The cautious approach adopted by all major constituency groups largely explains forest certification's relatively late arrival in Australia compared to our other comparator countries. It was only in 1999, following a series of important market and certification developments, that business and government were finally persuaded of certification's utility. At that point, and building on the experience of North America and Europe, they quickly mobilised national institutions to develop a competitor scheme to the FSC's.

Emergence of forest certification

Forest certification in Australia was spearheaded by two individuals with very different backgrounds: Tim Cadman and Leonie Van Der Maesen. Cadman was a founding member of the Native Forest Network (NFN) which formed in 1992. NFN was a global organisation with a substantial membership base in the US. It was at a conference of forest activists and biologists hosted by NFN in 1992 in the US that Cadman first learnt of the FSC initiative. Later the same year, he was consulted by Grant Rosoman of Greenpeace New Zealand about the role certification might play in Australia. Cadman noted that it could have a limited role if it were restricted to plantations. However, it would be unable to play a role with regard to Australia's native forests because several ECSOs had clear 'no native forest logging' policies. However, at this time the draft FSC P&Cs contained no principle or criteria covering tree plantations as their inclusion in the scheme was proving controversial. Thus the preliminary analysis conducted at this time suggested that forest certification would play a very limited role in Australia, although Cadman continued to be intrigued by its potential.

In the same year as Cadman was being consulted on the potential role of the FSC in Australia, a Dutch academic, Leonie Van Der Maesen, was investigating the sustainability of Western Australia's Jarrah, Karri and Tingle forests which were managed by the state agency, the Department of Conservation and Land Management. Van Der Maesen was a member of several networks, including FoE-International, which had branches in Western Australia as well as being on an advisory council to the Dutch government in Holland. On a visit to Canberra, Van Der Maesen met Bob Brown of the Australian Green party who encouraged her to visit Tasmania. It was on her visit to Tasmania that she met Cadman, who was deeply involved in a dispute over a wilderness area with high conservation value forests known as Jackie's Marsh. Cadman and Van Der Maesen quickly discovered that they shared a common interest in FSC certification and agreed to work together to introduce it into Australia.

Circumstances were hardly propitious, however. In the early 1990s, Australia was undergoing wrenching changes with respect to forest policy as it prepared for and dealt with the aftermath of the 1992 UNCED meeting. Several of the forest processes were heavily contested by industry and environmental networks, especially the Resource Assessment Commission's assessment of the forest industry (Economou 1996). Matters culminated in a dispute over the ALP's decision to expand the woodchip industry on the one hand and 'lock up' more forests in protected areas on the other. The first decision saw environmental groups take to the streets in protest, while the second culminated in a blockade of the federal parliament buildings by aggrieved log-truck drivers supported by the wider industry (Lane 1999). In the confrontational, winner-takes-all forestry politics of mid-1990s Australia, the market-based instrument of forest certification appeared rather insipid and had few backers. TWS, which was leading the campaign to establish parks and protected areas, feared that certification would undermine its campaign against native forest logging at a time when it was gaining momentum.

With Australia's environmental movement unpersuaded of the merits of certification, the vacuum might have been filled by business or government, but it was not. The forest industry was certainly concerned about the emergence of the FSC internationally and monitored the domestic context to see whether Australian ECSOs were pursuing it. However, they also thought that certification could be quickly and cheaply introduced into the country by adapting the emerging ISO 14000 EMS to the forest industry. With this in mind, Australian representatives to the ISO endorsed the establishment of an ISO forestry standard. ISO proceeded to establish an informal study group on the idea of an ISO forestry standard which met in November 1995 and again in February 2006, with participants from Canada, Australia and New Zealand (Elliott 1999, 17). Their move, opposed by environmental groups, encountered considerable opposition from within ISO. Ultimately the study group recommended that a 'bridging document' be prepared to guide forest companies in the application of ISO 14001 to their sector and the initiative petered out.

As the ISO initiative got underway, bureaucrats based in DAFF commenced work on a set of Australian regional 'criteria and indicators' for sustainable forest management as part of Australia's contribution to the Montreal Process. In a series of meetings held in quick succession between 1994 and 1995, Australia's forest bureaucrats networked with their international peers to draw up a set of seven criteria and 67 indicators designed to facilitate national reporting on 'sustainable forest management'. There were clear synergies between this C&I exercise and forest certification, since C&Is could be adapted to different national contexts and appended as 'performance' standards to national certification systems.

By mid-1995, Australia's forest policy network was approaching certification from two complementary ends. Industry was exploring the possibility of adapting the ISO 14000 standard to deal specifically with forestry, while DAFF officials – and the wider forest science community – were developing C&Is for sustainable forest management. Moreover, as a full participant in the post-UNCED institutions, Australia was also engaged in the Intergovernmental Panel on Forests, a multilateral forum that was established to continue the international dialogue on forestry. At its first meeting in 1995, the IPF set out a programme of work that included, *inter alia* 'the issue of the voluntary certification and labelling of forest products so as to contribute to a better understanding of the role of voluntary certification with regard to the sustainable management of forests including the impact of certification on development' (IPF 1995, 6). In an effort to consolidate work already done, build awareness in Australia concerning forest certification and fulfil commitments made to the UN, Australia undertook to host a major conference on voluntary certification and labelling in Brisbane, Queensland in 1996.

The Brisbane Conference was an important event in the development of certification and labelling in Australia. Not only was the conference attended by a large number of international experts[4] but it passed a resolution concluding that 'certification and labelling are potentially useful tools among many others to promote sustainable forest management' (DPIE 1996, xiii). Nonetheless a cautionary note was also sounded. The resolution observed that more evaluation of certification and labelling was required, including 'the scientific basis for defining and measuring sustainable forest management; governance and credibility of certification schemes; the roles of governments and international institutions/organisations; consistency with international agreements; harmonization/mutual recognition between schemes; the trade impacts of certification and labelling; and the role of environmental, economic and social objectives in achieving sustainable forest management' (DPIE 1996, xiii). In short, while there were more questions than answers, certification was endorsed at the Brisbane Conference by the international community as an important policy tool. Perhaps more importantly, it assisted Cadman and Van Der Maesen to make the case for forest certification in an Australian context within the ECSO community.

Slow progress

Despite the apparent success of the Brisbane Conference, it proved difficult to progress forest certification in Australia on either an FSC or a national scheme. One reason for this was that major constituency groups were fully engaged in negotiating regional forest agreements at the state and commonwealth level. Little spare capacity was available, therefore, to pursue forest certification. Hope also existed, at least in some quarters, that successfully negotiated RFAs could resolve Australia's 'war in the woods', obviating the need for the state to embrace the FSC or develop its own forest certification system. As noted in Chapter 4, however, the RFA negotiations exacerbated many forestry–environment conflicts because their ultimate purpose was to extract the federal government from local state forest conflicts. The compromises arrived at, often by bureaucratic fiat, ended up satisfying neither of the main protagonists.

Following the Brisbane Conference, Cadman and Van Der Maesen worked assiduously to gain the environmental movement's approval of the FSC with respect to certifying management practices in tree plantations. However, they continued to encounter ongoing scepticism regarding the FSC's worth in Australia. Their efforts were hindered by a major setback aimed at demonstrating the FSC's merits. In 1997, the North East Forest Alliance, a community-based group concerned about forest practices in northeast New South Wales, worked with Boral, a major forest products company, to trial FSC certification in their region. However, this collaboration proved disastrous when contractors hired by Boral marked trees where there was known koala habitat and then proceeded to log them (Russell 2003). In an extended report on forest certification in 1998, therefore, Cadman noted that 'the road to certification is still a long way off for Australia' and that 'conservation groups are not exactly rushing forward with proposals'. He traced this reluctance to the desire of the majority of conservation groups for a complete cessation of logging of native forests and a transition to tree plantations and scepticism regarding the practice of ecosystem-based forestry which was emerging in North America.

During this period, industry was also preoccupied with the RFA process and was unconvinced it required a national forest certification system. Instead, it continued to pursue ISO. NAFI's Warren Lang noted in March 1998 that ISO had approved the report of the Working Group on Forestry 'by 31 votes to nil, with three absentions' (NAFI 1998) and that 'the use of the generic standard (ISO 14000) in conjunction with Montreal Process Criteria and Indicators ... will enable Australia's forest managers, if they so choose, to have their policies and practices audited against recognised international yardsticks' (NAFI 1998). In this article, Lang endorsed the broader industry view that urgent action was not required. This view had emerged the previous year following a study tour on certification by the Queensland Timber Board. Tour participants reported that they had seen 'very little evidence of any volume of the world timber trades as being certified' leading them to resolve 'to monitor ... developments in C&L around the world, and to advise its members accordingly' (*NAFI Newsletter* 1997).

Despite being busy managing the RFA processes, officials at DAFF's International Forestry Division remained the most interested in forest certification. This was a consequence of their ongoing involvement in the Montreal Process and international forestry negotiations at the IFF. The former had turned its attention to implementing national C&Is and DAFF had established a Montreal Process Implementation Group to adapt the international C&Is to the Australian context, a process that reduced the number of indicators from 67 to 44. As this work progressed, it became clear that Australia could do what Canada had done and append its national C&Is to either an ISO or an Australian forestry standard to make it appear a more 'performance-based' approach. Meanwhile, the IFF continued to debate the role of certification and labelling for sustainable forestry management with the report of its third session calling for more practical experience in the use of forest certification, a request that was being rapidly pursued in North America and Europe (IFF 1999, 9).

Notwithstanding the breakthrough meeting at Brisbane in 1996 in introducing forest certification to Australia, little progress on forest certification occurred in Australia over the subsequent three years at a time when there were major developments elsewhere. The explanation for this state of affairs lies in ECSO and industry scepticism concerning certification's suitability for Australia. ECSOs were suspicious, because they viewed certification as a threat to no-native-logging campaigns and because they doubted the feasibility of practising ecoforestry in Australia. Industry was disinterested because demand for certified timber was limited, especially in Asian markets. Only government officials appeared to consider forest certification important, but in the absence of industry interest, they were unable to move the agenda forward.

The Australian Forestry Standard (AFS)

In 1997–8, the Asian currency crisis highlighted the forest industry's dependence on a single region and with markets drying up, forced it to examine opportunities elsewhere. In exploring alternative markets, especially in North America and Europe, the industry noted that while significant opportunities existed, clients in these regions were seeking some form of certification. Since it was also clear that ISO certification did not provide the requisite assurances, the industry finally began to consider a national scheme for Australia. It was in this context that a series of consultations took place in early 1999 which culminated in a stakeholder workshop hosted by DAFF in August the same year. The purpose of the August workshop was to develop an 'action agenda' for forestry and it identified six broad themes to deal with 'impediments' to industry performance. Following the workshop, six focus groups were established to further elaborate each theme, with representation drawn largely from industry associations. A central concern emerging from these focus groups was the credibility of Australia's forest operations and products and recognition of the 'growing demand for assurances' that forest products are 'sourced from sustainably managed forests' (DAFF 2000, 4). To deal with this marketing impediment, the first two of the Action Agenda's 12 strategic imperatives focused on the development of a national forestry standard (Strategic Imperative 1) and its mutual recognised by other schemes (Strategic Imperative 2).

The first strategic imperative was 'Develop and implement an Australian Forestry Standard (AFS)'. The 'driver' identified to accomplish this was an AFS Steering Committee (AFS-SC) composed of representatives from DAFF, NAFI, AFG, CFMEU, the Standing Committee on Forestry and the Standing Committee on Conservation. The Standing Committee on Forestry was a committee of the Ministerial Council on Forestry, Fisheries and

Aquaculture composed of forestry bureaucrats from state and commonwealth governments. The Standing Committee on Conservation was a committee of the Australian and New Zealand Environment Conservation Council, composed of environmental bureaucrats from state and commonwealth governments. To achieve its objectives under this strategic imperative, the Action Agenda identified several activities: (a) develop AFS via Standards Australia; (b) have AFS endorsed by the Forest Products Council and the MCFFA; and (c) trial and implement the AFS and develop a marketing and communications strategy to promote it (DAFF 2000, 11).

The second, and complementary strategic imperative, was to 'promote international acceptance of environmental and product certification schemes'. The main driver of this strategic objective was DAFF, which was given responsibility to 'explore and develop a voluntary mechanism for international cooperation on forest certification', to 'develop mutual recognition agreements (MRAs) with other forest practice certification schemes' and to 'lead international discussion with relevant involvement of Department of Foreign Affairs and Trade (DFAT)' (DAFF 2000, 11).

Action to implement both strategic imperatives began immediately. With respect to the first, the AFS-SC was established in 1999 as 'a partnership of the Commonwealth and State Governments, National Association of Forest Industries, Plantation Timber Association of Australia, Australian Forest Growers and the Australian Council of Trade Unions' (Standards Australia 2003, 2). Table 5.1 provides a breakdown of the membership of the AFS-SC,

Table 5.1 Membership of the Australian Forestry Standard Steering Committee, 1999–2003

Group	Representative
Primary Industry Ministerial Council (PIMC) (through the Primary Industry Standing Committee and the Forestry and Forest Products Committee) (3 members)	Hans Drielsma (Chair), Mike Bullen (State) and Rob Rawson (or alternate) (Commonwealth)
Natural Resource Management Ministerial Council (NRMMC) (through the Natural Resource Management Steering Committee) (1 member)	Max Kitchell (or alternate)
National Association of Forest Industries (NAFI) (1 member)	Warren Lang (or alternate)
Plantation Timber Association of Australia (PTAA) (1 member)	Richard Stanton (or alternate)
Australian Forest Growers (AFG) (1 member)	Peter Taylor (or alternate)
Australian Council of Trade Unions (1 member)	Michael O'Connor (or alternate)

Source: Standards Australia 2003.

which was composed of high-level representatives from national industry associations. The AFS-SC set up a Technical Reference Committee (TRC) of 19 members to draft the standard. AFS-SC characterised the TRC as being composed of 'independent professional and scientific experts, forest owners and processors, community and consumer interests and regulatory or controlling bodies' (Standards Australia 2003, 8). While the AFS-TRC was more representative than the AFS-SC in terms of stakeholder representation, producer groups dominated with less representation from environmental, social and indigenous interests (Table 5.2).

The AFS-TRC met between 2000 and 2002 to negotiate the content of the AFS. It was agreed at the outset that the AFS would be developed as a 'national standard' requiring Standards Australia processes to be followed. It was also agreed that it would be based on a set of national criteria and indicators derived from the Montreal Process. The draft standard put forward in September 2001 consisted of eight criteria; following public comments and further refinement, an interim AFS with nine criteria was subsequently published by Standards Australia in February 2003 (the additional criterion involved maintaining forests contributions to carbon cycles). Standards Australia issued the AFS as an 'interim' standard because it deemed it to lack the 'wider representation of interests' required. Thus in the preface to the Interim Standard, Standards Australia stated that the 'date of expiry for comment is two years

Table 5.2 Membership of the Australian Forestry Standard Technical Steering Committee, 1999–2003

Group	Representative	Participation
Chair	Hans Drielsma	Present
PISC via FFPC ˙		Present
NRMCC		Presents
PTAA		Present
AFG		Present
ACTU		Present
IFA		Present
National Furnishing Industry of Australia		Present
Timber Merchants Association		Present
Royal Australian Institute of Architects		Resigned
Consumers' Federation of Australia		Did not participate?
Forest scientist		Present
Environmental scientist		Present
WWF-Australia		Resigned
Tim Cadman (for ENGOs)		Resigned
ATSIC		Did not participate?
Victoria Forest Harvesting and Cartage Council		Present

Source: Standards Australia 2003.

after publication, at which time this Interim Standard will be confirmed, with-drawn or revised in the light of public comment, or published as an Australian Standard' (Standards Australia 2003, 2). The purpose of the two-year consulta-tion period provided by the Interim Standard process was to 'secure the partic-ipation of a wider representation of those interests concerned with sustainable forest management' (Standards Australia 2003, 2).

In fact, it proved extremely difficult for the AFS-SC to convince Standards Australia that it has secured the necessary 'wider representation'. This was because ECSOs launch a savage attack on the Interim AFS in a November 2005 publication titled *Certifying the Incredible*. ESCOs had two broad criti-cisms related to (a) the process used to develop the AFS, and (b) its substan-tive content. From a process perspective, ESCOs claimed that the AFS-SC was 'completely dominated by forestry interests' (TWS 2005b), that the AFS-TRC was similarly dominated, that voting arrangements within the AFS-TRC did not offset industry interests, that the chair was biased and that the procedures were poorly implemented (FERN 2004; TWS 2005b). On the sub-stantive side, ECSOs claimed that the standard would certify land clearing, broad-scale clearfelling, pesticide use, GMO use and the destruction of old-growth forests. It would also legitimate poorly structured and implemented forest practices codes and public consultation procedures.

The AFS-SC issued a detailed rebuttal of these ECSO claims in February 2006 (AFS 2006), prompting a rejoinder by TWS on behalf of ECSOs later the same month. This war of words and the failure of AFS-SC to attract the necessary 'wider representation' on the TRC continued through 2006 and into 2007. By then, however, the AFS-SC had managed to recruit the Ecological Society of Australia into the TRC and its members subsequently approved the revised Interim Standard following some modifications. This additional approval from the ecological community convinced Standards Australia to approve the standard resulting in its release as a full standard in August 2007.

The FSC

The development of the FSC in Australia followed an equally tortuous path to that of the AFS. In the late 1990s, Cadman and Van Der Maesen continued to seek the Australian environmental movement's endorsement of the FSC. The major forum in which to achieve this endorsement was the National Forest Summit, which met annually to discuss campaigns and network. At NFS meetings in 1999 and 2000, Cadman and Van Der Maesen proposed that the FSC be endorsed, but it was only at the April 2001 Summit in Hepburn Springs, Victoria that they finally achieved their goal. At this summit, approximately 35 ECSOs representing the vast majority of Australia's active forest campaigners endorsed in principle a proposal 'to investigate an FSC-type certification process for plantations and agroforests'

(Young and Cadman 2001). It was also agreed to establish several working groups to progress forest certification including definition of terms and a guide to processes. In taking these decisions, however, conservationists were motivated mainly by external events. The development of a national standard under the AFS and the decision by a major plantation company, Hancock Victoria Plantations (HVP), to seek FSC certification meant it was no longer feasible to postpone a decision on the FSC.

The HVP decision was of concern to some ECSOs who viewed it as a pre-emptive attempt to establish forest certification in Australia without the involvement of the conservation movement. Cadman, on behalf of the NFN, liaised with HVP, SmartWood and the FSC-IC to establish the terms and conditions under which the Australian environmental movement would support the initiative. Cadman made the case to the ECSO community at subsequent National Forest Summits that the international context was changing and that support was moving away from preservationist positions towards supporting alternative and more ecological forms of forest management backed by credible certification systems.

In 2001, Cadman resigned as NFN coordinator and took up a position as FSC Contact Person in Australia. His new role was to encourage the establishment of an the FSC 'National Initiative' and several efforts were made in the ensuing five years to establish one. At a meeting in Goolabri, near Canberra, in 2002, discussions occurred among a group of stakeholders that included members of the forest industry, forest workers and environmentalists. No agreement could be reached, however, on the form that the FSC would take and the meeting ended in some disarray. In a report of the meeting, Cummine and Wettenhall (2002, 9) note that 'despite desperate last minute attempts to salvage the situation, the conference did not agree ... to form the official FSC working group. ... All that was agreed was for nominated participants from each chamber to convene another meeting to attempt to resolve the still unresolved issues'.

More optimistically, Cadman reported the meeting induced several plantation companies to consider FSC certification (Cadman 2003). Given the difficulty of moving the FSC forward as an official national initiative, it now appeared that the best way of promoting the FSC in Australia would be to have Australian plantation companies certified under the interim standards of certifying bodies such as SmartWood and Soil Association Woodmark. Thus HVP received FSC certification in February 2004 under interim standards developed by SmartWood; and subsequently, a number of other plantation companies were certified including Integrated Tree Cropping (May 2004), TimberCorp (October 2004), Albany Forest Plantation Company (October 2004), Rewards Group (February 2006) and Hansol (March 2006).

In February 2003, Heiko Leidecker, the executive director of FSC, visited Australia and a small and more intimate meeting of those sympathetic to the FSC occurred in Melbourne. Given stakeholder sensitivities, Leidecker met with members of each of the FSC's three chambers separately. Considerable enthusiasm among a small group of like-minded individuals emerged at this time and they began to work with Cadman to establish an Australian national initiative. At a meeting in Melbourne in July 2004, the group resolved that those present constituted an FSC Interim Board and agreed to oversee the organisation's formal incorporation and to organise elections to establish an inaugural FSC board. While this constituted progress of a kind, the FSC Interim Board lacked resources and a year later little progress had been made. Meanwhile, Cadman had resigned as FSC Contact Person in July 2005 to pursue his doctoral studies. Not only was the FSC Interim Board not making much progress it had also lost its key motivator.

Circumstances improved, however, when Michael Spencer returned in August 2005 from the FSC-IC where he had been working as a marketing manager. Spencer had high-level contacts within Australia among environmental activists and industry and was committed to getting an Australian National Initiative off the ground. In discussions with several interim board members, a new strategy to establish the FSC in Australia was launched. The new strategy consisted of establishing an interim board of high-level individuals drawn from each of the FSC's three constituencies. Spencer also worked with a legal firm to finalise the FSC's articles of association. Finally, in March 2006, a new, six-member FSC Australian Interim Board was established with the environmental chamber represented by TWS and ACF, the economic chamber by TimberCorp and Spicers, and the social chamber by Doctors for Forests and Wombat Community Forest.

Analysing Australia's response to forest certification

As the above account makes clear, certification in Australia occurred along two separate tracks. The native forest industry collaborated with the government to develop the AFS while the plantation sector collaborated with environmentalists to promote the FSC. The native forest sector in Australia includes both public and private actors. Public actors include state forest agencies such as Forestry Tasmania, Forests New South Wales and Victoria's Department of Environment and Sustainability (including VicForests). Private actors with interests in native forest logging are grouped federally within the NAFI and locally within their own states through organisations such as VAFI and FIAT. Moreover, the corporatisation of public forest agencies in the 1990s meant that there was a close alignment between 'public' agencies of the state and the private sector, not only because both are charged with making a profit to return to shareholders but also because in many cases they are business partners. Thus, for example, the private Tasmanian company, Gunns Limited, signed a 20-year wood supply agreement with Forestry Tasmania in 2007, locking both parties into a long-term commercial arrangement.

An important institutional vehicle for bringing 'public' forestry officials together with regard to forest certification was the Ministerial Council on Forestry, Fisheries and Aquaculture. When the functions of the various ministerial councils were reviewed in 1999, the PIMC took over from the MCFFA as the major vehicle for mediating intergovernmental relations in the forestry sector. The MCFFA was managed by DAFF which, following the changed circumstances confronting the forest industry after 1997–8 Asian currency crisis, sponsored the major stakeholder workshop to identify strategic imperatives with respect to forestry. Hans Drielsma, the current general manager of Forestry at Forestry Tasmania, and formerly the general manager of State Forests New South Wales, chaired the AFS process.

After 2002, there was another forum where 'public' and 'private' interests were able to jointly pursue forest certification in Australia. This was the Forest and Wood Products Council which was another outcome of the 2001 stakeholder workshop. Since then, the council has played an important role in lobbying the government and private sector bodies on forest matters, including the promotion of the AFS. It achieved a long-standing campaign objective in 2009 when it persuaded the Green Building Council of Australia to expand its recognition of certified timber beyond the FSC to include the AFS.

Not only has the strong public–private triadic forest policy network that operates in Australia heavily endorsed the AFS scheme but the Commonwealth government's overt involvement in the scheme's development has meant that it became enmeshed in the certification wars. In 2006 then Minister of Agriculture, Fisheries and Forestry, Eric Abetz, was drawn into a dispute over whether the AFS provided a guarantee of legality and sustainability. The dispute arose because the UK government was assessing certification schemes under its FLEGT programme to determine which ones provided assurance of legality and sustainability. Australian ECSOs lobbied the agency conducting the assessment, the Central Point for Expertise on Timber (CPET), to deny accreditation to the AFS. The Australian forest industry launched a counterattack,impugning the structure and operation of the FSC (NAFI 2006). Abetz implied that the FSC was run by WWF and Greenpeace, while the opposition spokesman, Malcolm Ferguson, in an extended statement to the parliament, came close to suggesting the FSC was a protection racket run by WWF which

> has a history of establishing buyer groups that effectively boycott timber products that are not FSC certified. Consequently, producers and suppliers are pressured to obtain FSC certification to maintain their businesses and

their market access. The FSC's business interests are effectively protected by the environmental NGOs, who have mounted a concerted attack over recent years on other certification schemes. The AFS is just one of these schemes.

(Ferguson 2006, 144)

Section summary

Forest certification arrived late in Australia because the environmental movement was very suspicious of the FSC and unwilling to back it. This reluctance created a vacuum which industry, albeit reluctantly, eventually filled. While unconvinced at the outset, industry and government became increasingly aware that certification would become a basic requirement of market entry. After 1999, it mobilised the necessary resources – availing itself of substantial public and private funding – to develop an industry-friendly, national certification scheme, utilising existing institutional arrangements. The Australian state played a major role not only in developing the AFS but also in attempting to undermine the FSC by calling into question its legitimacy and objectivity. The scheme has been vilified by spokesmen from both major parties and, at times, with the full authority of the state. In short, in its determination of the 'public interest', Commonwealth and state governments sided very closely with the private sector. Ironically, some of the major proponents of AFS have recently discovered that it provides rather weak cover in deteriorating international markets. Both Forestry Tasmania and Gunns Limited have recently announced their intention to seek FSC certification, despite their overt past hostility, because important Japanese and Swedish companies are demanding it (ABC 2010b).

Fisheries certification in Australia

Fisheries management is fraught with challenges and uncertainties, often conflict ridden and, as noted in Chapter 4, Australian fisheries are no different. Many Australian fisheries are fully fished, with some overfished. This provides the context for interaction and conflict. Conflicts include those that are directly fishery based between regulators and the regulated as well as those between commercial fishers and recreational interests. Other conflicts arise from the challenges of accommodating different values and interests related to Australian indigenous peoples and the environmental movement. The increasing level of external scrutiny (from civil society organisations and government agencies outside the fisheries portfolio) over Australian fisheries is another factor in influencing the politics of fisheries management. It is this political and administrative environment that has shaped the development of fisheries certification in Australia, and which has shaped the impact of the MSC.

Certification engages with and expands this politics. External scrutiny of fisheries invites a range of responses from support to scepticism to opposition and in turn creates its own politics. Analysis of certification in Australian fisheries highlights these dynamics, and the critical impact of key institutions and stakeholders in shaping the debate. At the same time this analysis illustrates the importance of non-state actors in developing and implementing market-based institutions such as certification. As noted in Chapter 2, different forms and types of certification have been developed. These variations can be identified in Australian fisheries. Australian fisheries have been subject to formal environmental assessment by the Commonwealth Department of Environment, Water, Heritage and the Arts (DEWHA). This 'strategic assessment' process, discussed in Chapter 4, developed core principles as the basis for assessing sustainability that are compatible with those of the MSC. Australian fisheries have also been able to engage with industry-based certification schemes such as 'Clean Green', developed by the Southern Rock Lobster Fisheries company linked to the framework of ISO 14001 EMS standard, and to interact with the MSC process, which had three fisheries certified as of March 2010.

Australian fishing politics

Although the Australian fishing industry is small on a world scale, it is exceedingly diverse and encompasses small inshore operators as well as integrated companies operating large vessels in the high seas well off the Australian coast. This diversity of interests helps explain the differing responses to external certification of fisheries in general and to the MSC in particular. The lack of a coherent industry voice or organisation reflects the federal basis to fisheries management, where local and state organisations have been the primary industry organisations. This fragmentation does not diminish the impact of fisheries politics. The idiom that 'all politics is local' emphasises the key issues in Australian fisheries: political interaction relates to impacts of regulation (reducing catches) and to the increase in 'outsiders' scrutiny of operations on individuals and communities. This politics has been overlain by significant and long-standing interactions between Commonwealth and state governments over fisheries. This created further differentiation of an already fragmented industry organisation. In the case of fishery issues that transcended state boundaries or occurred in waters that had two or more states' fleets operating, the problems increased. The fragmentation of the industry continues to provide problems for government in trying to find an industry voice, and often government has been forced to develop institutional structures to facilitate government–industry interaction. Increased use of economic instruments, chiefly through the introduction of ITQs and statutory-based fishing rights have, however, been matched by more formal engagement with industry over management of fisheries.

The Australian fishing policy network

As noted the state facilitated if not drove industry organisation in the period 1960–90 with the Commonwealth influence matching an increase in its direct responsibilities over the past half century (Haward 1995). This has given a particular orientation to government–industry relations, with the organisation of interests very different to that experienced in the forestry sector. Fisheries provided a classic clientelist pattern, with groups developing close links with government agencies but without necessarily developing strong and enduring institutional structures to mediate policy ideas or policy differences.

Australian fisheries have traditionally been coastal or offshore operations, involving small-scale owner-operators. Industry organisation reflected a similar focus, with fishers organised around an association with their home-port. Shifts to fishery bodies and a segmenting of the industry occurred in the 1970s and 1980s, following the increasing active involvement of the Commonwealth government in fisheries licensing. Commonwealth licence holders' interests were supported by the development of the Australian Fishing Industry Council (AFIC) in 1968. AFIC provided a vehicle for Commonwealth fishers but operated alongside long-standing state-based industry organisations organised on regions or, from the 1980s, increasingly along fishery or gear lines. AFIC also provided a means for the Commonwealth to legitimate its activity as a fishery manager (Herr and Davis 1982; Harrison 1991) following the development of the first Commonwealth legislation in the previous decade. Initial opposition from the states continued with state officials critical of increasing intervention and action by Commonwealth officials, despite, or perhaps because of, increased formalisation of intergovernmental institutions in the 1960s (see Haward 1989; François 1984).

While the Commonwealth government actively supported the develop-ment of what became the ASIC in the mid-1980s, in an attempt to provide a national voice for industry, client relationships can wax and wane with lim-its to government support. The Commonwealth, for example, did not seek to intervene after the failure of the ASIC. Instead relevant interests reconsti-tuted around the development of the Commonwealth Fisheries Association, an organisation representing those fishers who fish for Commonwealth-managed fisheries. Other fishers returned to their sectoral or locality-based industry bodies.

While fisheries issues in Australia are generally considered 'low politics', particularly when compared with forestry, they have their moments of high drama. Despite this, the policy network has, in Pross's terms, focused on deliberations within the 'sub-government' (Pross 1992), with the 'attentive public' involved as issues escalate or disputes emerge. Civil society involvement in fisheries policy has increased, in part a response to the post-UNCED agenda, with 'public attention drawn to the issues by the scarcity of fish, high prices in markets or retail outlets or publicity and conflict surrounding closures or restrictions in different fisheries' (Haward 2009). The size and scale of recreational fisheries and the interests of Australian indigenous peoples also provide important elements of the policy network (Haward 2009).

Environmental organisations have become incorporated in management arrangements and advisory committees, and campaigns over fishing have increased in number and strength. Over the past two decades Australian industry has become more aware of its need to address environmental issues in catch and post-harvest operations. It has been increasingly vocal over concerns that imported fish do not necessarily meet standards of quality and safety and may not have been caught under the same conditions as those facing domestic fishers. Increased civil society concern over the 'state of fish stocks' has been promoted by and/or reflected in higher profile campaigns over specific fisheries, most notably attempts by the Humane Society International to have the Southern Bluefin Tuna, declared an endangered species. This campaign was rejected by the Australian Government, as was an attempt to have Patagonian Toothfish listed as a 'threatened species' under the Environment Protection Biodiversity Conservation Act.

Environmental groups, and broader civil society in Australia, have been more actively concerned over whaling. Australia ended its whaling operation in the early 1980s and moved to a staunchly pro-whale conservation or preservation position in regional and international forums. Opposition to commercial whaling led to a moratorium on catches being introduced through the International Whaling Commission that took effect in 1986. Australia has been active in opposing any resumption of commercial whaling and 'has been pushing to extend the status quo, and limit current Japanese scientific whaling programmes and lethal whale research' (Haward and Vince 2008, 44–5). It is arguable that the focus of NGO campaigns on whaling means that less attention has been given by such groups to fisheries.

Food safety and quality

Concern with seafood safety has been longstanding. As the trade in fish products and fisheries production from aquaculture has increased markedly in recent years there are growing concerns over the quality of fisheries products. These concerns are heightened in cases of illness or death caused by contaminated fisheries products and because most fish trade is from the global South to the developed North. While developed countries may have sophisticated quality assurance programmes in place, many European consumers in particular are increasingly concerned over the quality of unprocessed or uncooked fish products imported from developing countries. Food quality and safety is a marketing issue in Europe.

Australian concerns over the quality and safety of seafood match clearly this international agenda. Industry and government established cooperative working arrangements to promote a safe and quality-based seafood industry in the 1990s. SeaQual, an initiative of the then ASIC and Australian governments, was an important early element in identifying policy needs and practice. SeaQual began as a partnership between the seafood industry and governments to increase profitability and sustainability through quality management. SeaQual is now managed by Seafood Services Australia and helps members of the seafood industry – including catchers, aquaculture operators, co-operatives, wholesalers, retailers, processors, importers and exporters – to meet their needs for seafood safety and quality (Seafood Services Australia 2008). This assistance is mainly through providing information and advice, and through providing an industry perspective when dealing with governments on quality issues, with technical guides prepared for aquaculture, seafood processing, seafood harvesting and seafood retail operations (Seafood Services Australia 2008).

An internationally recognised industry standard is the Hazard Analysis Critical Control Point (HACCP) principles and methodology, a system that takes a proactive approach to food safety with mandatory measures to prevent food safety hazards, rather than trying to find and correct problems after they emerge. Preventative measures are applied at control points throughout the seafood handling process, such as point of receipt, storage conditions, cooking and post-cooking preparation and shipping and/or transportation. The HACCP programme establishes a record-keeping system and regularly monitors critical control points, thus pinpointing potential problems before they occur (Bache, Haward and Dovers 2001). While HACCP does not address sustainability of the fishery, it has emerged as an important certification tool for export fisheries.

Certification: Clean Green

Clean Green, seen as MSC's 'main Australian rival' (*Mercury* 2008), is an industry-driven certification programme that has originated with the Southern Rock Lobster industry (fishers from South Australia, Tasmania and Victoria operating under the ISO 14001 EMS. EMS approaches have been seen by a number of industry groups as a means of supporting their claims for sustainable practices, and a mechanism to meet Australian government's (both national and state/territory) commitments to ecologically sustainable development (ESD) in fisheries management. This approach to certification is managed by the Joint Accreditation System of Australia and New Zealand (JAS-ANZ). JAS-ANZ is the 'government-appointed accreditation body for Australia and New Zealand responsible for providing accreditation of conformity assessment bodies in the fields of certification and inspection' (JAS-ANZ 2008). The Clean Green Australian Southern Rock Lobster Certification Program

> is an industry product certification program owned by Southern Rocklobster Limited. The program employs the Clean Green Southern Rocklobster Standard as the certification standard and covers the complete 'pot-to-plate' rock lobster supply chain.
>
> (JAS-ANZ 2008a)

Development of the MSC in Australia

The Australian fishing industry has generally accepted the need to improve its environmental credentials, particularly in relation to bycatch and discards. Australian fishers have been active in developing environmentally friendly fishing gear including work on turtle-exclusion devices in prawn fisheries (Bergin and Haward 1994) and the development of mitigation devices and practices to avoid incidental bycatch of seabirds such as albatross and petrels in longline fisheries (Haward, Bergin and Hall 1998; Hall and Haward 2001). Despite this technical work, the development of external certification systems, either using the MSC as a standard, or the EMS certification under ISO 14001 or even self-referencing systems, has been relatively slow. While Australian governments have been reluctant to formally endorse the MSC as a standard, it nonetheless has had a significant impact within Australia. As noted in Chapter 4, MSC principles and criteria have influenced the format of the Australian government's strategic assessment applied to Australian fisheries, although the extent to which this assessment has set a rigorous standard has been questioned.

As of March 2010, Australia had three MSC-certified fisheries (Western Rock Lobster, Australian Mackerel Ice Fish and the South Australian Lakes and Coorong fishery). Australian industry undertook the first fishery assessment under the MSC process with the certification of the Western Rock Lobster Fishery with this initiative 'driven by the vision of Mr Murray France a western Australian Rock-Lobster processor' (Rogers, Gould and McCallum 2003, 103). The Western Australian Fishing Industry Council, 'the principal industry body in fisheries in Western Australia' became the 'client of record' (Phillips, Ward and Chaffee 2003, 98). The Western Rock Lobster assessment was 'recognised as a *de facto* test case for the MSC programme' (Phillips, Ward and Chaffee 2003, 101). This provided both certifiers, the MSC and industry, valuable lessons in 'operationalising' the MSC principles and criteria (Phillips, Ward and Chaffee 2003). The Western Rock Lobster assessment was also critical for the MSC as it was the first large and complex commercial fishery to pass assessment. Prior to the Western Rock Lobster assessment, the fisheries that had undergone the MSC process were small in scale.

The Western Rock Lobster fishery is both the world's largest rock lobster fishery and Australia's most valuable single species wild fishery (Phillips, Ward and Chaffee 2003). Its current status is problematic, however, as 2008 and 2009 saw significantly reduced recruitment which has lead to major reductions in quota allocations to individual fishers. The Western Rock Lobster fishery has had longstanding management arrangements under Western Australian state fisheries legislation and regulations (under an OCS arrangement) that have established a sophisticated management regime based on a mix of input controls. These include biological controls such as the protection of spawning females, gear and seasonal restrictions and number and deployment of pots as well as compliance measures such as marking of pots, floats, bags and crates of catch (Phillips, Ward and Chaffee 2003).

The Rock Lobster Fishery was certified in March 2000, after a 15-month certification process (MSC 2007a). Reassessment under the MSC process began in September 2004. In early 2005 two environmental civil society groups, the Conservation Council of Western Australia (CCWA) and TWS opposed re-certification, although one of the grounds of their objections related to provision of no-take zones, an objection that was rejected by then WA Minister for Fisheries, who 'accused the conservation groups of sabotaging one of the best managed fisheries in the world' (ABC 2005). A certificate of achievement was awarded in December 2006, following review of corrective action requests and a determination by the certifying body that the fishery should be certified and closure of the 21-day period for objections on 6 November 2006. The fishery has been re-certified for a further five years, subject to annual audits.

The significant reductions in recruitment to the fishery in 2008 and 2009 saw lower catches and concerns over the future of the fishery as MSC certified. This led to an MSC special audit, the first time this process has been used in an MSC-certified fishery, that 'involved a complete rescore of Principle 1, addressing stock issues in the fishery' (SCS 2010, 1). Scientific Certification Systems found that 'after careful review ... the fishery continues to meet the MSC standard and so products from the fishery remain endorsed by the MSC and continue to carry the MSC logo' (SCS 2010, 1).

In contrast to the Western Rock Lobster fishery, the Australian Mackerel Ice Fish fishery is a relatively new fishery developed in the late 1990s. According to the Australian Antarctic Division 'Mackerel icefish was once the most abundant species found near shore in waters less than 400m [with] declines in population sizes in the 1970s and 1980s linked to overfishing' (AAD 2008). The fishery occurs in the area under the application of the Convention on the Conservation of Antarctic Marine Living Resources, with CCAMLR Conservation Measures, as well as Australian government regulations important in shaping management. Stocks of icefish (*Champsocephalus gunnari*) are believed to undergo large natural variations in abundance, and commercial fishing for the species is restricted to periods of high abundance estimated from pre-recruit trawl surveys (CCAMLR 2008). The fishery had a Total Allowable Catch (TAC) in 2002–3 of 2980 tonnes (MSC 2007b). This fishery is subject to strict controls under a fishery management plan that sees annual adjustments in catches. In 2006–7, for example, the fishery had a TAC of 42 tonnes, and as a consequence no fishing was undertaken; in 2007–8, on the other hand, the TAC was 220 tonnes.

The proponents of this fishery sought MSC certification to help market a new and relatively unknown product into Europe (Exel 2006). Addressing sustainability of the fishery through external certification directly addressed any concerns over catches from a fishery in the Southern Ocean. The proponents' experience and familiarisation with, and support for, the MSC was also crucial. Murray France was a partner and executive director of the Kailis and France Group, a group including Austral Fisheries and New Fisheries, companies that were active in sponsoring and supporting the process for certification of the Australian Mackerel Ice Fish fishery. The assessment of this fishery began in August 2003 and certification was awarded on 31 March 2006, with the fishery undergoing its first annual surveillance.

The third Australian MSC-certified fishery, the South Australian Lakes and Coorong fishery in the ecologically sensitive Murray River area, completed the MSC assessment in June 2008. This fishery, established for 154 years 'uses manually operated gear to reduce bycatch, control fishing effort and produces the highest possible quality seafood' (FRDC 2008, 20). The community-based fishery involves 37 operators with 19 types of gear and focuses on scale fish, invertebrates and sharks, skates and rays, with the exception of the white shark (MSC 2007). This factor differentiates it from many other MSC-certified fisheries. The Lakes and Coorong fishery is also run as a cooperative voluntary organisation, an association rather than a corporation, a factor that affected the certification process. The president of the Southern Fisheries Association, the applicant for certification, recognised that 'the biggest hurdle is that we don't have a paid executive, everything is done on a voluntary basis' (FRDC 2008, 20). The fishery operates in state waters off the Coorong and associated lakes at the mouth of the Murray River. The main products are mulloway, golden perch, yellow-eyed mullet and cockle, and are sold in Australian domestic markets (MSC 2007).

The fishery has been active in ensuring it maintains a high environmental performance, with work in the late 1990s centred on utilising EMS approaches to management and seeing compliance with external standards such as relevant ISO standards and supporting HACCP certification from processors. The participants in the fishery recognised in the mid-1990s the need to verify the sustainability of the fishery (FRDC 2008). This work included developing an environmental management plan for the fishery (Southern Fishermen's Association 1998). This plan was developed over two years in what was claimed at the time as a world's first application of an EMS on a whole-of-fishery basis (Baker and Pierce 1998). The motivations of the Southern Fishermen's Association were clearly stated in the opening paragraph of their environmental management plan:

> Broadly speaking, society as a whole does not perceive the commercial fishing industry and its participants as environmentally responsible. In fact the contrary is often the case with members of the public viewing the industry as a symptom and cause of significant damage and degradation to the aquatic environment.
>
> (Southern Fishermen's Association 1998, 2)

The Southern Fishermen's Association noted that the environmental management plan enabled work towards external certification to occur (Southern Fishermen's Association 1998).

The Lakes and Coorong fishery's MSC assessment began in July 2004, with draft performance indicators and scoring guideposts released in March 2005. Prior to a period for public comments called in late June 2005, final performance indicators and scoring guideposts were publicised. Client and peer review took place in late 2006 and early 2007. In June 2007 a draft of the public review of the assessment report was in progress (MSC 2007). MSC certification was achieved in June 2008. The cost of the assessment and certification was 'about $250,000' (FRDC 2008). The Southern Fishermen's Association received funding from WWF Australia, the Packard Foundation of the US and the Australian government's Department of Environment and Heritage to offset the cost of the certification process.

Australia's engagement with the MSC is significant in other dimensions. Australians have been active in key MSC institutions. Murray France has been a longstanding, founding member of the MSC board. Dr Keith Sainsbury, a fisheries scientist formerly with Australian premier government science institution CSIRO, is currently vice chair of the MSC board and chair of MSC Technical Advisory Board, and Annie Jarrett is member of the MSC board and co-chair of the MSC Stakeholder Council. Australian fisheries scientists have served on certification panels or have provided technical advice and support to the MSC certification process. Dr Alistair Hobday, also with CSIRO and formerly with University of Tasmania, has worked on risk assessment models central to the application of the MSC standard to data-poor fisheries. Dr Tony Smith, a fisheries scientist at CSIRO had a major role in the assessment and re-assessment of the Western Rock Lobster fishery.

As noted earlier, while the MSC certification for the Western Rock Lobster fishery was part of an industry campaign to ensure increased market penetration into Europe, the ready support for the MSC by key industry leaders in Western Australia was also central to the certification process. The MSC, too, benefited greatly by having this fishery nominated for assessment. As the first large-scale, export-oriented fishery to be assessed, it became a valued test case for operationalising the MSC's principles and criteria. This process was also enhanced by the availability of data and information and the support provided by Western Australian fisheries managers to the industry body seeking assessment.

Australia is also the location of an MSC regional office and Duncan Leadbitter, the initial local representative, had been active in promoting eco-labelling and the MSC throughout Australasia and the Asia-Pacific, leading to an increased presence of the MSC in Japan. One important and direct result was the first certification of the MSC fisheries in Japan that occurred in 2008 with important impact of extending the reach of the MSC into one of the world's major fisheries markets. At the same time this work of the MSC's Asia-Pacific group based initially in Australia, has done much to extend the MSC's presence and reduce the perception of the 'Euro-centric' orientation of the organisation.

Analysis of the MSC in Australia

The Australian public is becoming increasingly aware of this nation's responsibility for management of the fishery resources through publicity over state of fish stocks in what inevitably are high-profile fisheries. The less public face of fisheries management – institutional arrangements, the impact of legislative and regulatory obligations on government officials (and increasingly on industry), the relationship between science and management and industry–government relations – is given little public attention.

Despite the early support for the MSC standard in the Western Rock Lobster fishery, driven clearly by the support from Murray France, the MSC experience in Australia is mixed. The MSC certification process and outcome for the Western Rock Lobster fishery was important for two reasons. As noted above it was the first 'test case' for the MSC assessment methodology and much was learnt in the process as applied to a large and commercially significant fishery. The second reason was the use of MSC certification to secure better access to Europe markets that had proved challenging to Australian producers in the past. With the downturn in exports to Asian markets in the aftermath of the Asian currency crisis of 1997–8, producers looked to Europe for markets. The emphasis on food safety and quality for seafood meant that European markets were active in asserting standards such as HACCP, and environmental certification provided some additional market leverage. The opportunity to use MSC certification and the product label as a marketing tool to leverage improved market access is a significant driver in both the Western Rock Lobster and Mackerel Ice Fish cases.

Both fisheries, too, were supported by commercial interests that were well aware of constraints and opportunities of export markets, particularly consumer sentiments in Europe. Key proponents for certification of these fisheries were key supporters of the MSC. The Western Rock Lobster fishery had spent considerable effort in developing market access into Europe, and into Asia. MSC certification was seen as key to ensure access to Europe. The Lakes and Coorong fishery, on the other hand, is a very different case: a small diversified community-based fishery, but with participants that had a high awareness of, and support for, measures that would enable them to show sustainability of the fishery.

Although these certifications are significant achievements, the MSC initiative has met with some scepticism from within the Australian fisheries sector. This scepticism has arisen in part due to a lack of clarity and transparency within the assessment process, and the cost of the process (that is, as noted, borne by the proponent). For many fishers benefits from certification are not clear, particularly if they are operating in domestic Australian markets where consumer pressure for certified fisheries is not as strong as in Europe or North America. The cost of MSC certification and the lack of direct and tangible benefits from this certification have been cited as reasons for industry-driven approaches such as the Southern Rock Lobster Clean Green programme (*Mercury* 2008). In contrast the Lakes and Coorong fishery indicates that support for internal industry-driven EMS type certification and the MSC are not mutually exclusive, although as noted above, unlike the Southern Rock Lobster industry, the former group chose MSC certification. Australian fisheries, too, have been subjected to increased level of environmental scrutiny from the Australian government's 'strategic assessment' processes.

The introduction of strategic assessments may, however, have hindered rather than helped the take-up of the MSC. There is no doubt that ecolabelling is a growing trend; however, the MSC label lacks visibility in the Australian market. Consumer behaviour, responding to, and seeking access to, wider and more detailed producer information is apparent in key markets. Individual and corporate consumers want some assurance that the world's fisheries are being managed sustainably. Scepticism over the MSC from the catch sector remains a significant factor. The Tasmanian Rock Lobster Fisherman's Association (TRLFA), operating in the second largest rock lobster fishery in Australia targeting the Southern Rock Lobster in waters around Tasmania, has rejected the MSC approach. The TRLFA argues that the EPBC strategic assessment and re-assessment programme on a five-yearly cycle is an effective arrangement that links sustainability of operations to export licenses.

Conclusion

Australia was a late adopter of forest certification and an early adopter of fisheries certification. Its late adoption of forest certification was due to a combination of environmentalist suspicion and industry disinterest.

With neither group pushing forest certification, state and Commonwealth governments remained on the sidelines monitoring its development internationally. In contrast, Australia adopted fisheries certification relatively early on in its development cycle with the Western Australian Rock Lobster fishery acting as a test case for MSC's methodology.

The timing of certification's arrival in Australia appears closely linked to the Asian currency crisis of 1997–8. In both the forestry and the fisheries sectors, overseas markets in Asia were severely impacted and industry sought markets in North America and Europe. When it did so, it discovered that these markets were populated with environmentally conscious consumers and retailers who were concerned to secure assurances that timber and fish were sourced from 'sustainably' managed locations. Some form of certification was required to obtain market entry into Europe and both the forestry and fisheries sectors adopted and developed schemes to promote their products.

Forest certification was much more contentious than fisheries certification in Australia, reflecting its great salience on the country's political agenda in the 1980s and 1990s. Whereas Australian governments were, at most, indifferent to the MSC, state and Commonwealth officials were actively hostile to the FSC. The FSC was perceived by industry and government officials as unnecessary, costly, intrusive and lacking legitimacy. Rather than work with it, they opted for a 'made-in-Australia' approach, building on the experiences of other countries in North America and Europe to develop the AFS.

Despite early suspicion, environmental support for the FSC now appears higher than for the MSC in Australia. In the 2000s, and following the establishment of AFS, the Australian environmental movement endorsed the establishment of the FSC with two high-profile organisations – TWS and ACF – representing environmental interests on the FSC-Australia board. It is still strongly supported by WWF Australia, although it has endured criticism by FoE-Australia over the ongoing certification of HVP in Victoria. The MSC continues to have the support of WWF-Australia, which provided assistance to fund the Lakes and Coorong Fishery. However, other environmental groups appear less supportive and some continue to criticise the organisation. TWS and the CCWA objected to the certification of the Western Rock Lobster Fisher in 2000; and in January 2010 the CCWA reiterated concerns over the fishery's re-certification (CCWA 2010). Greenpeace International released a briefing in June 2009 that was also highly critical of the MSC's 'deficiencies', citing Australia's Mackerel Ice Fishery and the Western Rock Lobster Fishery in examples.

6

Forest and Fisheries Certification in Canada

Canadian forest and fisheries policy networks were early adopters of certification. At both the federal and provincial levels, certification presented policy networks with significant strategic threats and opportunities. With respect to forest certification, the FSC was perceived as a direct threat to network interests which immediately responded by establishing an alternative, national forest certification scheme. Matters were different with respect to fisheries certification, with the MSC endorsed and promoted albeit not especially enthusiastically. In this chapter, we examine the development of forestry and fisheries certification in Canada, focusing on the actors involved in promoting and blocking it. We highlight the peculiarity of the trajectory of the FSC and the MSC schemes in the country. While the FSC received almost no support from industry and government at the outset, by 2010 it was surprisingly well placed, accounting for over 30 per cent of total area certified and growing faster than other schemes. In contrast, the MSC, which had been relatively well received at the outset, was coming under increasing criticism for the complexity of its certification processes and the environmental and social weakness of its standard. To understand these dynamics, we first examine the evolution of the FSC, followed by the MSC.

Forest certification in Canada

Canadian forestry politics

In comparison to Australia, forest certification arrived early in Canada, with initiatives commencing in the early 1990s by the FSC and two competitor schemes. These were the CSA scheme promoted by the CPPA and the SFI promoted by the AFPA (see Chapter 3). The early arrival of certification in Canada reflects the critical role that forestry plays in the country's national and provincial economies and the industry's dependence on environmentally sensitive markets in the US and Europe. As a

173

consequence of this dependence Canada's national and sub-national policy networks were deeply engaged in shaping the meaning of the emerging concept of 'sustainable forest management' through the post-UNCED forestry institutions including, especially, the Montreal Process (Gale and Cadman forthcoming).

The Canadian state's response to forest certification was fundamentally conditioned by the policy objective of maintaining a viable, productive forest sector that generated profits, jobs and revenues. Since the FSC approach to certification threatened these goals, forest policy networks located in forestry-dependent provinces rallied against it, notably in BC, Nova Scotia and, following a failed venture to obtain FSC certification, New Brunswick. Unlike in Australia, however, Canada's forest policy networks chose not to confront the FSC directly, in part because some companies were experimenting with the FSC and because at least one large province – Ontario – thought it might be able to qualify for FSC certification. Instead, the strategy adopted was to declare to be officially 'neutral' with respect to forest certification schemes, while favouring CSA and SFI over the FSC.

Forest certification in Canada pitted a powerful, but loosely coordinated environmental movement that was committed to the FSC against an equally powerful state and corporate sector that was committed to CSA or SFI. The intensity of the resulting 'forest certification war' (Humphreys 2006) belies the 'neutrality' of the language through which it was conducted. In this chapter, we first describe the key actors involved in promoting the FSC and competitor schemes and then describe the strategy and tactics adopted by both groups to champion their preferred schemes at federal and provincial levels and in key overseas markets. While at the outset it appeared that the FSC scheme would be swamped by the rapid development of CSA and SFI and their ability to certify vast areas of forests quickly, the FSC gained ground in the 2000s, especially following the development of the 2004 Standard for Boreal forests and the 2005 FSC-BC Regional Standard. By the end of certification's second decade in Canada, the FSC had made up substantial ground on its rivals, inviting sceptics to reappraise the scheme's standards, governance structures and worth.

Conflict over how forests should be managed emerged in Canada in the 1970s, driving a wedge between urban and rural communities on the one hand and between a rapidly developing environmental movement and a traditional resource-based industry on the other. The rawness of the struggle was evident in the community outrage that accompanied the blanket aerial pesticide spraying of Nova Scotia's forests to control a spruce budworm outbreak in the 1970s and in the blockades and mass arrests that occurred in Temagami, Ontario in 1989 and Clayoquot Sound, BC in 1993. It was in this context of forest management-related conflict that certification emerged in the 1990s, presenting Canada's traditional and new social movements with new strategic challenges and options. In this section, we outline the structure of these social movements focusing initially at the federal level and then examining the situation in the province of BC.

Environmental coalition

Federal level   The Canadian environmental coalition is a diverse and frag-mented group that lacks a strong presence at the federal level despite the existence of a national coordinating body. Federal environmental politics is dominated by the 'big three' ECSOs, Greenpeace Canada, Sierra Club Canada and the WWF-Canada. Each has played an important role in Canadian forest politics and contributed to building support for an ecosys-tem-based approach to forest management. Each, however, is differently structured, protects its organisational brand and campaigns separately. The longest-lived of the national organisations is WWF-Canada, a foundation with a self-appointed board which was established in 1967 to 'protect spe-cies, maintain healthy ecosystems, and create a future where humans live in harmony with nature' (WWF-Canada 2008). WWF Canada perceives itself as adopting a science-based, collaborative approach to biodiversity protection, forming strong alliances with governments and industry, from which it receives financial support.[1] However, the majority of WWF-Canada's support derives from individual donations (almost 66%). In 2008, only 7.2 per cent was contributed by industry and only 2.9 per cent by government on total revenues of $26.7 million (WWF-Canada 2008, 55–6). WWF-Canada was a founding member of the FSC in 1993 and of the Canadian Working Group (WWF-Canada 2008, 30). It works directly with companies to encourage them to obtain FSC certification. For example, in 2003 it announced that Domtar had agreed to become FSC certified (WWF-Canada 2003). A key link between forest certification and WWF is its work on High Conservation Value Forests; it has developed a tool to assist forest managers to identify and protect HCVFs within their management units (WWF-Canada 2008b).

In contrast to WWF Canada's corporate image, Greenpeace Canada eschews donations from government and business (Greenpeace 2007, 2). The organisation states that 'in order to remain independent, Greenpeace does not solicit donations from corporations or governments, but relies on individual donors to fund its environmental campaigns' (Greenpeace 2008). Greenpeace Canada claims to have over 100,000 'members', although these are actually supporters with no membership rights and who cannot vote. Structurally, Greenpeace Canada resembles WWF-Canada and consists of a small board of directors that 'determines the priorities and annual budget' with day-to-day responsibility devolved to an executive director (Greenpeace 2008). Greenpeace's approach to campaigning is based on its understanding of media politics. Zelco notes that when Greenpeace was founded in 1971 as the 'Don't Make a Wave Committee', its initiators were media-savvy individuals who had learnt the importance of a 'well-formulated media strategy'. Where earlier campaigners 'had naively assumed that the free and unfettered US media would accurately and fairly report their protest', Greenpeacers 'would need to develop strategies to ensure that the media could not ignore its protest' (Zelco 2004).[2] Like WWF-Canada, Greenpeace has adopted an aggressive and successful fundraising strategy built around donations, bequests and campaign grants. In 2007, it reported total donor contributions of $8.6 million, constituting about 80 per cent of total revenues of $10.7 million. Greenpeace reports that it spent about $3.8 million on fundraising, finance and administration costs which, at 36 per cent of its total budget, is quite high. Greenpeace Canada played a major role in the forest politics of BC and is further discussed below in that context.

The Sierra Club Canada (SCC) is the third major ECSO operating at the national level. Founded in 1969 in BC as a branch of the US Sierra Club (Sierra Club of BC 2008, a number of semi-autonomous groups were subsequently established across Canada before being incorporated as the Sierra Club Canada in 1992. According to its website, the SCC is 'a member-based organization that empowers people to protect, restore and enjoy a healthy and safe planet' (Sierra Club Canada 2009). The organisation has some 12,000 Canadian members, with the majority based in BC (Lalonde 2007). The SCC's preferred approach to environmental activism is conventional political lobbying via letter-writing campaigns and delegations coupled with legal action. With respect to the latter, for example, its 2007 annual report notes that it won 'two landmark Federal Court decisions affecting tar sands development: one which stopped approval of the Kearl tar sands mine to assess its climate change impact; and another which established the government's authority to withdraw an authorization for the project because of a faulty environmental assessment' (Sierra Club Canada 2007, 6). To fund these activities, SCC raises money from a variety of sources. Total revenues for 2007 were recorded at $2.946 million, with administrative overheads (under facilities and office, staff administration, communication and education and membership support) collectively estimated to be approximately $0.849 million, approximately 29 per cent of total revenues.[3]

Canada's environmental groups are united at the national level under the Canadian Environmental Network (RCEN).[4] RCEN was established in 1977 with the objective of supporting and strengthening civil society's input into environmental decision making by the country's federal environmental department, Environment Canada (RCEN 2008a). Its mandate has expanded since then and according to its 2007–8 annual report, RCEN now concentrates 'on helping environmental proponents communicate and share ideas with each other, gain access to important and timely information, and participate in consultations on environmental issues and policy development' (RCEN 2008b). Based in Ottawa, Ontario, RCEN provides networking facilities to over 600 Canadian ECSOs across the country ranging from the smallest and most local groups (e.g. the Environmental Education Association of the Yukon) to large ECSOs like the Sierra Club-Ontario Chapter and Nature Quebec. Important networking functions include the establishment and operation of issue-based caucuses in areas such as agriculture, atmosphere and energy, environmental planning and assessment, forests and fisheries and oceans. RCEN notes that its national caucuses 'offer a forum for ENGOs interested in similar issues to share information and work together to maximise the use of their collective expertise and benefit from the efficiencies of scale through collaboration' (RCEN 2008b, 7). Funding for networking and outreach activities comes largely from Environment Canada, with which RCEN has a contribution agreement. In the financial year 2007–8, Environment Canada provided grants and contracts worth $833,000, representing 78 per cent of a total budget of $1.069 million. Given its financial dependence on the federal government, it is unsurprising that RCEN is explicitly a 'non-partisan, non-advocacy organization that does not take positions on environmental issues' (RCEN 2008b, 4). It is thus very different in character and function to the Australian Conservation Foundation, which explicitly engages in advocacy work.

DeMarco (2005) articulates some dissatisfaction with the effectiveness of Canada's environmental movement at the federal level. Citing a 1993 paper by Paul Griss, DeMarco argues that little has changed in the intervening decade and that 'environmental groups of all stripes are weathering some serious and unexpected storms at present ... including fiscal constraints and changing decision making processes in both the public and private sectors' (Griss 1993). Building on *The Death of Environmentalism* debate in the US,[5] DeMarco questions whether the Canadian environmental movement is 'articulating a vision commensurate with the crisis', whether the focus on technical policy fixes is appropriate and, notably, whether leaders are not 'failing to question their basic assumptions and solutions' (DeMarco 2005, 4). Whatever the shortcomings of the RCEN, it is important to note that it is really a network of networks. It brings together under one national umbrella the provincial environmental networks that are spread across the country from the British Columbia Environmental Network in the west to the Newfoundland and Labrador Environmental Network in the east. RCEN is run by a national council of 28 volunteer members representing regional networks, aboriginal peoples, and the francophone and youth communities.

Provincial level   Given the relative weakness of the Canadian federal environmental movement, and the constitutional responsibility of the provinces for much environmental, forestry and protected area policy, forest politics in Canada occurs mainly at the sub-national level. In each province a regional environmental network has formed to coordinate action directed at provincial governments. One of the most established of these sub-national networks is the British Columbian Environmental Network (BCEN), which has been in operation since 1981. Although highly effective in the 1980s and early 1990s, the BCEN declined in importance after 1993 and provincial environmental groups interested in advancing the cause of the FSC in the province bypassed it to organise directly among themselves. These ECSOs included the Ecoforestry Institute Society, ForestEthics, Friends of Clayoquot Sound, Greenpeace, Kootenay Centre for Forestry Alternatives, Sierra Club of BC, Silva Forest Foundation and the Western Canada Wilderness Committee (WCWC).[6]

Of these groups, one of the most important from the point of view of forest discourse was the Silva Forest Foundation, based in BC's Slocan Valley, near the town of Nelson. Founded in 1992 and led by a charismatic forester, Herb Hammond, SFF pioneered an ecosystem-based approach to forestry that constituted a direct challenge to the dominant industrial forestry paradigm then being practised. On his land, and in a series of books and articles, Hammond promoted a new vision of forestry grounded in the emerging discipline of Conservation Biology that focused on what to leave behind in the forest rather than on what to take. According to SFF:

> High levels of social economic health are maintained by protecting ecosystems and natural capital, which are the foundation for societies and economies. SFF believes that the primary concern of forest use must be the protection, maintenance, and, where necessary, restoration of fully functioning forests for the welfare of all beings and the whole forest. Ecosystem character (how a natural forest functions) and condition (how human activities have impacted on forest function) form the context within which social and economic criteria are designed and adjusted.
>
> (SFF 1999, 1)[7]

SFF cooperated with the Ecoforestry Institute Society to write an early report outlining forest certification's potential in BC. The organisation was also represented at the FSC's founding meeting in Toronto in 1993. In 1994, SFF joined the Pacific Certification Council, a regional network of certifiers from California, Oregon and BC which subsequently drafted an early set of principles, criteria and indicators for a regional standard covering these regions (SFF 1999, 1). While PCC ultimately disintegrated due to lack of funding, SFF employed its standard to conduct one of the earliest certifications in BC when it certified Rod Blake's woodlot operation near Williams Lake in 1999 (Brewer 1999). In 2000, SFF became the first Canadian company to receive FSC forest management certification accreditation.

Another BC ECSO involved in introducing the FSC to BC was the Friends of Clayoquot Sound (FOCS). Established in 1979 and based in the picturesque town of Tofino on the west coast of Vancouver Island, FOCS's overarching goals are 'to preserve natural biological diversity and ecosystem dynamics, concentrating on Clayoquot Sound's ocean and forest ... [and] to stimulate creation of a conservation-based society, with a corresponding conservation-based economy' (FOCS 2009a). To achieve these goals, FOCS has undertaken a range of actions over the past 30 years including a blockade of Macmillan Bloedel's logging operations on Meares Island (1984/85), large-scale blockade of logging in Clayoquot Sound in 1993 that saw 856 people arrested and the co-sponsoring, with the Sierra Club and Greenpeace, of the 'Markets Initiative' designed to shift Canadian companies out of reliance on paper from 'ancient and endangered forests' (FOCS 2009b).

A third example of the diversity of an environmental group engaged in forest politics in BC is the WCWC. WCWC was established to 'preserve Canada's natural biodiversity in the face of growing industrial development' (WCWC 2008, 5). To effect the vision WCWC works to protect native biodiversity in wildlands, safeguard wildlife, defend public lands, secure wild salmon and ensure health communities. This work occurs largely through devolved chapters based in Victoria, Vancouver and Winnipeg. WCWC has been active on Vancouver Island, especially during the Clayoquot Sound dispute, and produced a strategy for Vancouver Island that envisages the large-scale protection of old-growth, high-conservation value forests. Claiming to have over 30,000 members in 2008, WCWC is run by an elected board of directors, many of whom have been there for decades. Its current treasurer, Alice Eaton, was a member of the board as far back as 1992 (WCWC 1992); and Bob Broughton, a director at large, has been on the board for more than 15 years. At the staff level, Joe Foy, WCWC's campaign director, has been with the organisation since 1984, becoming its first paid staff person in 1987. In addition, Andrea Reimer, the organisation's executive director, has been a staff member since 1992 and Matt Jong, the comptroller, since 1998. In 2008, WCWC reported revenues of $2.260 million, with donations accounting for 64 per cent of the total and memberships, just over 18 per cent (WCWC 2008, 14). WCWC has supported forest certification through its endorsement of Iisaak Forest Company, a First-Nation's run company that logs timber on Vancouver Island. Iisaak Forest Company is certified by the FSC, although a report by WCWC in 2006 sounded a warning note by stating that any effort by Iisaak to move to a more high-volume, low-value business model would likely lose WCWC's support (WCWC 2006).

The groups identified above as active in BC have their counterparts in other provinces and territories. In Ontario, the Ontario Environment Network brings together a huge number of provincial groups. Some of them such as Canadian Parks and Wilderness Society and WWF-Ontario worked with Tembec to introduce FSC certification to the Gordon Cousin's forest. In Nova Scotia, the Nova Scotian Environmental Network was founded in 1991 and acted as a coordinating forum for a large number of local ECSOs. With respect to forest certification, a large number of individuals representing a wide variety of groups were represented on the Technical Standards Writing Committee in the late 1990s engaged in writing a draft FSC-Maritimes Regional Standard. These individuals represented a wide diversity of Nova Scotia organisations including the Ecology Action Centre, the Falls Brook Centre and Genuine Progress Indicator-Atlantic.

Forest industry coalition

The structural power of the Canadian forest industry derives from the role that companies and workers play in the nation's economy. While it is only recently that sectoral interests have combined to form a single national association, in the 1990s powerful associations operating in the softwood lumber and pulp and paper sectors capably defended the perceived interests of their members. We commence with an analysis of the major federal business and union organisations, the Forest Products Association of Canada (FPAC) and the United Steelworkers of America. We then examine organisational arrangements at the provincial level using the case of BC to illustrate the triadic policy network that existed between the Confederation of Forest Industries (COFI), the BC Forest Alliance and local branches of the Industrial, Wood and Allied workers of Canada (IWA).

Federal level In the 1990s, business interests were represented at the national level by the CPPA and COFI. This division of labour between the pulp and paper and softwood lumber sectors was partly geographical, because pulp and paper production were dominant in Canada's central and eastern provinces while in western Canada, softwood lumber production prevailed. The globalisation of the industry during the 1990s rendered this sectoral and geographic split increasingly anachronistic as Canadian producers encountered a range of similar industrial difficulties related to the US trade barriers (e.g. the softwood lumber dispute), competition from low-cost producers in Latin America and Asia, opportunities for new markets in China and Europe and criticism from an increasingly transnationalised and media-savvy environmental movement. Increasingly, industry leaders realised that their message was not cutting through in Ottawa. For example, an editorial in the *Logging and Sawmilling Journal* states that the

> industry itself has to get its act together. Too often it speaks with two accents that sound more like argument than debate. Eastern and western Canada are two different industries. But if they want Paul Martin to listen, he needs to be able to hear at least a coherent voice. Until then, forestry will simply continue to be the big contributor to trade balances – and the federal government will be content to let it be so.
>
> (Clarke 2001, 3)

To achieve better outcomes at the federal level, the various forest industry bodies formed the FPAC in late 2001. FPAC brought together most of the major brand-name timber and pulp and paper companies operating in Canada. In the pulp and paper sector, these included AbitibiBowater, Alberta-Pacific Forest Industries, Kruger Inc and SFK Pulp; in the timber sector, they included Tembec Inc., Tolko Industries and West Fraser Timber Company (FPAC 2009). FPAC was established not only to ensure the industry's voice was heard in Ottawa but also internationally. In a press release issued at the time it was established, Descarries notes, that the 'new association, with an office in Brussels, will present a strong, clear voice on behalf of the entire forest sector, from coast to coast' (2001, 410). In setting up the European office, FPAC's aim was to advertise Canada's world leadership role 'in adopting third party audits to ensure sustainable forestry' (Descarries 2001, 410).

The major union organising forest product workers in Canada was the IWA. Originally standing for the International Woodworkers of America, IWA-Canada broke away from its larger US body in 1987 and established itself as a Canadian sectoral union (Wilson 1998, 41). As elsewhere, unions confronted very hostile conditions in Canada through the 1990s as a consequence of neoliberal policies including privatisation and contracting out in part driven by Canada's increasing integration into the US economy via the Canada-US Free Trade Agreement (1987) and the North American Free Trade Agreement (1993). Industrial consolidation across borders was one consequence of these changes and unions underwent rationalisation on a continental scale as well (Bickerton and Stinson 2008). In 2004, IWA-Canada members agreed to merge with the United Steelworkers of America which operates in three districts across the country: Western Canada, Quebec, and Ontario and Atlantic Canada (USWA 2009).

At the federal level, Canadian forestry unions fought side-by-side with industry to combat US protectionism and environmental regulation. From the union's perspective, US government action to limit softwood lumber imports from Canada and environmental claims of unsustainability both threatened jobs. In 2006, a union spokesperson, Kim Pollock, launched a broadside against the Conservative government's deal to resolve the softwood lumber dispute. Arguing that Canada was close to winning the case at the North American Free Trade and WTO tribunals, Pollock argued that the deal might be good for Canadian and US governments and corporations, but 'it does nothing for workers and their communities' (2006, 1). Pollock had also been outspoken about environmentalists in the past, arguing they were anti-employment (Pollock 1996).

Provincial level Reflecting Canada's devolved federal system, the forest industry is especially well organised at the provincial level. Its influence over policy has been noted in several jurisdictions, most especially in BC, where several commentators have described the triadic, corporatist arrangements that linked industry associations and government in the 1980s and 1990s. Wilson (1998, 24) noted that the BC forest sector has been dominated by some very large companies, and that 80 per cent of the annual allowable cut was in the hands of just 25 companies, with ten controlling almost 60 per cent of the cut. Many of these were household names in the 1990s and included Macmillan Bloedel, Slocan Forest Products, Canadian Forest Products (Canfor), West Fraser Mills and International Forest Products (Interfor). These companies not only shared similar interests but also were interconnected via a range of joint ownerships, minority shareholdings and common personal relationships (Wilson 1998, 29).

To ensure it was able to influence BC forest policy, the BC forest industry established COFI in 1960 as an 'association of associations' that included the BC Lumber Manufacturers' Association, the Plywood Manufacturers' Association, the Canadian Pulp and Paper Association (BC Division), and the Consolidated Red Cedar Shingle Association (Wilson 1998, 32). COFI provided services to members and engaged in policy and lobbying work. In the 1970s and 1980s, its members maintained close ties to the governing Social Credit party. For example, Mike Apsey, COFI's president and CEO in 1984 moved between industry and government with relative ease during these decades.

> In the years leading up to his taking of that post, Apsey engaged in a classic bit of elite hopping. After a stint on the government-appointed advisory committee that translated the recommendations of the Pearse Royal Commission into the new Forest Act of 1978, Apsey left his COFI vice-president position to become deputy minister of forests in mid-1978. His return to COFI after six years in that post raised a few eyebrows, especially since his years as deputy minister had been highlighted by government moves to increase the availability of tree farm licences, downsize the ministry, and delegate more forest management responsibility to the industry'.
>
> (Wilson 1998, 33)

COFI's relative power declined somewhat when the Social Credit party lost the 1991 election and the New Democratic Party took over. Moreover, with the BC industry targeted by European environmental groups, some industry actors wanted immediate action that COFI was unable to deliver. In early 1991, a small group of major BC companies established the BC Forest Alliance following advice from the major advertising agency Burson-Marsteller (Wilson 1998, 37). Set up as a 'grassroots' membership body that eventually attracted over 4000 members, the BC Forest Alliance was effectively, if aggressively, run by its first chairman, Jack Munro. The chief focus of its work was to debunk the 'misinformation' campaign run by environmentalists both within BC and in the US and Europe. It launched several campaigns with a multimillion dollar budget to achieve this goal. Following some initial success, the BC Forest Alliance was unable to sustain its early momentum. Viewed by Cashore, Auld and Newsom (2004) as an important player in BC's forest policy network in the mid-1990s, it had folded a decade later.

BC governments are lobbied by both industry and unions over forest policy arrangements. When it comes to environmental issues, both groups have often been closely aligned. The IWA advocated on behalf of the New Democratic Party in 1991 – and today its members continue to campaign against the incumbent Liberal Party (see, for example, Pollock 2009) and promote the NDP (e.g. BCNDP 2009). Wilson (1998, 263) notes that in the early 1990s, the 'IWA' and 'Green' factions of the NDP clashed over forest policy. This led to several bitter disputes over protected areas, annual allowable cut and silvicultural practices. The IWA was determined that any worker that lost their job as a result of any reforms would be compensated, an arrangement subsequently translated into policy by government with the establishment of Forest Renewal BC (FRBC). According to Wilson (1998, 273), 'following intense bargaining among IWA officials, NDP greens, and key cabinet ministers', the premier was able to avert a nasty debate at the 1994 party convention by promising to ensure continued employment in the forests by establishing the FRBC.

Forest certification in Canada

Forest certification created a range of opportunities and threats for actors engaged in forestry in the 1990s in Canada. Canadian environmental organisations – especially the large, federally based associations – were very sympathetic to the FSC, having played an important role in its establishment. With vast amounts of old-growth, high-conservation value forests, these groups were seeking ecosystem-based forestry rather than a cessation of native forest logging as in Australia. In contrast, industry was very concerned about the FSC and the costs it thought would be associated with adhering to its standard. The industry thus mobilised quickly to avoid the FSC becoming the monopoly provider of certification in Canada and enlisted the Canadian Standards Association to develop a competitor scheme that could be recognised by the Standards Council of Canada, a statutory body. With environmentalists backing the FSC scheme and companies, unions and governments backing the CSA, Canada experienced a series of 'certification wars' through the 1990s and 2000s. Cashore, Auld and Newsom (2004) argue that these certification wars were beneficial because they created a dynamic of 'converting' and 'conformance' behaviour among proponents of competing schemes leading to improvements in the structure and operation of all schemes. While this is true, the evidence suggests that industry schemes have been under more pressure to convert and conform to the FSC standard than vice versa. Despite a surge in CSA and SFI certification in Canada in the early 2000s, therefore, companies found it necessary to also certify to the FSC leading to a steady increase in FSC certifications in the past three years.

Emergence of forest certification

Certification emerged in Canada in the early 1990s as Canadian policymakers wrestled with three major threats to the country's forest industry. The 1990 global recession had induced a downturn in housing starts in the US and in demand for pulp and paper around the world. Canada's forest industry was in a slump, experiencing a profit squeeze and industry was looking for relief. Second, Canada's 1986 softwood lumber agreement with the US was proving extremely costly. The agreement required Canada to phase in commercially based stumpage fees, but as such matters fell under provincial jurisdiction they were unevenly addressed. In 1991, Canada withdrew unilaterally from the SLA arguing that provincial governments had met the requirements by increasing stumpage rates. This was disputed by the US which, under pressure from the Coalition for Fair Lumber Imports, launched a Section 301 investigation to determine whether Canada continued to subsidise its forest industry (Gorte and Grimmett 2006, 2).

The Canadian forest industry was also under siege from environmentalists who were objecting to the lack of parks and protected areas and to a range of silvicultural practices including, notably, large-scale clearcutting. Given the extent of old growth temperate rainforests in BC, the conflict was especially acute there, and the BC government was savagely criticised by domestically and internationally based environmental groups over its 'liquidation and conversion' forest policies (Wilson 1998; Tollefson, Gale and Haley 2008). In response to these environmental criticisms, the federal and provincial governments launched an 'image' programme to improve the perception of the industry in foreign markets. Part of the image programme involved substantive changes such as the adoption by NDP Premier Mike Harcourt of a Protected Areas Strategy for BC to increase the number of parks and protected areas from eight to 12 per cent of the province's total land area (Tollefson, Gale and Haley 2008, 66). Another part of the image programme aimed to combat negative perceptions in overseas markets. In 1991, as noted earlier, the BC Forest Alliance established a presence in Brussels to directly combat negative environmentalist messages (Elliott 1999, 196) and in 1992, Mike Harcourt undertook a tour of Europe to reassure purchasers of BC lumber, pulp and paper that it was coming from 'sustainable sources'.

As the challenges proliferated, it became increasingly clear to Canada's forest policy network that a multi-pronged strategy was required to reposition the industry to meet the sustainability challenge. The 1992 National Forest Strategy was designed to achieve this objective and set out nine 'strategic directions' the country needed to take to implement its vision of sustainable forest management (see Chapter 4). In paragraphs 4.13 and 4.17, oblique references are made to labelling and certification systems. Paragraph 4.13 states: 'By 1995, industry and governments will develop and put into operation a means of identifying and promoting Canadian forest products that reflect our commitment to sustainable forests and environmentally sound technologies.' Likewise, paragraph 4.17 states: 'By 1994, forest-based industry associations will adopt self-regulating codes of environmental practice'. However, while these paragraphs signal an awareness of the need to utilise market mechanisms and codes to publicise the sustainability of timber products, a full-fledged third-party certification and labelling scheme is not yet being suggested.

Canada's and BC's international image programme suffered a dramatic reversal in 1993 when the Harcourt government decided to permit logging in Clayoquot Sound, resulting in a joint environmental-First Nation's blockade of Macmillan Bloedel's operations and the mass arrest of over 800 people. The public relations disaster resonated through America and Europe, undoing the gains previously made. Canada's claims to be managing its forests sustainably were subjected to a daily forensic media analysis that highlighted the gap between rhetoric and reality. At the same time, Canadian and industry representatives watched with alarm as efforts to establish the FSC appeared to be succeeding with the announcement of an inaugural assembly in October 1993. The fear of the industry, which was now strategising globally at the newly constituted International Forest Industry Roundtable (IFIR), was that the FSC would 'corner the market' on certification unless other options were available. To prevent this, and to regain control of the forest certification agenda, the CPPA announced its intention to establish a forest certification scheme in October 1993. Subsequently, it set up the Canadian Sustainable Forestry Certification Coalition to oversee the implementation of certification in Canada and contracted with the CSA to develop the standard (Elliott 1999, 297; Clancy 1998, 112).

While Canada's forest policy network was frantically trying to manage US opposition to its subsidised softwood lumber imports and European outrage over old-growth logging, the Canadian environmental movement was becoming increasingly engaged in forest certification initiatives. The Silva Forest Foundation began working on a forest certification standard for BC in 1993 (Hammond 1995a) and led the country when it certified a government-sponsored forest operation in Vernon, BC in 1995 to its standard (Hammond 1995b). SFF was a founding member of the newly formed Pacific Certification Council, which united five organisations in the Pacific Northwest to develop a common certification standard for the region (Smith 1995). SFF was also involved in early discussions to establish the FSC, although it was not a member of FSC's Interim Steering Committee. SFF did attend the FSC's founding assembly in Toronto, however, and was part of the faction expressing deep reservations over the decision to extend membership to industrial forestry operators and retailers.

Despite SFF's early involvement in certification and labelling, Canadian ECSOs were less involved in the founding of the FSC than might be expected. In Synnott's (2005) outline of the history of the FSC, the influence of American and British ECSOs is emphasised – notably FoE-UK and the Ecological Trading Company in the UK and the WARP in the US. While 16 Canadian participants attended the FSC's founding assembly in Toronto in October 1993, overall numbers favoured the US (37 participants) and representatives from Britain appeared to have played an important role in proceedings in promoting industrial forestry representation (Chris Elliott, WWF-International) and in resisting it (Simon Counsell, FoE-International). In part the relative absence of Canadian ECSO influence represents the fact that the FSC was being promoted by WWF-International on the one hand and US charitable foundations on the other. It thus took the Canadian environmental movement a few years to begin to implement the FSC in Canada. As in Australia, the vacuum was filled early by a coalition of industry and government promoting a nationally based forest certification scheme.

Establishment of forest certification

The FSC-IC was established in October 1993 and its first executive director, Tim Synnott, was appointed in April 1994. Subsequently, Canada began to figure more centrally in the organisation's strategic plans. In 1996, Marcelo Levy was appointed as Canada's FSC 'contact person' whose role, like Cadman's in Australia, was to promote the FSC to Canadian groups and facilitate the setting up of a Canadian national initiative. Levy's appointment enabled him to facilitate the development of regional standards across Canada. He encouraged the Maritime provinces to establish a standard-setting process and an FSC Maritimes group met for the first time in April 1996 at Truro to develop an 'Acadian' forestry standard for Nova Scotia, New Brunswick, Newfoundland and Prince Edward Island. In the same year, a regional standards' group formed in BC under the auspices of the Ecoforestry Institute Society (EIS). Coordinated by Lara Beckett (nee Lamport) and working mainly through email, it pulled together the components of a first draft of a standard for BC over the ensuing two years. A year later, in central Canada, the Great Lakes St. Lawrence regional standards committee was established at a meeting in Parry Sound. Subsequently, a small eight-person steering committee was elected, which developed a draft standard which was presented to the public in May 1998 (Wood 2000, 10).

There can be little doubt that FSC members at all levels underestimated the time, expertise and resources required to develop regional standards. Unlike governmental processes discussed below, in the mid-1990s the FSC had no template for standards development and, as of 1996, no endorsed national standards to build on. While the FSC's 10 principles and 56 criteria provided high-level guidance, standards development involved turning these into practical, performance-based, auditable actions. Ideally, standards development would have been supported by the publication of a standards development manual and facilitated by paid staff with expertise in standards writing. The FSC-IC had very limited funds at its disposal, however, and the scarcity of resources meant that most standards were developed by busy but committed individuals on a voluntary basis. Nowhere was this truer than in the Maritimes, where a volunteer Technical Standards Writing Committee devoted countless hours preparing, considering and revising the draft standard in an effort to reach consensus (Clancy 1998, 115; Tollefson, Gale and Haley 2008, 104).

The FSC's difficulties in resourcing standards' development were not experienced by the Canadian forest policy network. Once it reached a decision to develop a competing national standard to the FSC's, the resources were deployed to achieve the outcome. In early 1994, the CSFCC moved quickly to sign a contract with the CSA and a first meeting of the CSA Technical Committee was held in July. The first meeting was chaired by Jacques Mercier, former dean of the Forestry Department at Laval University, and consisted of over 24 individuals selected from several CSA stakeholder categories including producers, environmentalists, academia, practitioners and government (Elliott 1999, 303). Two non-negotiable decisions were taken prior to the first TC meeting. First, it was agreed that the CSA standard would be based on the ISO 14000 EMS approach, not the performance approach adopted by the FSC. Second, it was agreed that, once developed, the CSA standard would be put forward at the ISO as the basis for the development of an international forestry standard (Elliott 1999, 303; Clancy 1998, 114–15). Organisers even managed to present a first draft of the CSA standard to the July TC meeting, based on input from several individuals including foresters at the Canadian Forest Service of NRCan (Elliott 1999, 303).

Policymakers viewed the CSA process as potentially compatible with the development of criteria and indicators for sustainable forest management then getting underway in Canada and through the Montreal Process. The development of C&Is was listed as a priority development in the 1992 National Forest Strategy and work to develop them for Canada began in 1993. Through the development of C&Is, policymakers aimed to demonstrate progress towards 'sustainable forest management'. Canada published its national C&Is in March 1995, well before the conclusion of the CSA standard negotiations in July 1996 (Elliott 1999, 303–4). It thus proved relatively straightforward to attach the CCFM C&Is to the CSA standard, giving the standard the appearance of being performance-based and deflecting criticism. The CSA standard was subsequently published in two documents: Z808, *A Sustainable Forest Management System: Guidance Document*; and Z809,

A Sustainable Forest Management System: Specifications Document. The standard that emerged did not have the support of major environmental groups, however. Both Elizabeth May of the Sierra Club Canada and Monte Hummel of WWF-Canada, though regularly listed as Technical Committee participants, had never attended any meetings (Elliott 1999, 305). And consultative meetings across the country to consider stakeholder views had been accompanied by protests outside and expressions of deep concern inside.

Given the environmental movement's commitment to the FSC, industry and governments appear to have taken the strategic view that such groups would never endorse the CSA. They determined to make ongoing efforts to engage with environmentalists to demonstrate 'good faith' while simultaneously refusing to alter the CSA's standard development process. In 1996, however, and in the knowledge that the CSA standard would soon be published, the forest policy network confronted another strategic question. What position should the country take on its CSA standard? Logically, there were three options: endorse it, ignore it or remain neutral. In mid-1996, a position paper drafted by Mike Fullerton of NRCan set out a framework for evaluating 'voluntary certification systems for sustainable forest management' (Elliott 1999, 35). The position paper argued that Canada should not be involved in forest certification since it was a market-driven mechanism for achieving sustainable forestry. However, it noted that government could offer logistical and other support to those schemes that met a list of criteria that included being consistent with Canada's national and provincial laws, not acting as a barrier to trade, being developed in an open and transparent manner, and so forth.

With the CSA standard close to being approved, Canada's forest policy network sought to have it recognised as an international forestry standard at June 1995 meeting of the ISO in Oslo. As noted in Chapter 5, this proposal received little backing from most delegations and environmentalists although it was strongly endorsed by Australia. Developing countries like Brazil were suspicious that the proposal might curb their capacity to convert forests to beef and palm oil plantations. Indonesia worried that its own international standard, the *Lembaga Ekolabel Indonesia*, might suffer in comparison to Canada's. The US was concerned that a specific standard for forestry would weaken the more general application of ISO 14000 series, which was supposed to be applicable to all industrial sectors. The Oslo meeting agreed to establish a working group to report on ways that ISO 14000 could be applied to the forest industry, and this group advised at a follow-up in June 1996 meeting that consensus could only be reached on developing a reference document setting out how ISO 14000 might be applied to the forest sector (Clancy 1998, 119; Elliott 1999, 310–12).

The efficiency of the development of the CSA standard forms a stark contrast to the stop-start nature of the FSC regional standard development in Canada. In BC, the draft standard that emanated from the EIS process was ultimately overtaken by an alternative process championed by a range of higher-profile ECSOs including Greenpeace, Sierra Club of BC and the WCWC. In late-1998, a newly established FSC-BC Interim Steering Committee hired two consultants to complete the first draft of the FSC-BC standard (D1). Its release in May 1999 was accompanied by the launching of a new standards development process managed by a formally structured and elected steering committee backed by donor finance.

The Standards Council of Canada endorsed the CSA's Sustainable Forest Management Standard in October 1996. In contrast, the FSC only succeeded in negotiating its first Canadian standard for the Maritimes in December 1999 (Tollefson, Gale and Haley 2008, 109), although Ontario's Haliburton Forest was certified by SmartWood in 1998 under its own industry-based standards. Where the CSA began to experience difficulties, however, was in implementation. This was somewhat surprising given that the CSA SFM standard was a management standard that contained high-level perform- ance components via the attachment of the CCFM's C&Is. Given the forest industry's experience with management standards, and familiarity with ISO 9000 and ISO 14000, implementation could have been expected to be fairly straightforward. However, managers implementing the CSA SFM standard encountered difficulties related to its new and somewhat oner- ous public participation requirements. These requirements had emerged over the course of negotiations and were significantly more detailed than those contained in the standard's first draft (Elliott 1999, 307). The public participation requirements acted as a break on the implementation of the CSA standard, as managers experimented with consultative processes in 'designated forest areas'. Thus while some 15 companies were reported to be undertaking trials of the CSA certification system in 1998, none was yet registered to the standard (Elliott 1999, 298).

Growth of forest certification

By the end 1999, only a handful of forest management operations had been certified in Canada (Metafore 2009). The lack of certified forest area belied the efforts being made by companies to have their lands certified under the SFI and CSA schemes. Thus, between 1999 and 2000, the area of certified forest in Canada grew from only 600,000 hectares to 9.7 million hectares and doubled again to 17 million hectares in 2001 (Metafore 2009). Growth accelerated rapidly in the first decade of the twenty-first century and by 2006 around 110 million hectares of forest in Canada were certified to one of the three standards (Metafore 2009). Of this, the FSC had certified only about 5 million hectares (about 4%), whereas SFI had certified approxi- mately 40 million (33%) and CSA about 75 million hectares (63%). Thus, in only five years, CSA and SFI schemes had leapt ahead of the FSC in terms of forest area and timber volume certified. In the battle to certify forest area, it was clear that SFI and CSA were winning. The sudden leap in certified area by SFI and CSA schemes can be explained by the observation that neither required major changes to industrial forest management practices. In addition, both were well resourced, being backed by their respective industry associations and by provincial and federal governments.

Of the two schemes, Canadian companies initially experienced more difficulty implementing the CSA standard, which contained more substantive and time-consuming community and stakeholder consultation arrangements than SFI. By 2000, though, a methodology for managing the public consultation arrangements had been developed. Consequently, CSA certification proceeded apace. Indeed, so quickly was the uptake by industry of CSA certification that in January 2002 the Forest Products Association of Canada was able to announce that its member companies had committed to having all their forestry operations certified under one of the three schemes by 2006 (FPAC 2007, 1).

Underlying Canada's official position of neutrality with respect to forest certification schemes was a clear preference for the CSA scheme. While there is a dearth of data, available evidence suggests that provincial and federal governments provided direct and indirect support to the CSA scheme, whereas little support was offered to FSC's. The Canadian Forest Service, for example, underwrote the development of the criteria and indicators on behalf of the CCFM. Individuals from CFS were also involved in developing early drafts of the CSA standard. There was little sympathy for FSC certification in BC with the Assistant Deputy Minister Don Wright saying he opposed any system that might work against the BC Government's policy objectives (quoted in Elliott 1999, 314). The FSC received somewhat more support in Ontario where a more diverse economy and public concern over Temagami created the conditions for more progressive forest policies under the Rae New Democratic Party government. Indeed, it appeared that a compromise between the FSC and the Ministry of Natural Resources was achieved in 2001 when the FSC's new executive director, Maharaj Muthoo, signed a memorandum of understanding to certify all Ontario Ministry of Natural Resources lands to the FSC standard. The announcement, however, sent shock waves through FSC-Canada and FSC-International as it appeared that a foundation stone of the FSC system – the development of national and regional standards by relevant and duly constituted groups – had been replaced with a centralised bargaining process between governments and the FSC-IC. As the protests grew, Muthoo's position became untenable and he resigned after barely six months in the position.

Despite becoming operational at the same time as CSA and SFI in Canada, the FSC clearly struggled to compete in the early years of certification's development. Across the country, regional standard development proceeded at a snail's pace due to lack of funds and the need to reach meaningful compromises. By 1999, only the FSC Maritimes regional standard had been endorsed by the FSC-IC. In BC, although the first draft of the FSC-BC regional standard was released the same year, it was sidelined by the decision of the new steering committee to have a Technical Advisory Committee prepare a second draft, subsequently published in March 2001. And after early enthusiasm in Ontario and Quebec, the Great Lakes St. Lawrence regional standard process ran out of steam and failed to produce results.

The FSC standard development in Canada improved in the early 2000s for several reasons. First, the FSC Canada board reasserted authority over the fragmented standards development process, most notably by taking responsibility for the development of a national standard for Boreal forests in 2001. This was a major initiative and promised the elaboration of a standard for fully three-quarters of Canada's forests. In BC, the newly reconstituted steering committee began to make substantial progress on standards development, assisted in part by donor support from the Rockefeller Brothers and the Tides Foundation. In Nova Scotia, while large industrial operators like J. D. Irving backed away from the FSC in favour of CSA and SFI, one small operation, Nagaya Forest Restoration, demonstrated it was possible to adopt the FSC Maritimes standard and obtained certification in 2002. Meanwhile, in January 2001, Tembec announced its intention to seek FSC certification for its Gordon Cousins Forest in Northern Ontario in partnership with WWF-Canada (WWF-Canada 2001). By early 2005, therefore, Canada had an endorsed FSC-Boreal standard and an FSC-BC standard, with the latter standard fully tested by Tembec on its Parsons Timber Forest Licence in BC's East Kootenay region.

Consolidation of forest certification

In 2005, the FSC began to regain some of the ground it had lost to SFI and CSA in the previous decade. The year before, Ontario decided to mandate certification for major licensees, a decision that took effect in 2007 (Lister 2009, 148). The decision followed several years of observing FSC and CSA standards development processes and was designed to 'gain market recognition for Ontario's forest products; confirm Ontario's high-quality legislative and regulatory framework by independently verifying SFM practices in the province; and accelerate the certification of Ontario's forests to ensure the Ontario forest industry remained competitive with neighbouring jurisdictions' (Lister 2009, 148). Ontario's decision – and that of New Brunswick's two years earlier in 2002 – suggested that certification was increasingly becoming a market requirement. With certification becoming mainstream, the FSC was well positioned to benefit. At the end of 2004, total FSC-certified forest area in Canada was about one million hectares. The following year, this jumped to about 8 million hectares and then quadrupled by end-2009 to over 32 million hectares, which represented about 22 per cent of the total certified area of almost 150 million hectares. Several reasons explain the FSC's sudden growth spurt, the most important being the finalisation of FSC-Canada's Boreal Standard. The FSC Boreal

Standard enabled a number of large, industrial companies such as Domtar, Tembec and AlPac to become certified to the FSC standard. While less prescriptive than the FSC-BC regional standard, the FSC-Boreal standard requires companies to undertake several ecosystem-based forestry requirements. These include managing the forest to maintain its pre-industrial condition, obtaining informed consent from First Nations communities and ensuring appropriate management of high-conservation value forests.

To enhance their international acceptability, SFI and CSA both sought and received endorsement from the PEFC in 2005. Both schemes were also recognised by the CPET as providing assurances of legality and sustainability with regard to imports of timber products into the UK (CPET 2009). However, the Minister for Environment, Food and Rural Affairs, Hillary Benn, recently announced the inclusion of social criteria in CPET's definition of sustainable forest management, requiring that 'the management of the forest must have full regard for the interests of indigenous peoples, local communities and forest workers' (CPET 2010). The requirement is likely to benefit the FSC, which has far more thoroughgoing social criteria built into its definition of sustainable forest management than schemes affiliated with the PEFC. In another setback for PEFC, in 2006 government, environmental and industry groups reached a compromise over forest management practices in the Great Bear Rainforest in BC. The deal, which was finalised in 2009 in the Coast Land Use Decision, reserves about 2.1 million hectares, while providing for ecosystem-based management on the remaining land (Sierra Club of BC 2009). In December 2009, a consortium of companies that included Interfor, BC Timber Sales and Western Forest Products announced they had succeeded in obtaining FSC certification on almost a million hectares in the region utilising the 2005 FSC-BC regional standard.

Canada's response to forest certification

Given the importance of forests to Canada's economy, governments have responded strategically to the issue of forest certification and labelling. When forest certification was first mooted in the early 1990s, the relatively closed forest policy networks operating at federal and provincial levels were sceptical of its potential. However, out of fear that the FSC would monopolise certification and labelling services, governments and industry mobilised to create an alternative, industry-friendly approach based on the ISO's EMS. Substantial resources were directed by industry into developing the CSA's scheme, while provincial and federal government forest departments provided a range of supportive services, including monitoring developments, contributing to CSA draft standards and developing criteria and indicators for sustainable forest management. Over time, however, it became apparent to many government and industry officials that the FSC certification might be required in European markets and that a policy of official neutrality was preferable to one of outright endorsement of CSA. An outright endorsement of the CSA scheme also ran the risk of upsetting producers in the US, where industry was certifying to the SFI standard.

Moreover, as industry began to certify to the FSC in the past decade, criticism of the scheme has become more muted and governments have sought to work out compromises to secure certification to the standard. One early indication of the increasing acceptability of the FSC's approach to standard setting occurred when the Central Coast Framework Agreement employed the same language to identify the technical experts for its Ecosystem-Based Management Information Team that was used by FSC-BC to recruit the technical experts that developed the FSC-BC regional standard (Government of BC 2001). More dramatically, however, is the recognition contained in BC's Great Bear Rainforest Settlement where the FSC certification provides important reassurances to environmentalists and First Nations that forest management practices on the non-reserved land base would be adequately protected. In another move signalling change, in late-2009 the Government of Nova Scotia funded the development of FSC group certification in the province (Government of Nova Scotia 2009).

Fisheries certification in Canada

Canadian engagement with the MSC illustrates a range of complexities as well as pragmatic considerations affecting the introduction of external certification schemes into large-scale commercial fisheries. At the outset, Canada lagged behind Australia and the UK in terms of MSC certification. While take-up of MSC certification was slow, there are now more fisheries certified (four) and under assessment than in Australia. As elsewhere, Canada's engagement with certification has led to considerable questioning over the role of the MSC and scepticism over its impact and purported benefits. In the Canadian case this questioning associated with the legitimacy of external scrutiny and certification of fisheries has been exacerbated by the structure of fisheries politics, which has subjected fisheries management to ongoing pressure and uncertainty.

Fisheries politics in Canada

Fisheries issues rank high on Canada's political agenda, well above the economic contribution the industry makes to the country and linked closely to social and cultural issues. The political salience of fisheries has encouraged considerable scrutiny of management and government performance and generated both push and pull factors with regard to external assessment and certification of fisheries. A number of civil society actors are engaged in the debate over fisheries sustainability including high-profile ECSOs such as the David Suzuki Foundation, Greenpeace and the Ecology Acton Centre.

Sustainable Seafood Canada is an umbrella organisation of like-minded groups that includes the Canadian Parks and Wilderness Society, the Living Oceans Society, the Sierra Club Canada (BC Chapter) and the David Suzuki Foundation (David Suzuki Foundation 2010). Sustainable Seafood Canada supports actions to limit over-fishing and campaigns to educate consumers on sustainable seafood. It also promotes its own 'Best Choice' list, which it notes may not contain fish from MSC-certified fisheries due to differences in assessment practices.

The fishing industry, too, is a major actor in campaigns over sustainability. These organisations range from local fisheries associations to major umbrella organisations such as the Fisheries Council of Canada that represents companies that process a majority of Canada's seafood products. Broad-based groups such as the Atlantic Fisheries Federation and the Pacific Salmon Foundation provide support to fisheries on, respectively, the East and West coasts. At the provincial levels industry organisations such as the BC Seafood Alliance are important in promoting fishing interests in the harvest and post-harvest sectors and increasingly addressing industry initiatives in sustainable practices. Labour organisations have had an important historical and social role in Canadian fishery politics. They are more visible in Canadian fisheries politics and in policy networks than in either Australia or the UK, organising to provide voices to fish plant workers and crew. The Fish Food and Allied Workers, originally the Fishermen's Union established in 1970 in Newfoundland, is based on the Atlantic Coast. The United Fishermen and Allied Workers Union, established in 1948, is based in BC. Both unions belong to the Canadian Auto Workers Union.

Canadian fisheries politics has been shaped by the collapse of the cod fishery in Atlantic Canada and questions over the state of the stocks in the Pacific salmon fisheries, the latter also introducing issues of First Nations rights and access.

> The collapse of the Atlantic cod fishery off the province of Newfoundland in the early 1990s was shattering to coastal communities and the Atlantic region's economy. ... The resultant 'cod crisis' also served as a catalyst for a reassessment of management tools and policies used to govern Canada's fisheries.
>
> (Haward and Vince 2008, 130)

The social costs of the collapse were immense and ensured that the politics of the cod fishery broadened to consider issues of regional development and social support to communities. Labour and environmental groups joined criticism of government, although from different perspectives. Government's failure to regulate the cod fishery raised questions about the basis of fisheries management and the need to look to new approaches to maintain stocks. The problems of managing cod epitomised the difficulties of 'managing straddling stocks, a key feature of a crisis it was experiencing in its fisheries in the late 1980s and early 1990s' (Haward and Vince 2008, 119). It also saw an entanglement of domestic and international politics with the arrest of foreign fishing vessels accused of plundering cod and other species. Environmental groups pointed out that domestic fishing practices, underreporting of catches and 'high grading' (discarding small fish when larger fish could be landed) all contributed to the failure of the cod fishery and should not be hidden by claims of excess foreign fishing.

While Canadian fishers and fisheries managers have recognised the significance of ecolabels and consumer concerns over sustainability, initial reluctance coupled with criticism of the MSC approach has meant that the scheme has not been readily endorsed. More specifically the MSC process has been criticised over its processes, most notably by the University of British Columbia's Fisheries Centre, a world-leading institute for fisheries assessment, and on sustainability indicators for fisheries. Criticism was also levelled at the MSC by industry in relation to the process of certification of Pacific salmon fishery. These issues were recognised by the MSC and appear to have been resolved in the current certification process underway for Pacific salmon. However, the experiences of this fishery clearly illustrate the problems that can arise in fisheries based on similar, if not the same, stock, yet subject to different governance arrangements. As PriceWaterhouseCoopers stated in their *State of the BC Seafood Industry Report*, 'a significant challenge for the BC seafood industry is how best to communicate the industry's quality, safety and sustainability' (PriceWaterhouseCoopers 2000, 70).

Development of the MSC in Canada

Government, industry and environmental civil society organisations in Canada, as in other countries, observed the development of the MSC initiative. The increasing interest in the application of ecolabels to fisheries products, particularly from other North American fisheries that were similar to or targeted similar markets, put pressure on Canadian fisheries. The announcement from MSC's founding partner Unilever that it would only source products from MSC-certified fisheries had a direct impact on Canadian salmon fisheries, as BC salmon was a major product for Unilever. Later a further driver for Canadian action on the MSC was the certification of US Alaskan salmon, a controversial process and outcome that did much to affect the MSC's initial standing with industry and fisheries scientists in Canada. The decision by Sainsbury's to support MSC-labelled product also increased pressure on Canadian Pacific salmon producers.

As the UK is a major market for canned salmon from BC, a major impetus for adopting MSC certification of Canadian Pacific salmon was the US Alaska salmon fishery which obtained certification in 2000. The Canadian fisheries policy network linking industry and government mobilised to ensure that Canadian salmon was not shut out of the UK and other markets. However, this effort did not include the American market, where there is an absence of demand. 'In the US, however, where consumer awareness of and demand for certified fish remains low, few retailers save organic grocery chains use the MSC label when they sell wild Alaskan salmon' (Searle et al. 2004, 7).

The MSC system of certification that utilises performance indicators has a number of 'competitors' in Canada, where a significant effort has been undertaken to develop sustainability indicators and apply them to fisheries management. The most significant initiatives are those by the University of British Columbia's Fisheries Centre. The Fisheries Centre has developed 'RapFish' and is also involved in the development of *Fish Source* (see Pitcher 2007). Rapfish is an approach that provides rapid appraisal of the relative status fisheries against a series of attributes (Potts 2006).

Alternative approaches to the MSC include those developed in Atlantic Canada by GPI Atlantic (Charles 2005) that are deliberately built around broader sustainability indices. 'GPI Atlantic is an independent, non-profit research and education organization committed to the development of the Genuine Progress Index (GPI) – a new measure of sustainability, well-being and quality of life' (GPI Atlantic 2008). ISO 9000/14000 standards have also been used in Canadian fisheries, with a number of aquaculture operations utilising and promoting the ISO standards. Canada has promoted development of integrated and harmonious standards related to seafood safety and quality in the Asia-Pacific Economic Cooperation forum, and has noted the proliferation of 'green labels', particularly in Europe.

In seeking to explore the impacts and effects of ecolabelling in general and the MSC in particular, DFO entered into discussions with Anthony Charles in 2001 (Charles 2005). Charles, a world-recognised authority on fisheries policy and management, worked with DFO to develop the ingredients of a study to analyse the options. This led subsequently to formulation by DFO of terms of reference for the study that was put out for tender. This tender was won by Canadian consulting firm Gardner Pinfold, which released their report in March 2004 (Gardner Pinfold 2004). The report was to provide the DFO

> ... with information about key trends, drivers, challenges and opportunities for seafood eco-labelling, both globally and domestically. This research will be used by DFO to help the department to further define its position regarding seafood eco-labelling, given its relationships with industry, its membership in international fora, and recent developments in the area of seafood eco-labelling.
>
> (Gardner Pinfold 2004, 1)

In an interesting and important comment bearing on Canadian industry attitudes to the MSC initiative, Gardner Pinfold noted some fisheries have sought MSC certification reluctantly. This may be true of the BC Salmon Fishery which is undergoing certification due to pressure from customers that are showing preference for product from the MSC certified Alaskan Salmon fisheries.

(Gardner Pinfold 2004, 14)

In relation to the first MSC certification in Canada, a process that commenced in 2001, Gardner Pinfold recognised that

[t]he BC Salmon fishery represents the first fishery to pursue certification following the certification of a direct competitor for the same fish species. There has been much discussion over the differences between the application of criteria in the Alaska and BC fisheries. The Alaskan salmon fishery was evaluated against 26 indicators whereas the adjacent BC salmon fishery is being evaluated on the basis of 47 indicators.

(Gardner Pinfold 2004, 16)

The differences in the processes used to assess these two salmon fisheries reflected a maturing of the MSC and broader experience in the certification process, as well as the organisation's commitment to continual improvement and learning (Howes 2005). This is an interesting and important case of the MSC in operation: not only were the two salmon fisheries similar but also the same certifier, SCS was responsible for both assessments. While there has been criticism that the BC salmon fishery was subjected to a different standard than the Alaskan fishery, others note that it was not a different standard but simply a different application of that standard through the increased number of indicators (Gardner Pinfold 2004, 1). Canadian stakeholders 'anticipate that Alaskan Salmon will also be subject to a more labour intensive assessment as they apply for certification renewal' (Gardner Pinfold 2004, 17).

As noted in Chapter 4, the DFO is the major management and regulatory agency for Canadian fisheries. Provincial governments have some delegated responsibilities, but all marine fisheries are managed by the DFO. This includes provision of stock assessments and related science. As a result the DFO holds significant data on stocks in addition to information on management procedures that is necessary for any MSC-type certification or assessment. As in Australia, DFO supports industry in seeking external certification but does not mandate such action. On the other hand the department is well aware of the consequences of an MSC assessment, that is, if a fishery fails an assessment there will be a presumption that government is failing in fisheries management. This reinforces the perception that the development of third-party external standards to assess sustainability of fisheries poses particular challenges for government. This point has been reiterated by Shelton and Sinclair (2008) who note that recent developments in Canadian fisheries policy, most notably the development of a 'harvest strategy framework' reflects commitments to sustainable fisheries management, and links a number of domestic and international initiatives that Canada has developed or to which it has been party. This policy framework will enable fisheries management in Canada 'to be better able to meet new ecocertification and ecolabelling standards' (2008, 2306). The authors note, however, that 'ultimate responsibility for fisheries management in Canada rests with the Minister for Fisheries and Oceans Canada and is not shifted to a third party such as the MSC' (Shelton and Sinclair 2008, 2308).

BC salmon

The BC salmon fishery 'is exploited by commercial fisherman, aboriginal fisherman and sport fishermen' (MSC 2002). The fishery is based on five species: Sockeye, Pink, Chum, Chinook and Coho salmon. These species are anadromous, spawning in fresh water, where juvenile fish spend up to a year before migrating to the Pacific Ocean. In addition to wild spawning stocks, salmon hatcheries have been used for over a century to bolster stocks of particular species. The salmon fishery also faces direct competition from developing aquaculture operations, a factor noted by the MSC in responding to the question of certification (Lincoln 2005). Thus farmed salmon increased in value from C$107 million in 1990 to C$309 million in 2002 (PriceWaterhouseCoopers 2000, 3). At the same time 'total wild landing of salmon declined by 80% during the decade [and] wholesale and export values for BC wild salmon products declined proportionately' (PriceWaterhouseCoopers 2000, 3). This led to significant restructuring of the industry and concomitant changes in processing and marketing arrangements (Brown 2005).

The assessment of the BC salmon fishery was the first case of a similar and/or competing fishery being certified under the MSC standard. Inevitably this could be expected to lead to concerns over varying interpretations of the MSC standard, particularly the application and assessment of the fishery under the MSC principles and criteria. Not surprisingly, the BC salmon industry tried to make sure that they were getting a process that was similar to that undertaken for Alaskan salmon. Despite industry efforts to engage Alaskan industry and civil society organisations, including using the same certifier, SCS, the process of assessing the BC salmon fishery was perceived to be highly cumbersome compared with that undertaken in the Alaskan salmon fishery.

The Alaska salmon fishery was certified as a single fishery whereas the initial proposal for the BC salmon fishery was that it be subdivided by species (Sockeye, Pink, Chum, Chinook and Coho salmon respectively) and further subdivided by river runs, leading to 40 stocks and 46/49 criteria. Canadian industry was concerned over 'creeping standards': each time the criteria were applied they became more stringent and they were being subjected to more stringent standards than those faced by Alaskan fisheries, their direct competitor in European markets. The initial focus was on Sockeye salmon, with Pink and Chum salmon undergoing assessment in 2008.

The British Columbia Salmon Marketing Council (BCSMC) was the applicant for the MSC process.

The British Columbia Salmon Marketing Council is an association formed in 1991 to represent the harvesters and processors of commercially caught BC wild salmon with a mandate to:

- benefit and promote the BC wild salmon industry;
- conduct research and educational programs for the development and promotion of commercially harvested BC wild salmon, and;
- communicate to national and international markets the quality, availability and value of BC wild salmon.

(BCSMC 2008)

SCS established an evaluation team comprising Chet Chaffee of SCS with 'Karl English, Vice-President, Western Operations, LGL Ltd, Sidney, BC; James Joseph, former director, Inter-American Tropical Tuna Commission, now an independent consultant; and Dana Schmidt, senior fisheries biologist, Golder Associates, Castlegar, BC' (SCS 2003). In June 2003 SCS published the evaluation team's guidelines on the certification, noting that 'the evaluation process will cover all commercially harvested salmon in BC that can be legally sold, including salmon caught by First Nations' for commercial sale' (SCS 2003, 2). The guidelines incorporated the specific performance indicators and scoring guidelines to be applied to the fishery. BCSMC responded to the guidelines in late November 2003 and a technical report was prepared with the assistance of Fisheries and Oceans Canada.

Following a long delay, in 2008 a change in certifier was negotiated with TAVEL Certification Inc replacing SCS in order to 'expedite the BC Sockeye Salmon MSC assessment' and 'complete the certification process initiated in 2001' (TAVEL 2008). TAVEL Certification announced that it would contract the same team members to complete the assessment and contract the peer reviewers identified by SCS (TAVEL 2008). While previous work would be used, the new certifier would ensure that the assessment team would re-score the fishery under 'the relevant performance indicators' recognising that considerable time had passed from the initial analysis and management changes had been implemented since the initial work had been undertaken. In January 2008 the MSC announced that Pink and Chum salmon fisheries had applied for assessment. TAVEL Certification was contracted to undertake the certification and indicated that it was expecting to complete the assessment by February 2009. The complexities of the assessment have meant that the process has taken longer than the certifiers had expected. A number of ECSOs have indicated concern with the salmon assessment, particularly in relation to fish populations considered to be vulnerable. These concerns included non-target species as well as the assessment of the primary stock. The David Suzuki Foundation was particularly concerned with the assessment of Sockeye salmon. 'BC sockeye salmon are currently ranked as "yellow" with some concerns by SeaChoice Canada. We recommend that sockeye should be consumed infrequently or when a green choice is not available' (David Suzuki Foundation 2010).

Canadian Northern Prawn Trawl

In August 2008 the Northern Prawn Trawl fishery was the first Canadian fishery to achieve MSC certification. This is the largest coldwater shrimp/prawn fishery in the world (MSC 2008) with a catch of 68,000 tonnes per year. The fishery targets the northern prawn or shrimp (*Pandalus borealis*), 'usually found in areas with soft, muddy sediment where temperatures range from 1–6°C and depths range from 150–600 metres' (MSC 2009a). The fishery uses bottom 'otter trawls with a minimum mesh size of 40 mm, and fitted with a Nordmore separator grate [to minimise bycatch of non-prawn species]. Prawn pass through the grate, but ground fish with swim bladders are directed upwards towards an exit triangle in the upper panel' (MSC 2009a). The management of bycatch is common to all prawn fisheries and has led to significant effort worldwide to combat problems of non-target species catch (Bache, Haward and Dovers 2001) and environmental damage to the seafloor and environment.

The prawn fishery around Newfoundland and Labrador increased in significance with the collapse of the cod fishery in the early 1990s. Derek Butler, the executive director of the Association of Seafood Producers noted 'the resource has changed – shellfish in the place of ground fish – and the landed values are historically high. In spite of that, we have kept processing and harvesting capacity too high, such that the value created is spread too thin' (*Globe and Mail*, 15 November 2007). Butler's organisation, the Association of Seafood Producers, 'is a not-for-profit corporation which represents the interests of seafood producers generally in the Province of Newfoundland & Labrador (NL), Canada' (ASP 2009). The association became the client of record for the MSC assessment of the Canadian Northern Prawn Trawl fishery, recognising the importance of external certification of sustainable practices for a European export-oriented fishery.

The assessment was contracted to Moody Marine, with the assessment team announced in January 2007. The team incorporated lead assessor Andrew Hough (Moody Marine, UK), Michaela Aschan (Professor in fisheries biology at the Norwegian College of Fishery Science), Colin Bannister (former head of the Shellfish Resource Group at CEFAS, Lowestoft) and Howard Powles (a fishery consultant and former director of Fisheries Science and of Biodiversity Science at the Canadian Department of Fisheries and Oceans Headquarters (Moody Marine 2007). The assessment process ran from October 2006 to August 2008, following a similar approach to other MSC fisheries assessments. A key aspect of the assessment process is, as noted in Chapter 3, stakeholder and public consultation and feedback at all stages of the process. In the Canadian Northern Prawn Trawl Fishery this involved consultation, distribution of material to fishers and site visits. It is noteworthy that no objections were received in relation to the initial assessment stage over the assessors of the fishery, nor any objections received over the 'public certification report' (MSC 2009a). SeaChoice, a Canadian national sustainable seafood programme did, however, reiterate its concerns over 'continual damage to underwater ecosystems from bottom trawling' following the release of the MSC certification for the fishery (SeaChoice 2008).

The achievement of the MSC certification for the Northern Prawn Trawl fishery gave impetus to the MSC in Canada (MSC 2008). The MSC also received support from the minister of Fisheries and Oceans, the Honourable Loyala Hearn, who congratulated the Association of Seafood Producers on their effort. Minister Hearn noted that 'government is only one part of the big picture. That's why it's so important to recognize efforts by private industry and independent third parties who share our commitment to conservation and sustainable use of the fishery' (DFO 2008). The Loblaw Group of companies, Canada's largest food retailer, noted with regard to this fishery certification that 'Canadians are increasingly aware of and concerned over sustainable fishing methods. We proudly announced offering MSC-certified fish in our stores earlier this summer and will have 15 MSC certified private label seafood items by year end' (MSC 2008).

Gulf of St Lawrence Northern Shrimp

This fishery targets the northern shrimp/prawn in the Gulf of St Lawrence, with the fishery based in the Canadian provinces of Quebec and New Brunswick. The fishery was certified as sustainable by the MSC in September 2008 following an assessment process carried out by TAVEL Certifications Inc for clients Association Québécoise de l'Industrie de la Pêche (Quebec), Produits Belle-Baie Ltée and L'Association Cooperative des Pecheurs de l'Ile Ltée (New Brunswick). 'The holders of this MSC certificate are five shrimp processors from Québec that are members of the Association of Quebec Seafood Producers (AQIP), and two New Brunswick-based shrimp processors' (MSC 2009). The assessment of the Gulf of St Lawrence Northern Shrimp fishery paralleled that of the Northern Prawn fishery, beginning in November 2006 and concluding in September 2008. This assessment involved a 'harmonised assessment' for an area that was targeted by Newfoundland- Labrador and Quebec-New Brunswick fishers, with both certifiers, TAVEL and Moody Marine collaborating on the assessment and reporting.

Canadian Pacific halibut

Pacific halibut (*Hippoglosus stenlolepis*) is found across the North Pacific Ocean from the Sea of Japan through the Bering Sea to Southern California (halibut fact sheet) and is fished commercially and as a sport fish. MSC certification for Canadian Pacific halibut is being sought by the Pacific Halibut Management Association which represents over 90 per cent of commercial halibut licence holders in BC (PHMA 2009). The association is active in ensuring a sustainable fishery, and through it 'halibut fishers have entered into formal co-management arrangements and assumed a greater role in the day-to-day operations of the fishery. This initiative is part of the commercial halibut industry's commitment to fish responsibly and be accountable for its actions, its fishery and its future' (PHMA 2009).

The fishery is managed through a longstanding international convention between Canada and the US. This convention established what was initially called the International Fisheries Commission in 1923 (IPHC 2009). The International Pacific Halibut Commission (IPHC) is a joint US/Canadian Fisheries management agency (MSC 2009) whose 'mandate is research on and management of the stocks of Pacific halibut ... within the Convention waters of both nations. ... The IPHC consists of three government-appointed commissioners for each country who serve their terms at the pleasure of the President of the United States and the Canadian government respectively' (IPHC 2009).

SCS appointed an assessment panel of Dr Rick Deriso (Inter American Tropical Tuna Commission) with expertise in stock assessment, Dr Robert Furness (University of Glasgow), ecosystem impacts and Dr Mike Shepard (independent consultant) to address management systems. Following development and completion of the performance indicators and scoring guideposts the draft report was completed. Peer reviews were publicised in June 2008 and a revised list of peer reviewers was notified at the end of October 2008, with comments to be directed to the certifier in November 2008. The fishery was certified by the MSC in September 2009. This certification was endorsed by the David Suzuki Foundation, after initial strong opposition when the fishery was first proposed. This opposition centred on concerns over bycatch and adequate monitoring of bycatch. Responses by the industry to these concerns were welcomed by the foundation, but they retained a qualified position: 'although the David Suzuki Foundation supports the certification of BC's halibut fishery, the sheer complexity of the marine ecosystem and inherent vulnerability of the species involved mean ongoing vigilance and re-evaluation is necessary' (David Suzuki Foundation 2010).

Analysis of the MSC in Canada

Canada has been at the forefront of work on use of market-based approaches and external certification of fisheries, particularly in relation to the assessment of fisheries sustainability. Key researchers with the Fisheries Centre,

Daniel Pauly and Tony Pitcher, had been involved in the development of the MSC. Pitcher and Pauly participated in the Bagshot MSC meeting (Pitcher 2001, see also Chapter 3), which led to the codification of the MSC principles and criteria (Sumaila, Pitcher and Pauly 2005, 4). However, the Fishery Centre, and Tony Pitcher are now significant critics of the MSC process.

Canada's engagement with the MSC highlights the influence and inter-relationships of internal and external drivers. Action by Alaskan fishers in the Pacific Salmon fishery to seek MSC certification led to Canadian industry's initial foray with the MSC. As discussed above, Canadian industry's concern over Alaska Salmon MSC certification process and distrust of DFO in relation to US/Canada issues appears to have coloured early experiences. The complexity of the Pacific Salmon fishery raised important challenges over the process, and how a multi-species anadromous fishery could be assessed. Given these factors it is no surprise that the first fisheries certified under the MSC were much smaller scale and without international or bilateral issues affecting their management.

The MSC was also criticised over its methodology. A critique by the University of British Columbia Fisheries Centre was published in an open letter to the MSC in the Fisheries Centre newsletter, *Fish Bytes*, in 2001. This letter attacked the lack of transparency in the scoring process in the first major fisheries assessed, the Alaskan salmon fishery and Western Australian rock lobster fishery (Pitcher 2001). The 'blanket approval of all species, localities and gear types for the Alaskan salmon fishery raises serious questions over the scoring process' (Pitcher 2001). Sumaila, Pitcher and Pauly criticised the fact that 'almost no information could be found on fisheries that failed accreditation' (Sumaila, Pitcher and Pauly 2005, 5). In response the MSC argued that changes to the certification process in 2002, which were codified in 2004 with the MSC's redrafted Fisheries Certification methodology, addressed the criticism admitting that 'the scoring process was indeed a black box in the early days, lacking transparency' (Howes 2005, 1). Subsequently, Sumaila accepted an invitation to serve on the MSC Stakeholder Council.

Concern, too, has been expressed over the 2008 Prawn fishery certification. SeaChoice and the David Suzuki Foundation have been critical of the 'minimal passing grade in the ecosystem category of the MSC's certification process for the fishery' (SeaChoice 2008). These environmental critiques of the implementation of the MSC standard in Canada are similar to those expressed in other countries: the standard (MSC's principles and criteria) is not applied at a high enough level. SeaChoice highlighted that several sub-criteria did not reach required 'pass' levels.

The Canadian government has provided indirect support to the MSC initiatives within Canada, providing scientific and research support and encouragement to industry bodies seeking certification. The Canadian government, unsurprisingly, clearly sees MSC certification as an adjunct to its management efforts. Gardner Pinfold noted that

> in Canada, industry cannot attain eco-label certification without DFO's assistance. Costs associated with supporting certification information requirements will vary across fisheries. DFO's initial experience with BC salmon is expected to be costly. It would be helpful for DFO to develop a standard approach to respond to requests for support from industry. *Ignoring independent certification bodies, such as the MSC, is not an option*
> (Gardner Pinfold 2004, 4, emphasis in original)

The presence of MSC-certified products is increasing in Canada. Major retailers such as Loblaws have provided direct support to the MSC label and encouraged more Canadian fisheries to seek MSC certification. The fishery sector has recognised the need for external validation of claims for sustainability in relation to export products. The Canadian experience emphasises the complexities in introducing both external certification and the MSC programme. Recent evidence on increase in fisheries certified and those under assessment suggests, however, that the MSC programme may have overcome earlier Canadian concerns and hostility. In Canada, as in Australia and the UK, the interactions between state, market and civil society provide important parameters for the development and ultimate success of certification and labelling schemes. ECSO activity and campaigns have increasingly focused on sustainable fisheries. This activity has centred on publicising 'red lists' of overfished species and the promotion of online guides to sustainable seafood. These guides include SeaChoice, State of the Catch, Ocean Wise and Seafood Watch, the latter a widely distributed publication of the Monterey Bay Aquarium. None of its member organisations have provided unequivocal support or endorsement of the MSC standard. The David Suzuki Foundation and Greenpeace have been critical of the MSC certification process and standard, with the former promoting alternative approaches to consumer education. The David Suzuki Foundation has engaged with the MSC process in key fisheries and was able to support the halibut certification when industry committed to strong measures to manage and monitor bycatch.

Scrutiny of the practices and management of the Canadian fishing industry by environmental groups will inevitably increase. This scrutiny will be directed at individual fisheries (and possibly towards individual fishers) and more broadly at the industry as a whole. Government's response is important and the development of Canada's harvest strategy, in a similar approach to that adopted in Australia, will better place it and the industry to address issues such as 'environmentally friendly gear', maintenance of habitats and reduction of bycatch. It is obvious that while existing management practices involve an increasing level of stakeholder participation and external 'oversight' of these practices, the increasing use of external certification provides great challenges for fisheries managers and industry alike. The challenges arise from a range of sources: the legislative base of much fisheries management and the responsibility of fisheries managers, problems of defining and involving all stakeholders and providing effective institutional arrangements for the management of fisheries.

Conclusion

While the certification of forests and fisheries have both been controversial in Canada, government support for the MSC has been more forthcoming than for the FSC. FSC certification was perceived as a threat by industry and governments alike, and it called forth an immediate response in the form of an alternative, national forestry standard: the CSA. In contrast, MSC certification was perceived, albeit reluctantly, as a requirement to access European markets for which a 'made-in-Canada' solution would not be feasible. Consequently, the FSC's development in Canada was fraught with difficulty as it encountered extensive hostility from federal and provincial policy networks who constantly and consistently promoted alternative certification schemes including CSA, SFI and ISO. In contrast, the MSC's development while certainly not unproblematic had fewer barriers placed in its way by government and industry.

Given the relative differences in the acceptability of the FSC and the MSC at the outset, it is somewhat surprising that today the FSC appears to be the more successful of the two schemes in the Canadian context. With over 35 million hectares of forest certified, about 8 million hectares undergoing certification and with demand for its products increasing in Europe, North America and Asia, the FSC appears to have 'cracked' the Canadian market. If this is indeed the case, then the FSC has succeeded in spite of, rather than because of, governmental and forest industry support in Canada. In contrast, although the MSC was initially well received by industry, governments and environmentalists in Canada, experience has not strengthened that support. The difficulties and anomalies that emerged in the certification of the Pacific Salmon fishery led to concerns being expressed not only by industry but also by environmentalists and fisheries scientists. The former were concerned about the burdensomeness and 'creeping' nature of the standard while the latter were concerned about its comparative weakness from an ecosystem-based management perspective.

Finally, in contrasting the Canadian case with the Australian case, we note that despite the evident disaffection of the Canadian forest policy network for the FSC, a more nuanced strategy was required in combating it than that adopted in Australia. The reason for the official policy of 'neutrality' with respect to the FSC in Canada in contrast to Australia's aggressive promotion of AFS relates to the structure and operation of the two countries' commodity chains. While Canadian producers are heavily dependent on US and European markets, Australian producers rely on Japanese and Chinese markets. For Canadian producers, this meant that there were several other national standards with which CSA would compete beyond the FSC standard; notably a battle between CSA and SFI would have been very counterproductive. In contrast, the Australian policy network has been under no such pressure until recently, enabling it to aggressively promote its own national standard domestically and internationally while vilifying the FSC standard as an environmental conspiracy. With regard to fisheries certification an increasing number of fisheries have been certified or are under certification. This indicates a growing acceptance of the process and standard by industry. The increasing number of fisheries under the MSC reflects the lack of a viable competitor standard, although the alternative sustainability guides have high visibility and impact. In contrast to Australian industry that is directed towards North Asian markets, the salience of the European market for Canadian exports is a factor supporting MSC certification.

7
Forest and Fisheries Certification in the UK

Of the three cases of forest and fisheries certification examined in this study, the UK's has been the most successful. Not only are all public forest lands in the UK now certified, but so too is a significant portion of private forest land. Moreover, this certification success has been achieved via a compromise between environmental, economic and social interests that has forestalled the emergence of an FSC competitor scheme. Thus, while there is a PEFC-UK initiative, it has been unable to generate much support in the UK, where brand recognition for the FSC's 'tick-tree' logo is widespread. Likewise, local fishers embraced the MSC model early, and have continued to support it. Retail chains especially indicated their support for 'eco labelled' fish products early, recognising the interests and concerns of highly informed consumers.

With respect to the MSC, ecolabelling and certification of fisheries, and more importantly fisheries products, has greater presence in the UK than in either Australia or Canada. Thus UK fisheries were 'early adopters of the MSC approach', with the Thames Herring fishery being one of three certification test cases 'conducted and supervised by the MSC in 1999–2000 – the others being the Western Australian Rock Lobster and the Alaska Salmon fisheries' (May et al. 2003, 23). In addition to the Thames Herring fishery, the Bury Inlet Cockle fishery and Southwest Handline Mackerel fishery received certification in 2001. While these fisheries were small in scale, the early adoption of the MSC standard provides interesting insights into the dynamics of fisheries certification processes.

The FSC-UK story is even more interesting with the government emerging as a strong supporter of an environmentally and socially acceptable forest standard. This support appeared in the late 1980s when the UK's international development arm, the Overseas Development Administration, backed a proposal for a feasibility study at the ITTO. Implicit support for certification and labelling also developed within DEFRA. And, although initially vigorously opposed to third-party certification, the UK Forestry Commission eventually endorsed it in the late 1990s and played a crucial role in brokering an industry–environment compromise. The shift to a 'multipurpose' forestry paradigm in the 1980s that we charted in Chapter 4 created the discursive context for the Forestry Commission to reprioritise its forestry agenda. This shift, together with the extreme dependence of the UK on timber imports from overseas, forced the commission to rethink its oppositional stance to FSC-style third-party certification.

As in our other two chapters, we commence with an account of the development of forest certification in the UK, focusing on the actors involved and its emergence, development and growth. Following an account of the role played by the UK's forest policy network in the emergence of the UKWAS, we turn our attention to fisheries certification and chart the emergence, development and growth of the MSC in the UK. The chapter concludes with a comparison of the UK's response to FSC and MSC certification and also highlights some of the main distinctions between this case and the Australian and Canadian case studies.

Forestry certification in the UK

UK forest politics

As outlined in Chapter 4, the general context of twentieth century forest politics in the UK was shaped by the dominance of the Forestry Commission and its 'home defence' argument that justified a policy of aggressive afforestation and reforestation on public and private lands. In prosecuting this policy, the commission both shaped and was shaped by an expanding forest policy network that came to depend on government for financial, technical and political support. The Forestry Commission defended the taxation arrangements for Britain's foresters, arguing that it was essential that the private sector had a financial incentive to grow trees. Similarly, the Woodland Grants Scheme was justified because it gave cash grants to landowners who dedicated a portion of their estates to wood production. But the Forestry Commission not only defended the interests of the broader forest industry in Westminster and Whitehall, it also provided a range of technical assistance measures to support tree growing. These included advice on species, location, spacing, fertilising and chemical use and, later on, on landscaping practices to reduce the visual impact of exotic monocultures. In summary, the Forestry Commission stood at the centre of a fractious forest policy network that sought to have its interests met through accommodative public policy.

For much of the twentieth century, and despite severe criticism from conservation, recreation and amenity interests, the UK forest policy network was able to secure its interests. Change began to occur in the 1960s, however, with the rise of an increasingly organised, financed, institutionalised and politically savvy environmental movement. At the core of the environmental policy network in the UK were WWF-UK and FoE-UK. These two organisations adopted 'insider' and 'outsider' strategies respectively, with WWF-UK associated with lobbying the establishment and FoE-UK reserving its right to use direct, non-violent action. This division of labour was not consciously agreed upon but reflected differences in conceptions of the environment that translated into alternative political styles. Importantly, however, both FoE-UK and WWF-UK were strong champions of forest certification and labelling, which they believed would help institutionalise sustainable forest management at home and abroad. This enabled them to present a unified position on it as a key policy solution to the problem of unsustainable logging at home and illegal logging overseas.

UK environmental movement

The two groups at the centre of the UK environmental policy community with respect to certification were FoE-UK and WWF-UK. FoE-UK is a grass roots environmental organisation with a devolved organisational structure. In the UK, there are a large number of semi-autonomous local groups spread around the country. FoE was launched in the UK in 1971 as a campaigning organisation. One of its first actions was to dump 1500 non-returnable bottles outside the headquarters of Schweppes, and it has maintained its reputation for dramatic media spectacle down the years.[1] Today, FoE-UK claims to have 230 local campaigning groups throughout England, Wales and Northern Ireland, with a separate and large presence in Scotland (FoE-UK 2008a, 3). It is hard to gauge membership numbers directly and no explicit information is provided on its website or in recent annual reports. Rootes and Miller provide data for FoE-UK membership between 1971 and 1998 that suggest significant rates of growth, with 1981 membership of about 18,000 growing to 114,000 by 1998 (Rootes and Miller 2000b, Table 7).

FoE-UK's policy agenda is wide ranging and includes a large number of very local actions. Indeed, its structure means that there is often a struggle between local and central campaigners over its policy and campaign agenda. According to its recent annual review, the 2008 campaigns focused on climate change, sustainable economics, monopolistic supermarket behaviour and environmental justice. While some of these topics are also pursued by the environmental mainstream, it is its environmental justice campaigns that often set apart FoE-UK from WWF-UK and other conservation groups. FoE-UK has launched a Rights and Justice Centre to provide 'free environmental legal advice' to UK citizens, while further afield its Action for Justice Programme has sought to empower Indonesian communities in their struggle against palm oil companies (FoE-UK 2008b, 6). To fund these activities, FoE-UK relies on funding from individual donations and 97 per cent of its total income of almost £4 million came from this source in 2008 (FoE-UK 2008a).[2]

In contrast to FoE-UK's devolved and 'outsider' mode of environmental campaigning, WWF-UK adopts a more corporatist approach. WWF-UK was founded simultaneously in Switzerland and the UK in 1961 and was motivated by ongoing decolonisation and 'fears that habitat destruction and hunting would soon bring about the extinction of much of Africa's wildlife' (WWF-UK 2009). Its founders included several notable individuals including biologist Julian Huxley and naturalist and painter Peter Scott. Scott designed the organisation's famous panda logo and for the next 20 years the organisation was best known for its efforts to raise money in the West to protect endangered species in developing countries such as the panda, tiger, blue whale and other 'charismatic megafauna'.

However, in 1980 and in conjunction with the IUCN and the United Nations Environment Programme, WWF-UK joined other national efforts and its international headquarters in launching the World Conservation Strategy. The strategy developed the concept of 'sustainable development' later popularised by the UN Commission on Sustainable Development (Brundtland Commission) and took a broader-based approach to conceptualising the relationship between conservation of endangered species, habitat protection and the wider social, economic and political context in which these goals had to occur, including 'the concept of living within the limits of the natural environment' (WWF-UK 2009). After 1992, Rootes observes, WWF-UK again 'widened its ambit to work with other NGOs to form a common agenda on development and the environment' and, following a member survey that suggested its supporters were more radicalised than its cautious board, fully embraced the sustainable development agenda (Rootes 2006, 771).

WWF-UK's preferred *modus operandi* has been to cooperate with governments and corporations, with the latter making a significant contribution to its income after 2000. Rootes reports that between 2000 and 2005 'the fastest-rising part of WWF-UK's income was that from corporate donations and sponsorships'. It rose from a miniscule 0.25 per cent of total income in 2000–1 to 15 per cent of total income in 2004–5 (Rootes 2006, 775). In its 2008 annual report, corporate donations were reported to be around 14 per cent of total income, up from about 12 per cent for 2007. Income from governmental grants for 2008 was reported to be almost 11 per cent of total income, slightly down on 2007 income from this category, which was reported as 12.5 per cent (WWF-UK 2008, 22). Notwithstanding this significant support from corporations and governments, the 2008 report notes that 'the core of our funding comes from concerned individuals who support us through cash gifts, monthly direct debits and thoughtful legacy donations' which 'account for about 57% of all income generated' (WWF 2008, 13).

FoE-UK and WWF-UK were the main environmental organisations lobbying for the adoption of certification and labelling of timber products in the 1980s and 1990s in the UK. However, a successful FSC process requires participation of social sector groups and this proved more difficult to obtain. In its last several annual reports, the FSC-UK lists the National Trust, the Forest Management Foundation and four individuals as members of its social chamber. Broader recreational and amenity interests are represented via an umbrella group called Wildlife and Countryside Link (WCL). WCL is listed as an organisation appointing representatives to the UKWAS Steering Group (UKWAS 2000, 8) and members of the WCL also explicitly endorse the FSC and UKWAS in a 'Manifesto' published in 2007 (WCL 2007). In the latter document it is stated that WCL 'supports additional voluntary · certification through the UK Woodland Assurance Scheme (UKWAS), which meets international Forest Stewardship Council "criteria and principles for sustainable forest management"' (WCL 2007, 5).

WCL began as Wildlife Link in 1980 and was 'set up by Lord Peter Melchett, whose position in the House of Lords convinced him that better coordination was needed between voluntary organisations with similar core objectives' (WCL 2009b). In 1990 it merged with Countryside Link and today brings together 37 organisations representing a diversity of interests focused on protecting the UK's wildlife, landscape and marine environments. It claims that its 37 member organisations 'collectively employ over 10,000 full-time staff, have the help of 170,000 volunteers and the support of over 8.3 million people in the UK' (WCL 2009a). WCL is responsible for 'coordinating groups of experts in collectively agreed policy areas such as biodiversity, farming and rural development, land use planning, water, whales and wildlife and trade' (WCL 2009c). FoE-UK and WWF-UK are both members of WCL, as are Greenpeace, the Royal Society for the Protection of Birds, the National Trust, the Woodland Trust, the Ramblers Association, the Campaign to Protect Rural England and the Royal Society for the Protection of Cruelty to Animals. It thus brings together a large number of groups with environmental, animal welfare, recreation and amenity interests in rural Britain. Given its diverse membership, it is significant that WCL supports UKWAS and the FSC. However, it did not play a particularly active role in the politics of certification in the UK in the 1980s and 1990s despite being very active regarding other more domestic forestry matters related to the Flow Country and to access to rural lands.

UK forest industry

Of our three forest policy networks, the UK's had undergone the most change in the past two decades. When certification was first mooted in the mid-1980s in the country, the forest policy network included three powerful business groups representing landowners, forest growers and timber importers and traders. Associations representing these interests separately and jointly lobbied the dominant governmental institution, the Forestry Commission. The commission was sympathetic to their concerns not only because it was instrumental in creating some of the associations but also because it shared the paradigm of sustained-yield forestry that dominated forestry thinking at the time and that justified the aggressive afforestation and reforestation policies from which all network members benefited.

A key interest in the British forest policy context is private landowners, some of the largest of which are members of Britain's nobility who own large estates of many thousands of hectares. Large landowners dominated forest policy in the early years of the twentieth century and the Acland Committee that recommended the establishment of the Forestry Commission included Lord Lovat, an important member of Britain's landed gentry with an estate in Scotland. Both he and Lord Clinton were among the first group of commissioners to be appointed to the Forestry Commission when it was constituted in 1919.[3] The interests of large private landowners were represented by the Country Landowners Association (CLA), an organisation founded in 1907 to protect private property rights and advance the cause of wealthy individuals.[4] For almost its entire existence, the association has been linked to privilege and the British upper class. Lancien (2007, 11–27) notes, however, that in 1989 only '421 large landowners had estates of more than 3000 acres, and the rest of its 40,000 members fall below that line', data that suggest that the CLA in fact represents a large number of smaller landowners.[5]

The CLA was an effective body during the first half of the twentieth century, ably assisted by its close connections to British nobility. English and Scottish lords were not only able to represent their interests through the CLA but also directly as some of them were Forestry Commission commissioners, while others were active in select committees of parliament on matters related to forestry, agriculture, recreation and rural affairs. However, when the Forestry Commission began to rethink its forest management paradigm in the 1980s and moved to embrace multiple-purpose forestry, the CLA's grip over public policy began to decline. This decline was both relative to other interests that the Forestry Commission found it necessary to engage with and absolute in that multiple-purpose forestry called into question the absolutist notion of private property rights.

Membership in the CLA overlapped with a more sectorally specific association representing the forest industry in England and Wales. In the 1970s, this body was known as the Timber Growers Organisation (TGO) and it responded aggressively to perceived threats to its members' interests, especially the treasury's multi-departmental cost-benefit study that called into question the economic rationale of Britain's 'domestic defence' forestry policy. Winter (1996, 293) cites Garthwaite that 'in the words of the Timber Growers' Organisation, it [the cost-benefit analysis study] plunged forestry into the greatest crisis in a generation'. In 1985 the TGO, now named Timber Growers UK (TGUK), spearheaded the establishment of a private code of conduct known as the Forestry and Woodland Code which 'provided guidance on how forestry could be practiced in harmony with nature and the community' (Tsouvalis 2000, 121).[6]

Winter, in a discussion of the UK's forest 'lobby', notes that the TGO and its Scottish equivalent were at the centre of interest representation in the early 1980s and became even more influential following their merger into TGUK in 1983. According to Winter's account, 'TGUK has never been a large body, with just 2654 members in 1983' which 'has been, and remains, dominated by large landowners, including the forest companies' (1996, 300). As noted in Chapter 4, there are only a handful of these large timber management companies in the UK, such as Fountains Forestry Ltd and UPM Tilhill. The interests of these companies were represented by TGUK whose chief executive Peter Wilson famously compared a WWF seminar advocating FSC certification to a Nazi rally (quoted in McNichol 1999, 18). Yet, notwithstanding the vicious rhetoric, many members of the TGUK endorsed third-party certification after 1997 and ended up being active participants in the negotiations to introduce the UKWAS. In a curious *volte face*, Peter Wilson went on to become the executive chairman of UKWAS.

A third major group engaged in influencing forest policy in the UK is the Timber Trade Federation (TTF). Founded in 1892, the TTF 'represents the timber industry in the UK' which in turn encompasses timber 'used for building, making furniture, packaging, transport, making tools, decorating and a host of other uses' (TTF 2009a). The TTF felt especially aggrieved in the early years of the forest certification debate as it viewed certification as costly, infeasible and discriminatory. In 1990 it launched its *Forests Forever Campaign* to lobby national and intergovernmental organisations on forest policy. Its December 1991 news release warned member nations attending the ITTO of 'the increasing pressure for external application of labelling and certification' arguing that, to obviate this dangerous development, 'producer countries should initiate, as soon as possible, their own plans for assurances about their timber exports' (TTF 1991). Early the following year, its comments were even more pointed, and it called on governments attending the ITTO to urgently consider 'the unilateral proposals originating from the USA and the UK for foreign voluntary audit of individual tropical timber logging companies' and 'the growing militancy of some environmental groups in the marketplace' (TTF 1992). Following the negotiation of UKWAS in 1997–8 the TTF subsequently became a champion of certification, especially with respect to its role in controlling illegal timber imports (TTF 2009b).

These three organisations – CLA, TGO and TTF – worked separately and together in the 1980s and 1990s to oppose forest certification in the UK context. Tomkins argued that in fact the core group engaged in lobbying on UK forest policy was quite small, that there was a range of overlapping interests among the various groups that make up the forest lobby, which disguises the fact that it has at its core a mere 150 people. ... The real heart of the lobby lies with the major landowners. They fill the ranks of the TGUK and CLA, and sit on the Regional Advisory Committees.

(Tomkins quoted in Winter 1996, 300)

The Regional Advisory Committees to which Tomkins refers were established under the 1967 Forestry Act to advise the Forestry Commission on the implementation of its forestry policy. They were, until restructured in 1985, heavily dominated by forestry interests with Tsouvalis noting that even in 1989 'each [RAC] committee consisted of eight or nine members selected by the FC [Forestry Commission], four of which (often including the chairman) represented forest industry interests', with the remainder representing agricultural, environmental, trade union and planning sectors (Tsouvalis 2000, 84).

Forest certification in the UK

Forest certification arrived very early in the UK and was fraught with conflict. Efforts to establish a certification scheme commenced in the mid-1980s before the FSC was formed and continued until a deal was finally agreed between industry and environmentalists in 1997–8. Over the period, the Forestry Commission moved from a position of hostility to scepticism to broker, eventually playing a highly constructive role in mediating an industry–environment compromise in the form of the UKWAS standard. Agreement on UKWAS – which is a UK national standard that is recognised as equivalent to the FSC-UK standard – enabled forest certification to develop rapidly in the UK and created few incentives for companies to certify to the emerging PEFC-UK alternative. Consequently, all certified forest land in the UK is certified to the UKWAS standard utilising FSC-accredited certifying bodies.

Establishment of forest certification

In 1985, Charles Secrett of the FoE-UK launched a rainforest campaign that highlighted the destructive role of the tropical timber industry to biodiversity and habitat loss in Africa, Asia and the Americas. FoE-UK published *Rainforests: Protecting the Planet's Richest Resource* and *Timber! An Investigation of the UK's Tropical Timber Trade* in the same year and then developed a list of suppliers who were dedicated to importing 'sustainable' tropical timber (Lamb 1996, 208). The FSC-UK then published its *Good Wood Guide* in 1986 'which listed sustainably produced tropical timber products and acceptable temperate alternatives endorsed by The National Association of Retail Furnishers' (Lamb 1996, 208). Subsequently Koy Thomson, another FoE-UK forest campaigner working with Secrett, opened discussions with

Tim Synnott of the Oxford Forest Institute in 1987 over the practicalities of certifying and labelling timber products and together they developed a feasibility study on the idea which was submitted to the ITTO. As noted in Chapter 3, the reaction of both developing countries and the industry at the ITTO meant that that body could not take the lead on forest certification and ECSOs like FoE-UK and WWF-UK were forced to look elsewhere for institutional support.

Initially, FoE-UK looked to its *Good Wood Guide* as the potential driver of consumer power. In addition, it also had some success in calls for the tropical timber trade to sign on to a code of conduct that it developed and which required a government-approved management plan, label of origin, a commitment from retailers to stock products that had these attributes and the creation of an industry fund to support long-term sustainable tropical forest use (FoE-UK 1988). The *Code of Conduct* and *Good Wood Guide* became, in turn, levers to mount a more radical campaign targeting retailers that were not conforming to either. Lamb recounts how

> a code of conduct for the trade was devised and the Timber Trade Association [Timber Trade Federation] was vigorously lobbied to accept it. ... These materials became a basis for local group campaigns aimed at motivating consumers to use their spending power to affect the retail marketplace ... some of the groups moved on to picket selected shops and flypost products such as furniture or toilet seats with small hazard warning stickers bearing slogans such as 'CAUTION! THIS PRODUCT KILLS PARROTS' or 'A MONKEY LOST ITS NUTS FOR THIS PRODUCT'.
>
> (1996, 132)

However, confronted with increasing difficulties over validating the legality and sustainability of timber products, FoE-UK discontinued its *Good Wood Guide* in 1990.

In an important escalation of this emerging campaign to inform consumers and curb consumption of tropical timber that was illegal and unsustainable, the FoE-UK and its member branches in Europe lobbied the European Parliament in 1988 to pass a motion to control tropical timber imports (Lamb 1996, 210). Increasingly concerned, the TTF's spokesperson cautioned importers against putting labels on products that would lay them 'open to unscrupulous people' (*Timber Trade Journal* 1989a, 1). Furthermore, importers were encouraged to view the demands for ecolabelling by FoE-UK as 'way ahead of practicality' and to doubt that 'FoE as a voluntary organisation [was] able to judge what is environmentally friendly' (*Timber Trade Journal* 1989a, 1). At the same time as FoE-UK was becoming more aggressive in relation to ecolabelling and timber boycotts, however, other conservation organisations began to doubt their utility. Jeffrey Sayer of the IUCN noted at a House of Commons Select Committee inquiry that 'we do think that by totally boycotting tropical timbers one can do a lot of harm, because they would simply be sold elsewhere at lower prices. If one wishes governments to put resources into managing natural forests, then there must be a financial incentive for them to do so' (Sayer 1989).

In the early 1990s, FoE-UK's campaign was complemented by WWF-UK, which launched a series of forest seminars that aimed 'to bring producers, importers, retailers together to discuss ways to work together to solve forest sustainability issues' (McNichol 1999, 16). McNichol observes that it was at one of these meetings that the concept of the WWF 95 Buyers Group was formulated between Francis Sullivan of WWF-UK and Alan Knight, the newly appointed environmental manager of the major do-it-yourself (DIY) store, B&Q. McNichol notes that in frustration at the perceived lack of understanding by boycotters of the complexities of the issues, 'these two hatched a plan within several months to form a WWF-sponsored group of businesses committed to purchasing an increasing share of timber products from well-managed forests and developing a credible third-party certification and labelling scheme to do so' (McNichol 1999, 16).

1991 was a watershed year in the development of certification and labelling both internationally and in the UK. As noted in Chapter 3, in late 1990 the WARP had concluded its San Francisco meeting by endorsing the idea of a global forest certification standard laying the foundation for the establishment of an interim board which met over 1991 to explore options. Meanwhile, WWF-UK formally announced the establishment of its WWF 95 Buyers Group in February, which committed 15 retail companies to the purchase of timber from well-managed forests by 1995 (Cashore et al. 2004, 139). Later in the year, WWF-UK issued a report that showed that the vast number of existing claims made by the timber trade could not be substantiated and contained 'a remarkable mix of fact, conjecture and allusion together with a smokescreen of diverting but wholly irrelevant information' (WWF report quoted in Bartley 2003).

The UK certification wars

In 1992, the Soil Association, a certification body in the UK that had developed widely used organic standards, established a Responsible Forestry Programme (Cashore et al. 2004, Table 5.2). The aim of the programme was to develop a 'Woodmark scheme' that would enable a label to be attached to forest products that met the scheme's requirements. The Woodmark scheme's development was supported by the ODA, which provided £90,000 over two years 'to support the development of forestry standards, technical documentation and an administrative system to handle certification procedures' (DFID 2009). The Woodmark certification scheme was launched in 1994 and later that year two British forests were certified to its new label. However, by this time, the FSC-AC had been established and support for certification schemes led by certifying bodies had waned. Instead, efforts were made to establish the FSC in the UK with the work done by the Soil Association Woodmark subsequently feeding into negotiations on the FSC-GB Standard.

To stimulate timber certification, and build on developments in the UK, Synnott of the FSC-AC appointed Hannah Scrase as the FSC's first-ever contact person in mid-1994. Scrase worked closely with members of the WWF 95+ Group to establish an FSC-UK Working Group (WWF/World Bank Alliance 2002a). In 1995, the FSC-UK was set up as a trust with a nine-member steering committee composed of two trustees from each FSC chamber and three additional members with technical expertise. The FSC-UK immediately established a standards group and a promotions group and tendered out the writing of the first draft of a standard to Pryor and Rickett Silviculture in July 1995. Ultimately, three drafts of an FSC-GB Standard (Northern Ireland was excluded) were developed and circulated for public input and supplemented by several workshops held around the UK (WWF/World Bank Alliance 2002a).[7]

To combat the perceived threat of FSC certification in the UK, the forest policy network decided to develop its own sustainable forestry standard (McNichol 1999, 17). Operating within the umbrella body the Forest Industry Council of Great Britain and in conjunction with the Forestry Commission, they launched the FICGB Woodmark label in March 1994. The Woodmark label – not to be confused with the Soil Association's scheme of the same name – supported a claim that 'the wood fibre contained in a product originated in forests managed in accordance with UK government policy and regulations' (WWF/World Bank Alliance 2002a). In a separate initiative in the mid-1990s, the Forestry Commission undertook to consolidate its sustainable forest management guidelines into an UK Forestry Standard. The first guidelines on water had been produced in the late 1980s and, as part of the UNCED forest process and in conjunction with the MCPFE, the UK government and the Forestry Commission worked to tailor European-level Helsinki Process guidelines for sustainable forestry management to the UK context. These various strands came together in early 1998 in the launch of the Forestry Commission's UK Forestry Standard, although by then discussions on merging the FSC-GB and UK Forestry Standard were well underway.

The depth of industry feeling at the pressure it was under was evident in Peter Wilson's 1996 'Nazi' comment noted earlier. By then it was evident that the industry was losing the battle for the 'hearts and minds' of UK consumers as outsider groups like FoE-UK continued to apply pressure via an aggressive 'Mahogany is Murder' campaign. Meanwhile, WWF's 95+ Buyers Group continued to expand in numbers. In early 1996, the first FSC-certified wood products began to appear in UK retail outlets and 'by May 1997, 5646 Chain of Custody certificates had been issued to actors in the supply chain' (Eden 2009, 16). The problem for UK forest product producers was that Poland, which contained large swathes of state-owned forests, received initial FSC certification in 1996 (Paschalis-Jakubowicz 2006, 245–6). With Sweden set to follow, it was becoming increasingly clear that a reasonable volume of FSC-certified forest products would be available to UK retailers in the coming years, enabling those within the growing WWF 95+ Buyers Group to meet their commitments to stock only certified forest products from 2000 onwards. In short, FoE-UK was reducing public demand for uncertified timber at the same time as WWF-UK was growing retailer demand for certified forest products.

The negotiation of UKWAS

The imminent widespread availability of certified forest imports was a major motivator for the UK-based forest policy network. David Bills, the director-general of the Forestry Commission noted at the time: 'The landscape in the UK changed when first Poland and then Sweden had significant areas certified and when FSC products from certified forests in Malaysia appeared in British stores' (Bills 2001, 2). These developments occasioned a profound rethink in the Forestry Commission's approach to certification. It was noted, for example, that the UK was heavily trade exposed as it depended on foreign imports for over 80 per cent of its total supply and that the WWF 95+ group of companies 'purchased from many different countries, had little market loyalty to UK suppliers and were prepared to switch to suppliers in other countries' (WWF/World Bank Alliance 2002b). It was in this context that a letter from the director general was published in a forestry industry newsletter openly canvassing the idea of a third-party certification scheme that might have FSC recognition. The article 'caused a storm in the private sector, which accused the Forestry Commission of breaking trust' (WWF/World Bank Alliance 2002b).

In pursuit of the certification idea, the Forestry Commission and the FSC representatives opened discussions in September 1997. At this point, the FSC-GB Standard was close to completion with the third draft scheduled to be released in November. The Forestry Commission was also planning to release its own UK Forestry Standard early the following year. The idea behind the meeting was to determine if the two drafts could be melded into a single document that would satisfy the FSC-UK and the FSC-AC. To that end, the Forestry Commission chaired a meeting of 'about 20 key decision takers and opinion formers in the WWF 95+ Group, WWF, FSC, FICGB and government' (WWF/World Bank Alliance 2002b). The purpose of the meeting was to focus on the general purpose of certification and its role in forestry rather than a discussion of specific schemes, criteria or processes. The outcome was an agreement to consider at a subsequent meeting the similarities and differences between the two draft schemes.

In preparation for the subsequent discussion, the Forestry Commission hired a consultant to prepare a report on the two draft standards. The report noted many areas of similarity but also recognised that the draft FSC-GB standard was 'more demanding in terms of impact on the cost of forest management', a perhaps backhand way of recognising it contained more environmental and social rigour than the draft UK Forestry Standard (WWF/ World Bank Alliance 2002b). Despite the fact that differences were identified, the subsequent meeting, which had now expanded to include about 50 representatives, concluded that the two standards were close enough 'that the participants should continue to work together to develop an "audit protocol" for third party certification that would be independent of the FSC but that could be endorsed by the FSC' (WWF/World Bank Alliance 2002b).

While agreement was reached to develop an audit protocol, some thorny issues remained to be negotiated. These included the issues of genetically modified trees, use of chemicals and restoration forestry for ancient seminatural woodlands (Bills 2001, 2). A slightly longer list of key issues was provided by the author of the WWF/World Bank Alliance review. According to this account, in addition to the above, the key issues were 'monitoring implementation of the plan and forest impact; choice and percentages of species in plantation forests; scale and rate of felling and regeneration; ... consultation; public access; application of the standard to small forests' (WWF/World Bank Alliance 2002b). To secure agreement on an audit protocol from this list of potentially deal-breaking issues, all parties adopted a pragmatic approach. Bills (2001, 2), for example, describes how the forest policy network accepted the requirement that genetically modified trees not be permitted in certified forest operations by recognising that they were many years away from being commercially available. Likewise, the 'social' aspects of the FSC standard were restricted to 'stakeholder consultation and public access to forest land for recreation use' while the thorny issue of chemicals was handled by agreeing on reducing use and phasing them out rather than an outright ban (WWF/World Bank Alliance 2002b).

Movement towards forest policy network acceptance of an audit protocol was facilitated by two developments. First, at an FICGB meeting in May 1998 the previously unified group threatened to fragment. Peter Wilson, chief executive of the Timber Growers Association (TGA), described how 'the UK sawmillers and processors, who had been supportive of the cross-industry opposition to certification, advised that they now needed certification to safeguard UK timber's market share. These were our direct customers and theirs was a view that could not be cast aside lightly' (quoted in Cashore et al. 2004, 149). However, even though it was now recognised by the industry that certification was a requirement, there was still significant resistance to the FSC within the UK forest policy network. The decision by the Forestry Commission to draft an audit protocol that was the equivalent of the FSC Standard in content but different in structure created a face-saving mechanism for the industry, which could maintain its dignity by noting it never accepted FSC certification while in practice accepting a standard that was the negotiated equivalent.

By September 1998, what had become known as the UKWAS Working Group met over four days to finalise the draft audit protocol. The outcome was agreement on a 'final' text which it planned to launch in late December/ early January. However, it emerged subsequently that small forest operators had not been sufficiently consulted and they loudly protested the text of the agreement. According to one account 'the process was almost derailed at this point' (WWF/World Bank Alliance 2002b). Following frantic negotiations with UKWAS participants, the Forestry Commission held bilateral negotiations with representatives of the small forest operators with a view to understanding their concerns. It was agreed within the UKWAS Working Group to view the September text as a draft subject to revision and a further UKWAS Working Group meeting was held in March 1999 to finalise its content. It was only then, at the March 1999 meeting, that a final text of the UKWAS standard was agreed. Although this met the concerns of most parties, there was a contingent of landowners and small forest operators that remained adamantly opposed to FSC certification and to the idea of its effective monopoly of the UKWAS Standard in the UK. This group eventually launched the UK PEFC initiative in 2000 (discussed later).

In mid-1999, the UKWAS standard was published and an effort made to make it widely known within the industry. At that point it was not yet officially recognised as an equivalent of FSC standard. To achieve this outcome, the FSC-UK Working Group coordinator, Hannah Scrase, developed a document that cross-referenced all the elements in the revised FSC-UK (formerly FSC-GB) standard and submitted the FSC-UK standard, the UKWAS standard and the cross-referencing document to FSC-AC for endorsement (WWF/World Bank Alliance 2002a). In October 1999, and following clarifications between the FSC-AC'S Accreditation Business Unit and Scrase, the FSC-AC endorsed the FSC-UK standard and recognised its equivalency to UKWAS. This enabled FSC-accredited certifying bodies in the UK to grant FSC certification and the right to use the 'tick-tree logo' to interested forest owners and managers.

Growth of forest certification

The UKWAS compromise achieved in 1999 became institutionalised remarkably quickly in 2000. In support of FSC certification, the Forestry Commission's forest management agency, Forest Enterprises, achieved FSC certification under the UKWAS standard in January 2000 following an assessment conducted in the latter half of 1999 (Bills 2001, 5). As Bills notes, 'Each of our 32 districts covering one million ha of intensively used and managed forests was assessed', a process that aided 'clarity of thought and a consistent approach across the organisation' (2001, 5). However, British government agencies were not only supporting the production pole of the wood products chain. In the summer of 2000, the UK Environmental Minister announced plans to tackle the problem of illegal timber imports by 'actively seeking to buy timber and timber products from sustainable and legal sources' and included mention of independent certification schemes such as the FSC's as providing legality and sustainability verification (FSC-UK 2001a).

With a considerable portion of state forest lands quickly certified to the UKWAS standard and qualifying for the FSC label, pressure grew on large private operators to follow suit. One private operator, The National Trust, had its lands certified in 2001 (FSC-UK 2001b). In 2002, Fountains Forestry, one of the large UK forest management companies, was designated an FSC Certified Resource Manager. This enabled it to offer FSC certification to the UKWAS standard to its clients. And by 2007, UPM Tilhill, another large forest management company, claimed that over 60,000 hectares of client-owned woodland was UKWAS certified under its Resource Manager Scheme, while over 100,000 hectares was similarly certified under its Group Certification Scheme. Moreover UPM Tilhill's total certified area constituted about 35 per cent of the total private certified area of the UK (UPM Tilhill 2007). Public and private certified lands have continued to expand through the decade and as of March 2009, the total woodland area certified was 1.283 million hectares which represented 45 per cent of total UK woodland area, varying from 31 per cent of England's woodlands to 74 per cent of Northern Ireland's (Forestry Commission 2009). Moreover, at that time all of the land certified was under the FSC scheme, with none recorded under PEFC, a situation that was only altered with the emergence of dual certification in 2010.

The absence of PEFC-certified woodland until recently is perhaps somewhat surprising, given the extensive areas certified under the scheme in our two other case studies. It can be explained, however, as a reaction to market arrangements in the UK. Once agreement was reached on the UKWAS national standard, and on its equivalency to the FSC UK standard, woodland managers had little incentive to adopt PEFC certification since the FSC brand carried more cachet in the marketplace, where it is widely regarded as the 'gold standard' of forest certification. Forest owners and managers, having made the decision to obtain certification, clearly desired to maximise the benefits obtained, and those accrued to displaying the FSC rather than the PEFC logo. The subsequent emergence of dual certification has enabled them to add the PEFC logo to their products at minimal cost.[8]

As previously noted, the UK Minister of the Environment endorsed certification in early 2001 as an important mechanism to control illegal and unsustainable imports of timber into the UK. Subsequently, the UK government established the CPET to undertake a desk study of several certification schemes to determine if they provided a credible assurance that timber certified under them was 'legal' and 'sustainable'. In November 2004, CPET announced its initial findings, concluding that while all five of the schemes it surveyed provided legality verification, only CSA and FSC provided sustainability assurance (CPET 2004). The report caused consternation within PEFC and the governments whose schemes it endorsed. Len Yull on behalf of PEFC-UK called the report biased because it was compiled by a consultancy company called ProForest 'which endorses FSC, alongside environmentalists Greenpeace and the World Wildlife Fund' (ContractJournal.Com 2004). The pressure applied by Yull and other members of the PEFC network eventually told and in a subsequent 2006 decision, four of the five schemes reviewed (FSC, PEFC, CSA and SFI) were deemed to provide assurances of both legality and sustainability. Only the Malaysian scheme, MTCC, was deemed to require more work. This time it was the environmentalists' turn to protest vigorously, with FERN, Greenpeace, WWF-UK and FoE-UK jointly issuing a press release that accused the UK government of 'green washing bad logging practices and directly contributing to the destruction of ancient forests' (FERN et al. 2006). CPET's decision was upheld following a revision of its methodology and subsequent review in 2008, although as noted previously, it has now revised its definition of 'sustainability' to include social criteria which may drive significant further change in the certification community in the coming years (CPET 2010).

State responses to forest certification in the UK

Unlike in Australia and Canada, the UK government played a major role in supporting FSC certification in the country. It did so through the major agency responsible for forestry, the Forestry Commission, which facilitated negotiations between the forest industry and the environmental movement during an intensive two-year period from 1997 to 1999. The commission expended considerable resources in support of this effort. An officer of the commission, Mike Garforth, was responsible for managing the negotiations and the commission covered the costs of the meetings, which were held in attractive locations to facilitate inter-group and cross-group discussions. Once the UKWAS compromise had been reached, the UK government took immediate steps to certify Forestry Commission lands. This not only legitimised the UKWAS standard but applied pressure to the private sector to become certified or lose business since domestic purchasers were able to access both foreign and domestic supplies of FSC-certified timber. Subsequently, therefore, private operators became certified to the UKWAS standard, expanding the volume of certified timber products significantly and further legitimising the standard.

The Forestry Commission's 'conversion' to certification occurred through the 1990s. Although it was under pressure from environmentalists, its position up until 1996 was to support the forest industry rather than back the FSC. Thus it supported industry efforts to develop a FICGB Woodmark label and subsequently began to develop its own UK Forestry Standard. Key elements that affected the commission's stance on forest certification were its participation in the UNCED and MCPFE negotiations on sustainable forest management and criteria and indicators which signalled a further discursive shift away from the single-purpose approach to forestry it had adopted throughout the twentieth century. The emergence of certified forest products elsewhere in Europe and the consequent split in the British forest industry between growers and processors over certification provided additional political space to reconsider its stance regarding the FSC. While circumstances were therefore propitious in 1997, it nonetheless required a great deal of skill on behalf of several key actors – Scrase of FSC, Wilson of FICGB and Garforth of the Forestry Commission – to broker a deal.

Fisheries certification in the UK

Fisheries certification is clearly influenced by the broader political economy of UK fisheries, as noted in Chapter 4. At the same time the MSC has had a greater reach and impact in the UK than in either of our comparator countries. The support for the MSC standard in the UK can partly be explained by the strong links to the MSC by UK interests and individuals from its inception and support for ecolabelled fisheries products from major supermarket chains Sainsbury's and Tesco. The activities of corporate entities such as Unilever – as noted in Chapter 3, one of the two founding partners of the MSC – cannot be underestimated as Unilever's commitment to source all its fish from sustainable stocks was a major trigger for increased attention to means by which sustainability could be assessed.

Fisheries politics in the UK

Despite sharing similar concerns to their colleagues in Australia and Canada, industry leaders in the UK were encouraged to support certification by the partnership model envisaged by the MSC. The opportunities for external certification of fisheries addressed challenges posed by increasing scrutiny of fishing operations by ECSOs concerned over the impacts of the EU's Common Fisheries Policy. Industry too was clearly conscious of the pressure to improve environmental performance. 'The fishing industry is acutely aware that consumers need to have confidence that the food they buy has been produced in a sustainable and environmentally responsible way' (NFFO 2010).

Fisheries politics in the UK has also been influenced by devolution resulting in increased intergovernmental interaction. Scotland (the location of a significant proportion of UK fisheries) in particular has engaged directly with fisheries management. DEFRA is the responsible British Government Ministry and coordinates policy responses on fisheries with the four constituent governments of the UK. Scotland, however, takes a more independent position on fisheries issues albeit within the constraints of the EU Common

Fisheries Policy. While the UK engages with the EU over the CFP, devolution has established formal intergovernmental interactions within the UK as part of fisheries policymaking.

The intergovernmental dimension is also reflected in the peak industry body, Seafish. Seafish, supported by all four UK governments and from a levy on industry, increases the range of programmes and initiatives available along with its general focus on industry promotion. The development of Seafish reflects the broader political economy of UK fisheries. Seafish was established in 1981 through the amalgamation of the Herring Industry Board and the White Fish Authority, organisations established and focused on management arrangements for UK fisheries before the UK's entry into the EU. Seafish 'works across all sectors of the seafood industry to promote good quality, sustainable seafood. Our research and projects are aimed at raising standards, improving efficiency and ensuring that our industry develops in a viable way' (Seafish 2010).

Seafish commissioned an early report that focused on developments in ecolabelling and the MSC (MacMullen 1998). This report indicated that the MSC did not have widespread support through the UK industry, and noted that while large retailer and processor interests were present, the catch sector had been virtually absent in discussion over the development of the MSC principles (MacMullen 1998, 24). This concern over the interests of the catch sector has continued.

Seafish notwithstanding, the UK industry is mostly organised around regional and sector-based organisations such as the National Federation of Fishermen's Organisations (NFFO), which represents fishers. Members join NFFO via membership of local organisations, of a producer organisation or as an individual. NFFO was established in 1977, incorporating a number of smaller, more locally oriented organisations in the lead up to the release of the Common Fisheries Policy. 'During these extended and difficult negotiations, the need for fishermen to speak with a clear, coherent voice at national and international levels became apparent and this has been the NFFO's guiding principle ever since' (NFFO 2010). NFFO promotes sustainable operations, publishing *Sustainability Initiatives in Fisheries* (NFFO 2010a). This includes the observation that

> [t]he more confrontational, media-focused, NGOs such as Greenpeace are difficult to have a meaningful or informed dialogue with, but collaborative work is underway with WWF and the Royal Society for the Protection of Birds/Birdlife International on a number of fronts. The fishing industry also works with the statutory conservation agencies, English Nature (Natural England) and JNCC, the Joint Nature Conservancy Council.
>
> (NFFO 2010a)

The Scottish Fishermen's Federation, formed in 1973 in the same year that UK joined Europe, provides a similar role for Scotland's fishers, acting as an umbrella representative body covering nine organisations. 'The Federation plays an active role in advancing the interests of Scottish fishermen at national and international levels by lobby government officials in Edinburgh, London and Brussels' (Scottish Fishermen's Federation 2010). The Federation has a commitment to sustainability embodied in its *Environmental Policy Statement*, yet this document does not mention the MSC or its standard. The Federation also works with the NFFO on matters of mutual interest, with a current joint initiative to ensure that fishing interests have a voice in current consultations over marine protected areas.

For the UK industry, certification provided opportunities to add value to its catch, which was declining. The UK's fishing fleet has undergone significant decline following the twin drivers of the 'ocean enclosures' through declarations of EEZs under the LOSC and the impact of the CFP. Traditional fishing ports declined with consequent impacts on industry and the community, and criticism of the impacts of the CFP have been strong. Successive UK governments have worked within the CFP, at times being strongly criticised by what is now a relatively small and marginalised industry sector. As well, certification created marketing options via the promotion of the sustainability of operations to increasingly environmentally conscious consumers. These consumers, educated by an increasingly sophisticated environment movement focusing on fisheries (Gray, Gray and Hague 1999), entered the marketplace with demands that were also recognised by the retail sector. Marketers and retailers also engaged early with the MSC, no doubt responding to Unilever's initial (and, as it turned out, overly ambitious) commitments to sourcing certified products. Fishing associations, on the other hand, were concerned that the MSC was 'unaccountable, undemocratic, lacking transparency and driven by sectoral interest' (MacMullen 1998, 26).

A final factor was clear concerns from the fishing industry that it needed to engage with the emerging environmental agenda, described as a third wave of environmentalism that moved away from confrontation and 'doomsaying' to focus on 'providing solutions in partnership with industry' (Gray, Gray and Hague 1999, 120). This led to an interesting pattern of interaction between various interests. While UK fishers shared similar suspicions of the 'green agenda', they decided to develop partnerships with ECSOs to achieve common objectives. The MSC is clearly an outcome of such partnerships.

The UK industry and ECSO engagement with the MSC has been high, despite the lack of endorsement from industry peak bodies. The MSC gained direct support from the UK government through funding and statements of support from ministers, but also gained endorsement from major political figures such as John Gummer, a former Conservative party minister for Agriculture, Fisheries and Food and secretary of state for the environment. Gummer was founding chairman of the MSC board which also included members of the British royal family, including HRH Prince Charles, Prince of

Wales. Prince Charles, in a speech at the Marine Stewardship Reception Gala Dinner at the former Billingsgate Fish Market in March 2004, noted that

> [i]n addition to good science and good regulation, we need a system that harnesses the power of the consumer and provides economic incentives to well-managed, sustainable fisheries. That is exactly what the Marine Stewardship Council does, and that is why I have been such a strong supporter of its work right from the start.
>
> (HRH Prince Charles 2004)

Elite and governmental support contributed to a high visibility of the MSC in the UK, visibility no doubt enhanced by the choice of London as the organisation's international headquarters.

Development of the MSC in the UK

As noted in Chapter 4, a UK fishery was one of the first 'test cases' for the MSC principles and criteria assessment process. While industry and govern-ment reacted cautiously to the MSC initiative in the period 1996–9, there was direct interest by some industry groups in working closely with the MSC, particularly with strong support for external assessment of fisheries by organisations such as Unilever. Government officials too saw benefits and lesson learning as the MSC rolled out its standard and developed its ecola-belling scheme. Seafish recognised that as good as the MSC promised to be, a range of alternative approaches were possible, including vessel accredita-tion schemes. With its focus on helping industry, it recognised the need to develop stocks and industry through benchmarking methods for fishermen to improve industry. This has led to a programme developed in conjunction with NFFO and its participating associations that recognises responsible fishing practices in UK fisheries. 'Participating vessels will be required to undergo an audit which will establish if they meet the required standards for responsible fishing. The scheme is intended to develop, promote and reward good practice including minimising discards, use of selective fishing gear, traceability and catch handling' (NFFO 2010).

The MSC was launched at a time when UK consumers were faced with food contamination scares and when as a result quality and safety of food was high on the public agenda. 'Concern over health and food safety has helped to raise interest in food labelling schemes in a number of sectors including fisheries' (Potts and Haward 2007, 91). In the UK most concern was directed at the consequences of humans eating meat contaminated with Bovine Spongiform Encephalitis (BSE), more commonly known as Mad-Cow Disease (MCD). At the same time public concern at the safety and scrutiny of imported fish products also increased. A much-publicised case of contaminated Bangladesh-sourced prawns in the Netherlands in 1996 (see Bache, Haward and Dovers 2001) led to calls for increased regulation of imported food and improved standards in processing. Emphasis on improved food quality and safety was linked to demands for labelling and introduction of HACCP systems in food (including seafood processing) facilities.

The introduction of HACCP systems and greater scrutiny on food safety and quality through labelling of products has driven consumer concerns over food safety, with approximately 75 per cent of all seafood consumed in the UK being imported. This work is undertaken with reference to the science-based assessments of the *Codex Alimentarius* created by the FAO and World Health Organisation in 1962. 'The *Codex Alimentarius* is a set of norms or principles or code of conduct on which countries can agree with a view to protecting the health of consumers and to ensure that procedures followed in food trade are fair' (Bache, Haward and Dovers 2001).

The UK government had no direct involvement in the development of the MSC but ministers and officials were interested in the progress of the initiative. Government officials recognised early on that in the UK, as in Australia and Canada, fisheries choosing to undergo MSC assessment needed access to catch and harvest records collected and maintained by management agencies. The MSC too recognised its assessment processes' data needs. In 2005 the UK government moved from active, but 'hands-off', support to give more direct finance to the organisation. As noted in Chapter 3, the MSC is dependent on income from licensing its logo and support from beneficiaries. However, it found that the pace of accreditation was insufficient to provide cash flows to fund the organisation. In response to a request from the MSC chief executive officer, DEFRA provided a once-off payment of a few thousand pounds for financial support. The payment was not made by the fisheries section of the department, however, but from another section.

As the MSC has developed and broadened its focus from certification of fisheries to downstream consumer-oriented activities, it has continued to engage with and benefit from programmes developed by DEFRA. An MSC programme, Fish and Kids, part-funded by DEFRA, reached 900,000 children and was replicated in Sweden (MSC 2008a).

> Fish & Kids is a new and exciting way to teach children about fishing, sustainability and the marine environment. Children also learn about healthy eating, food labelling, shopping and consumer choice, and how to prepare simple, nutritious recipes using sustainable seafood.
>
> (MSC website 12 August 2009)

In July 2009 the MSC announced that it had gained £400,000 in new funding from DEFRA's 'Greener Living Fund' to develop the 'MSC on the Menu' programme (MSC 2009c). This latter initiative was designed to enable the organisation to work closely with the catering industry and restaurants interested in MSC certification. Gaining industry support for MSC products on menus provides opportunities to reach consumers: 'When people are eating out they are captive audiences with time and inclination to read about the seafood they are being offered. The materials we develop as part of MSC on the Menu will provide a positive and engaging way to communicate to them about certified sustainable fish' (Stewart 2009).

Retail outlets are important actors in the development of fisheries certification in the UK. While the retail chain responded to Unilever's initial engagement with the MSC, supermarkets have continued to be a major driver for certification of fisheries. Marks and Spencer joined other leading retailers in 2007 when it announced that all fish would come from 'MSC or other independently certified sources by 2012' (MSC 2007c). The major UK supermarket chains are significant players in the marketing and promotion of fish, described as a 'supermarket seafood revolution' by an Australian observer, to much greater degree than occurs in Australia (FRDC 2006, 37–38). The growth in MSC-labelled products in UK supermarkets has been impressive, from over 80 MSC-labelled fish and seafood products in 2007 to over 300 in 2009 (MSC 2007; MSC 2009). The active engagement of consumers, either through supermarket point of sale purchasing or through initiatives involving restaurants and catering firms is interesting and is a clear point of difference between the UK and either Australia or Canada.

The MSC has also benefited from the work of research and scientific groups based in the UK. MRAG, a research and consultancy company, 'spun off' from research programmes at Imperial College, has undertaken a major evaluation of the MSC. MRAG scientists have also been involved in the South Georgia Toothfish assessment. As noted in Chapter 3, MSC certification relies upon scientific analysis and assessments undertaken by external research groups; in this case MRAG work is replicated by the engagement of Australia's CSIRO scientists and the work of Canadian DFO scientists in MSC assessments, in their respective countries.

Certified fisheries

As of March 2010 12 UK fisheries have achieved certification (Figure 7.1), one fishery is pending final approval, with three of these fisheries having undergone re-certification under the MSC assessment model. Fifteen fisheries are currently under assessment. While many of these fisheries are small in scale, and some very localised, the UK experience indicates a steady roll-out and acceptance of the MSC standard. With the exception of the South Georgia Patagonian toothfish and North East Atlantic Mackerel fisheries, the fisheries have low annual catches, and are small and localised in their area of operations. Small landings place a premium on price, a factor that is a major driver for certification. The South Georgia Patagonian toothfish fishery is interesting, as this certified fishery is located within a larger, regionally managed fishery. The toothfish fishery, a high value fishery, has been the target of IUU fishing (Haward 2004). MSC certification provides enhanced certification, over and above the arrangements established by CCAMLR to certify catches, and is clearly useful in sophisticated and aware markets that were the subject of campaigns by ECSOs over the

---

- The South Wales Burry Inlet Cockle (*Cerastoderma edule*) fishery was first certified in April 2001, and recertified in February 2007. This fishery, with an annual catch of 3500 tonnes, involves raking and sieving of cockles (shellfish).

- The Hastings Dover Sole (*Solea solea*) Trawl and Gillnet fishery was certified as sustainable on 9th July 2009, with the two fishery types (Demersal trawl and Gillnet) subject to separate assessments. The fishery has an annual catch of 137 tonnes, with a majority of catch exported to France.

- The Hastings Fleet Dover sole fishery, using Trammel nets set by shore-launched small boats, was certified as sustainable in September 2005. The fishery has an annual catch of 72 tonnes.

- The Hastings Pelagic Herring and Mackerel (*Clupea harengus and Scomber scombrus*) fishery, with an annual catch of 10 tonnes catch by boats under 10 metres launched from the beach at Hastings.

- The Scottish Loch Torridon Nephrops (*Nephrops norvegicus*) Creel Fishery, target the commonly known Norway Lobster. The fishery has an annual catch of 120 tonnes and achieved initial certification in January 2003, and recertified in July 2008.

- North-Eastern Sea Fisheries Committee Sea Bass fishery located in north east England, using intertidal fixed gill-nets with a 7-tonne annual catch was certified as sustainable on 3 December 2007.

- Scottish Pelagic Sustainability Group North Sea Herring (*Clupea harengus*) was certified as sustainable on 9 July 2008.

- Scottish Pelagic Sustainability Group Western Component of North East Atlantic Mackerel (*Scomber scombrus*) fishery was certified as sustainable on 21 January 2009. This fishery had a 2009 catch of 140,000 tonnes using mid-water trawl gear.

- South Georgia Patagonian toothfish (*Dissostichus eleginoides*) longline fishery was certified as sustainable in March 2004. This fishery is located in South Georgia and the plateau to the west around Shag Rocks in the Southern Ocean. It has an annual catch of 3500 tonnes, and is managed with the arrangements set by the Commission for the Conservation of Antarctic Marine Living Resources (CCAMLR), see Haward (2004).

- South West Handline Mackerel Mackerel (*Scomber scombrus*) fishery is around 150 small vessels under 10 metres in length located off the south west coast of England targeting mackerel using handlines.

- Stornoway Nephrops Trawl (*Nephrops norvegicus*) certified fishery of 17,676 tonnes per annum, is undertaken using an otter trawl fishing gear. The fishing gear is designed to reduce environmental impact.

Figure 7.1   MSC-certified fisheries in the UK

problem of IUU catches of toothfish. In this case, as with the Australian certified mackerel Icefish (also caught in the Southern Ocean), MSC certification was sought to enhance market acceptance.

Analysis of the MSC in the UK

The MSC has a higher profile and more support in the UK than in either Australia or Canada. As noted above, the MSC gained early support which has continued over the past decade. Interestingly, formal and overt governmental support is much clearer in the UK than in either Australia or Canada. The UK government has provided financial support to the MSC, and has recognised its programmes and processes in a number of fisheries strategy documents. At the same time elite support has been gained for the MSC (particularly in relation to high-profile politicians' work on MSC board and Royal family presence at MSC events) helping to publicise the role of certification in fisheries and the MSC.

While the factors identified above, including the location of the MSC's headquarters in London, explain the impact of the organisation in the UK, Gray et al. point out the MSC also 'built on collaboration and partnerships between government, conservationists, industry and the public' (Gray, Gray and Hague 1999, 134). Such partnerships were part and parcel of new forms of governance whereby network governance was emerging in place of state-centric bureaucratic 'old politics'. Gray et al. note that while some see 'the state as being side-stepped by an alliance between ENGOs and economic actors [and where] consumer sovereignty is taking over from state sovereignty ... organisers of the MSC saw it less as a displacer of governments than a parallel force' (Gray, Gray and Hague 1999, 134).

The MSC appears to be more strongly embedded in the UK fisheries policy environment than in either Australia or Canada, with a clearer form of network governance emerging, albeit with the state continuing to play a major role. While ecolabels provide consumers with readily accessible information on sustainability, the external assessment of fisheries relies upon data from, but also provides feedback to, management organisations that in turn can improve management performance (Haward 2009). The MSC's success in the UK is also in part due to its work 'downstream', with the post-harvest sectors, addressing processing and marketing of fisheries products. UK consumers have been exposed to issues of food safety, most notably with BSE-infected beef. Demands for greener and cleaner foods occurred at the same time that the MSC was developing its programme. Targeting supply chains and consumers with the support of major retail outlets is a major feature of the MSC's activities in the UK. The growth in number of MSC-labelled sea-food products available to consumers reflects the observation that 'labelling will increase in the domestic market with consumer demands providing a major driver for market behaviour' (Haward 2009).

The UK's engagement with the MSC reflects the position of fisheries certification firmly within the agenda of the policy network with close links between the MSC and DEFRA. While the support for the MSC is high in the UK – bolstered by government commitments in the 2004 *Net Benefits* report to seek certification of all its fisheries under the MSC or similar standard by 2015 and the recent strategic document *Fisheries 2027* – this support is not uncontested or universal. The UK industry, while clearly recognising the benefits of external scrutiny of its operations has indicated wariness over the MSC process. This wariness in the main centres on the costs of certification as opposed to the relative benefits gained by the catch sector compared to the retail sector.

Government, too, despite statement such as those found within *Net Benefits* has steered a middle ground, working within the European Union process for fisheries ecolabelling (with the EU a key participant on behalf of its member states in the FAO discussions), but also recognising that sustainability of fisheries is a key issue, and that while the MSC has a role to play it is not a universal process. Thus the interaction over the MSC with regard to fisheries certification in the UK addresses many of the same issues faced in Australia and Canada, and may be a harbinger of future activities in those states.

Conclusion

FSC and MSC certification both became established in the UK with substantial government support. In contrast to Australia and Canada, the FSC was endorsed in the UK, although indirectly via the UKWAS standard which was a 'national standard', that was the negotiated equivalent of the FSC-UK standard. Once UKWAS was negotiated – a process that only took about 18 months once a decision was reached to do so – FSC certification spread rapidly throughout the UK, led by the Forestry Commission and joined subsequently by the private sector. A key motivation for UK adoption of the UKWAS compromise was its heavy dependence on the import of timber products from elsewhere in Europe, the strong commitment via the WWF 95+ Buyers Group to source certified forest products and the pressure applied by ECSOs to raise public awareness concerning the problems of illegal and unsustainable timber imports. While the UK forest policy network's initial response to FSC certification was hostile, resulting in the eventual establishment of a PEFC-UK group, PEFC failed to gain traction in the UK because by the time it was established, UKWAS had already been negotiated and there was little incentive for forest managers to obtain PEFC certification to UKWAS when they could obtain FSC certification instead.

Likewise, MSC certification was supported by the UK state, with financial backing received from DEFRA to support its organisation and marketing work. UK government agencies, like their counterparts in Australia and Canada, were also involved in the certification of fisheries through their role in providing data on the state of the fisheries. Government support for certification and the MSC standard is not matched by similar commitments from industry. While the UK has a significant number of certified fisheries, industry bodies do not actively endorse the MSC standard. Key industry documents do not mention the MSC, instead committing to sustainability programmes or providing alternative certification initiatives such as the Seafish vessel certification programme and industry commitment to best practice and codes of conduct. While there is no directly competing standard (as in forestry), the situation in UK fisheries emphasises that, as in Australia and Canada, the MSC standard is not universally supported.

8
Comparative Analysis of State Responses to the FSC and the MSC

In Chapter 2, we argued that to understand state responses to FSC and MSC certification schemes it was necessary to disaggregate the state to the sectoral level and investigate the structure, operation and evolution of policy networks. In studying such policy networks, we noted the need to be mindful of how they were shaped by the ecology of the resource, by shifts in management discourses and by the feedback actors received through the commodity chain. In Chapters 4 to 7, we outlined the ecological, policy, discursive and market influences of the forestry and fisheries policy networks and charted the emergence, growth and establishment of the FSC and the MSC in Australia, Canada and the UK. In this chapter, we systematically examine the relationship between policy networks and the reception they gave to the FSC and the MSC in our comparator countries. We commence with Australia and proceed to analyse Canada and the UK.

Australia's response to forestry and fisheries certification

Table 8.1 sets out the basic elements of our analysis. The table depicts the structure of the policy networks in operation in the forest and fisheries sectors in the 1990s and the ecological, commodity chain and discursive arrangements that underpinned their respective operation (columns 1–4). In column five, we set out schematically how each policy network responded to FSC and MSC certification as these emerged on the national agenda in the early 1990s (FSC) and later in the decade (MSC). In column 6 we identify the factors that exercised a significant influence on each policy network in the 2000s and consider whether these shifts induced a change in strategy or structure of the policy network (column 7). We present our understanding of the current status of the FSC and the MSC in the country in column 8.

Structure of the forest and fisheries policy networks

The first important difference we observe between the forest and fisheries policy networks at the outset of certification in the 1990s is that

Table 8.1 Policy network evaluation – Australia

National jurisdiction	Sub-national jurisdiction	Policy network type at outset of certification	Crucial elements influencing policy network at outset	Initial status	Shifts in crucial elements influencing policy network over past decade	Induced shifts in policy network	Current status
Australia – FSC		Triadic policy network composed of NAFI, CFMEU and DAFF; institutionalised via MCFFA (1990s) and PIMC (2000s).	EC: unique, endangered, endemic habitat & species; CC: Asian dependent; MD: SFM as sustained-yield plus.	Certification viewed as unnecessary but developments monitored; later viewed as useful and national AFS scheme developed; FSC scheme resisted.	EC: enhanced public perception of unique, endemic, habitat and species; CC: Asian buyers demand FSC; MD: public concern over failure of SFM; General: GFC-induced crisis in forest industry.	Triadic policy network fragmenting; environmentalist, Aboriginal and community interests becoming more influential.	FSC becoming accepted as legitimate; NAFI joins FSC-Australia; government websites; crisis in forest sector as companies go into administration.
	Tasmania	Triadic policy network composed of FIAT, CFMEU-Tasmania and Forestry Tasmania; institutionalised via FFIC.	EC: unique, endangered endemic habitat and species; CC: highly Asia dependent; MD: SFM as sustained-yield plus institutionalised via Tasmanian RFA.	Certification viewed as unnecessary at outset; later, Forestry Tasmania and Gunns early adopters of AFS; FSC viewed as threat to industry.	EC: enhanced public concern over environment increases; CC: Asian buyers demand FSC; Sodra demands Gunns obtain FSC for joint venture on pulp mill to proceed; GFC-induced crisis in local forest industry.	Triadic policy network fragmenting; environmental, Aboriginal and community interests becoming more influential; residual hostility to 'greenies' remains.	Forestry Tasmania, Gunns and Norske Skog explore FSC certification; Forest Enterprises Australia goes into receivership.

Australia – MSCw	Clientelistic policy network links state-based industry associations with Commonwealth managed by AFMA; weak national industry association as efforts to organise peak body fail with AFIC becoming ASIC which later wound up.	EC: Fisheries at least fully developed if not over exploited and under increased environmental scrutiny from EPBC Act 1999. CC: dependence on Asian markets affected by 1997–8 currency crisis driving search for new markets; marketing of new product (Mackerel Ice Fish); MD: sustained-yield giving way to ESBM.	Australia early adopter of MSC certification (Western Rock Lobster fishery); driven by influential individual (Murray France); but little follow-up; industry suspicious and only sporadically supportive; support from individual scientists; establishment of MSC regional office and active representation.	EC: increased concern over status of some fisheries; CC: Pragmatic adoption of MSC related to specific circumstances of individual 'clients of record'; WRLF seeks access EU markets; MIF seeks marketing advantage; LCF seeks to demonstrate achievements of SFA; MD: Strategic assessments of fisheries; Ministerial Direction rebuilding fisheries; introduction of harvest strategy approach.	Clientelistic fisheries policy network still intact; one AFMA commissioner (Keith Sainsbury) also MSC Chair of Technical Board; Increased recognition of need for external review of management and industry performance.	MSC viewed pragmatically for tangible benefits delivered; not widely supported by state, industry or environmentalists; WRLF recertification resulting in legitimacy crisis; MSC faces important test in proposal to certify Antarctic Krill fishery; benefits of MSC not seen by industry, which is suspicious of MSC agenda; Criticism of costs of certification; Growing support for industry based certification – e.g. 'Clean Green'.

Note: EC=ecological concerns; CC=commodity chain; MD=management discourse.

whereas the forestry policy network was strongly triadic and incorporated government, business and unions in a single network, the fisheries network was weakly clientelistic and managed for Commonwealth fisheries by the Australian Fisheries Management Authority and by state agencies for the fisheries under their jurisdiction. This difference in policy network structure appears to have made it easier for the fisheries policy network to experiment pragmatically with the MSC, whereas the forestry policy network was opposed outright to the FSC. In the forestry sector at the Commonwealth level, the Department of Agriculture, Fisheries and Forestry was officially responsible for forest policy but managed matters in strong collaboration of the National Association of Forest Industries, which itself represented private forest interests (such as those of Norske Skog, Gunnersens Limited, and Gunns Limited) as well as state forest interests (such as Forestry Tasmania and VicForests). Also represented on NAFI were unionised forest workers of the CFMEU and TCA, the industry-backed 'grassroots' forestry body that was founded to represent the interests of timber-dependent regions.

Australia's tight-knit national forest policy network had its counterparts at the state level at this time. In Tasmania, for example, a similarly organised triadic policy network was in operation. It was composed of the newly corporatised Forestry Tasmania, powerful local companies such as Gunns Limited, the Tasmanian branches of the CFMEU and TCA. Key institutional arrangements cementing the forestry policy network locally were the FIAT and FFIC. The latter served as a location for both determining industry needs and ensuring their translation into state policy. FFIC, initially the state's peak body to represent industry interests to government later took on the additional role of an industry council, to provide advice to government on policy matters, further cementing this triadic relationship.

The tight, triadic structure of the forest policy network contrasts vividly with the loose, clientelistic arrangements in the fisheries sector. Despite the best efforts of the AFMA to create a strong industry peak body, no coherent industry voice emerged. The early effort to establish an Australian Fisheries Industry Council was unsuccessful and a subsequent effort to set up a countrywide Australian Seafood Industry Council also failed. Instead, regional fisheries interests dominated over national interests. While AFMA made extensive efforts to respond to these interests – thus making the network clientelistic rather than bureaucratic – the absence of a coherent industry voice meant there was no single view on the value of certification in general or the MSC in particular. This created opportunities in different fisheries to experiment with the MSC, unlike the blanket opposition that greeted the FSC in the forestry sector.

Initial network reactions to the FSC and the MSC

The differences in network structure in the fisheries and forestry sector resulted in varied reactions to the proposed introduction of FSC and MSC certification. In the forestry sector, the forest policy network was dismissive of certification in general and the FSC in particular. Crucial elements influencing its position were the view that enough had been done – especially following the signing of Regional Forestry Agreements in the latter half of the 1990s – to appease environmental and wilderness interests and that the new discourse of sustainable forest management, embedded in the RFA compromises, struck an appropriate balance between preservation and use of Australia's forests. The network's view that Australia's forests were being managed sustainably interacted with the knowledge that customers at home and abroad were not demanding certification and that the environmental insensitivity of Asian markets contrasted strongly to the market conditions confronting producers in Canada and Europe. Moreover, since environmental groups were divided over the merits of certification in Australia, domestic pressure to respond to an Australian FSC initiative was also lacking.

The forest policy network's initial negative reaction to certification and the FSC altered in the late 1990s following the Asian currency crisis. With traditional markets in a slump in Japan, Korea and China, the industry sought access to North American and European markets where forest certification was well advanced. It was quickly apparent that Australia's forest products could be at a competitive disadvantage if they were unable to provide some authoritative certificate of sustainability, and the network moved quickly after 1999 to establish the AFS, which was endorsed as an interim standard by Standards Australia in 2002. The decision by the forest policy network to adopt a national standards approach rather than consider the FSC (or an UKWAS-style compromise) was motivated by two considerations. First, it was convinced that overseas markets would be satisfied with a national certification scheme, which could be harmonised with other emerging national schemes elsewhere such as CSA and SFI. Second, it viewed the FSC as extremely costly to implement in the Australian context, especially in the native hardwood sector where companies continued to practise widespread clear-cutting of native forests.

Where the forest policy network viewed native forests as basically well managed and capable of providing timber into the future, the general perception in the 1990s in the fisheries policy network was that many of Australia's fisheries were at least fully developed and that there was a significant danger of over-exploitation. The perception existed that current management practices were not achieving desired objectives and that new governance mechanisms – including certification – should be considered. As the network pondered these issues in the late 1990s, the Asian currency crisis forced it to reassess its dependence on Asian markets and to develop strategies to export a larger volume of product to North America and Europe. It was in this context that some regional members of the weak and fragmented clientelistic fisheries network considered whether the adoption of MSC certification in the discrete context of the Western Rock Lobster fishery might provide it with a marketing advantage in its ongoing efforts to maintain and grow market share in Europe.

Evolution of network reactions to the FSC and the MSC

The evolution of the forestry and fisheries policy network responses to FSC and MSC certification schemes followed very different trajectories in the first decade of the twenty-first century. Despite open and onging hostility by the forest policy network to the FSC through the decade, justifying Humphreys description of struggles between the FSC and national schemes as 'certification wars' (2006), by the decade's end both state and private sector network members were seeking FSC certification. In contrast, the early promise of more widespread adoption of the MSC signalled by the certification of the Western Rock Lobster fishery had not materialised. Instead, by decade's end the fisheries policy network was still undecided about the value of MSC certification, with some regional fisheries bodies actively opposing it in favour of alternative, localised ecolabels. To understand these peculiar trajectories, we need to examine the ecological, commodity chain and discursive changes that occurred in both sectors in the early years of the twenty-first century.

With respect to ecological concerns, heightened public awareness of the consequences of global warming and climate change generated very different opportunities for the forestry and fisheries sectors. In the forestry sector, it was evident that forests could play an important positive role in mitigating carbon emissions either by preventing deforestation in the first place or by encouraging afforestation and reforestation. Growing public awareness of the importance of tackling climate change coupled with a realisation of forests' key contribution to the effort created additional pressure to rein in illegal and unsustainable logging and forest practices. In the early 2000s, this pressure translated into policies to end deforestation, further improve forest management and limit the importation of illegal timber. Certification and labelling now appeared in a new light and to offer a significant mechanism to achieve many of these objectives, providing the standard set was sufficiently rigorous. It was in this context that the FSC began to gain ground during the decade in Australia. It was promoted by a united and strongly supportive environmental movement that was able to reach out globally to sister-organisations around the world.

In contrast, the implications of climate change for the fisheries sector were less salient. While sea level rise and warming waters threatened to displace and/or destroy existing fisheries stocks, no evident role existed for fishers to mitigate this outcome. Thus the ecological perceptions of the fisheries policy network remained similar to those of the 1990s: fisheries were fully exploited, action was required to manage them, but disagreement existed on how this should be done. While the problem of IUU fishing did gain ground during the 2000s, potentially strengthening demands for fisheries certification, by the end of the decade there was still little demand for certified fisheries products in Australia. In part this was because the environmental movement did not fully endorse the MSC due to the weakness of its certification standard and processes, while many in the industry remained suspicious or, in the case of the Tasmanian Rock Lobster fishery, downright hostile.

These alterations in the perception of the ecological importance of forestry and fisheries interacted with further shifts in the commodity chain as a consequence of the global financial crisis. While the 'Great Recession' affected both industries significantly, the Australian forest industry has been particularly badly hit, in part because it was over-leveraged through the setting up and running of managed investment schemes. Companies that were unable to reign in their levels of debt quickly in 2008 went under in 2009. These included several high-profile companies such as Timbercorp, Great Southern Plantations and, most recently, Forest Enterprises Australia. With the cost of borrowing soaring and demand and prices sinking, all companies in the sector got caught in a scissors' crisis – increasing debt and declining revenues. The generalised decline in demand for forest products induced by the GFC interacted with the steady rise of FSC-certified forest products available in the market following the certification of Canada's boreal forests after 2004. By mid-2009, consumers of forest products in Japan, North America and Europe were suddenly able to access considerable quantities of FSC-certified products and, in a buyers market, began to prefer the FSC. Japanese buyers visited Tasmania the same year to tell Forestry Tasmania that they now required FSC-certified products to maintain contracts, sending shockwaves through the state and national forest policy network.

Finally, in relation to natural resource management, the first decade of the 2000s has seen some strengthening of the ecosystem-based management discourse at the expense of conventional, sustainable natural resource management paradigms built around the concept of sustained yield. In the fisheries sector, this discourse has been in place for some time, but the political-economic costs of implementing it have given rise to considerable local resistance in specific fisheries. While a strengthened MSC could undoubtedly contribute to the implementation of ESBM, the wider environmental movement has not championed the scheme, leaving WWF to do so without widespread national support. In contrast, the discourse of ESBM has been largely absent in the forestry sector in Australia, where forestry has been dominated by the discourse of wilderness protection on the one hand and sustained-yield forestry on the other. With the FSC in the ascendant, however, it is now conceivable that the discourse of ESBM will develop apace as environmental, forestry and social interests seek to find compromises that will enable limited native forest logging within an acceptable environmental and social framework.

The changes outlined above have had a differential effect on Australia's forestry and fisheries policy networks. In the forestry sector, there are signs of a shift in the structure of the policy network with the closed, triadic network opening up to new interests and becoming more pluralistic. While it is still too early to determine if the more pluralistic policy network will be sustained, the ecological, economic and discursive pressures outlined above show no signs of dissipating suggesting that an FSC-brokered compromise will eventuate in Australia in the second decade of the twenty-first century. To fully cement the move to a more pluralistic forest policy network, governance arrangements at the corporate, state and national levels will require modification. In contrast, there are few signs of a fundamental structural shift in the fragmented clientelistic fisheries policy network. AFMA remains a key actor within the network, but as '*primus inter pares*' lacks the dominance required to impose its demands on the sector. State agencies and strong regional fisheries associations continue to seek their own local solutions which may include pragmatic support for the MSC on the one hand as with the Mackerel Ice Fishery, or outright competition to develop an alternative label as in the case of the Tasmanian Rock Lobster's 'Clean Green' ecolabel.

Canada's response to forestry and fisheries certification

The Canadian response to forest and fisheries certification shows some interesting contrasts to Australia's. While Canada's forest policy network was concerned initially about the costs of the FSC, its overall policy response was more muted and neutral in contrast to Australia's aggressive promotion of AFS. Moreover, as an early mover on certification, Canada was able to gain significant experience in three alternative approaches before Australia had finalised its draft national standard. The development of an FSC regional standard for boreal forests in 2004 created opportunities for industry to engage with the standard that were seized. In fisheries, a bureaucratic policy network managed by the DFO led to early, pragmatic support for the MSC in response to MSC's certification of the US Alaskan salmon fishery. However, DFO's declining authority following the collapse of the cod fishery, the rise of alternative, more locally based management arrangements and problems with MSC's certification system prevented DFO from garnering more widespread support. At the same time, however, the requirements for certification (fisheries data and management information) required and led to DFO support for fisheries nominating for assessment (see Foley 2010). In this section, we set out the response of Canada's forestry and fisheries policy networks to the FSC and the MSC. As in the Australian section, the basic elements of our analysis are summarised in Table 8.2.

Structure of the forestry and fisheries policy networks

Federally, the Canadian forest policy network of the 1990s is most appropriately characterised as clientelistic with business and government being the dominant partners. While forest workers were organised by the IWA, their influence is much greater at the provincial level. Across the country, it was the peak industry body, the Canadian Pulp and Paper Association, that co-managed forest policy in conjunction with foresters operating within the Canadian Forest Service and inter-provincially through the Canadian Council of Forest Ministers.

The federal network had its counterpart at the provincial level, although at the sub-state level the influence of forest workers and timber communities was stronger, giving rise to a triadic policy network in forestry-dependent provinces like BC. In BC in the early 1990s, the Confederation of Forest Industries collaborated closely with the BC Ministry of Forestry and IWA locals to promote the logging of old-growth forests. Actors representing other interests, including environmentalists, First Nations and non-timber forest interests were excluded from the discussions, leading to high profile and internationally important conflicts such as those that erupted over Clayoquot Sound in 1993–4.

In contrast, the Canadian fisheries industry was managed until recently by a bureaucratic network centred in Ottawa at the Department of Fisheries and Oceans. While DFO had links to an industry peak body, the Fisheries Council of Canada, as in Australia, was a weak organisation that struggled to represent the national fishing interest and was internally divided into strong regional industry associations on the East and West coasts. DFO's dominance waned in the 1990s, however, when the collapse of the Atlantic cod fishery undermined its authority and regionally based fisheries organisations explored alternative, more locally driven solutions. Unlike Canada's forestry policy network, but similar to Australia's fisheries network, the Canadian fisheries network adopted a pragmatic response to the MSC, viewing it as potentially valuable as a marketing tool in a limited number of fisheries.

Initial network reactions to the FSC and the MSC

Canada's clientelistic forest policy network reacted swiftly to the perceived threat of the FSC. The CPPA quickly created the Canadian Sustainable

Table 8.2 Policy network evaluation – Canada

National jurisdiction	Sub-national jurisdiction	Policy network type at outset of certification	Crucial elements influencing policy network at outset	Initial status	Shifts in crucial elements influencing policy network over past decade	Induced shifts in policy network type	Current status
Canada-FSC		Business-Government Corporatist policy network between CPPA and CCFM/CFS.	**EC:** substantial volumes of old-growth forests; **CC:** Economic dependence of Canada on forest industry; dependence on US market; AFPAs development of SFI; **MD:** SFM as sustained-yield plus.	Forest policy network mobilises to develop FSC-competitor scheme (CSA); initial disquiet over FSC gives way to policy of 'neutrality' among certification schemes after 1996; practical support granted to CSA scheme in its development, trialling and endorsement phases; little tangible support granted to FSC until present.	**EC:** Great Bear Rainforest settlement; **CC:** Increased FSC products available consequent on industry certifying to FSC-Canada Boreal and BC regional standards; **MD:** Discursive shift to 'ecosystem-based forestry'; general influence of climate change discourse.	Pluralisation of forest network includes previously excluded groups (especially First Nations and Environmentalists); *Vision for Canada's Forests* also focuses on transformation of industry and climate change.	FSC increasingly accepted across the country; large area of forests certified to FSC, especially in BC, Quebec and Ontario.

| Canada-BC | Triadic Network links COFI, IWA and MOF; other actors excluded. | EC: large areas of native, primary, coastal rainforests; lack of treaty settlements with First Nations; CC: dependence on softwood industry and on US markets; MD: SFM as sustained-yield plus encapsulated in Forest Practices Code. | COFI joins CSFCC; initial hostility of MOF to FSC replaced by official policy of neutrality after 1996; tangible support to companies to trial certification (including FSC); large area of BC certified to CSA standard. | EC: Great Bear Rainforest Settlement; mountain pine beetle destruction of BC's forests; CC: GFC creates drop in demand for softwood lumber exports; markets sought in Asia, including China; MD: Discursive shift to 'ecosystem-based forestry'; general influence of climate change discourse. | Forest policy network becoming more pluralistic, and opening up especially to First Nations involvement and community forestry. | FSC established in province; key role in 'resolving' the Great Bear Rainforest dispute; also playing a role in safeguarding management in community forestry tenures (i.e. Harrap-Proctor). |

(continued)

Table 8.2 Continued

| Canada-MSC | Bureaucratic policy network dominated by DFO but linked to industry via Fisheries Council of Canada; Conflicts over fisheries management. | EC: collapse of Atlantic cod fishery; concern over Pacific salmon fishery; CC: Market access to European markets, especially for salmon; MSC certification of US Alaskan fishery; MD: ESBM advocated but not practised. | MSC adopted to maintain access to EU Pacific salmon market; industry enthusiasm for MSC declines following experience of Pacific Salmon assessment; scientific criticism of MSC grows (e.g. UBCFC) due to lack of transparency; environmental criticism grows over weak MSC standard which approves all applicant fisheries; Government neutral on MSC but concerned that Canadian fisheries required to meet tougher test/higher bar than US fisheries. | EC: Legacy of cod fishery collapse continues to affect perceived legitimacy of DFO and 'science-based' fisheries management; CC: poor experience with MSC Pacific salmon certification; decline in support from scientific and environmental groups (e.g. CPAWS, David Suzuki); Reorganisation of MSC salmon certification gains support from industry; MD: Development of harvest strategy approach to management has an ecosystem orientation. | Bureaucratic fisheries policy network in decline, but still strong push for central role for government; network becoming more pluralistic, incorporating eastern and western interests more directly; alternative, locally based, community initiatives on the rise; increasing use and criticism of MSC. | MSC continuing to certify fisheries; retailers (e.g. Loblaws) beginning to drive MSC certification domestically; but scientific and environmental criticism mounting; successful certifications acknowledged by government but criticism of lack of action on bycatch in shrimp fisheries etc. |

Forestry Certification Coalition, which funded the development of a competing standard to the FSC's through the CSA. The swiftness of Canada's response to the FSC stands in considerable contrast to Australia's wait-and-see approach. It reflected the fact that environmental organisations were aggressively backing the FSC in North America, while the US industry was developing its own SFI. In this context, industry perceived the need to either obtain FSC or SFI certification or develop its own scheme. The perceived costs of obtaining the FSC were viewed as too high given the industry's dependence on old-growth forests and in the context of unsettled land claims from First Nations groups, especially in BC. While multinational timber companies operating in the US and Canada would likely qualify for SFI, purely Canadian-based companies might not and network members did not want to adopt an SFM standard set by the US industry. For these reasons – and because the policy network viewed itself as on the path to practising SFM following the release of the 1992 National Forest Strategy document, *Sustainable Forests: A Canadian Commitment* – the network moved rapidly to establish its own scheme.

In the fisheries sector, the bureaucratic policy network orchestrated by DFO spotted an early opportunity to pragmatically employ the MSC to address a strategic marketing problem with regard to Pacific salmon. The problem arose following the September 2000 certification of the US Alaskan Salmon fishery to the MSC standard, which placed pressure on Canadian Pacific salmon fishers to follow suit since products from both fisheries competed for market share in Europe. The British Columbia Salmon Marketing Council's decision to seek MSC certification for the Canadian Salmon fishery was backed by DFO and other industry bodies despite misgivings over the implications for state sovereignty and the acceptability of external scrutiny of Canada's fisheries management practices. Environmental groups and First Nations were less enthusiastic, as were members of the University of British Columbia's Fisheries Centre. The latter had developed a strong critique of MSC's certification of the US Alaskan Salmon fishery and feared the process to certify the Canadian Salmon fishery would be equally non-transparent and problematic.

In Canada, as in Australia, the state responded differentially to the FSC and the MSC. Perceived as a threat to profits, jobs and revenues, the clientelistic forest policy network mobilised en masse against the FSC and established an alternative, national competitor scheme under the CSA. In contrast, the bureaucratic fisheries network, while uneasy about the implications of certification and labelling schemes for national sovereignty, perceived the MSC as a pragmatic opportunity to enhance the marketability of Pacific salmon. These divergent reactions resulted in one network resisting the FSC and the other embracing the MSC. At the outset, then, the FSC was under considerable pressure in Canada to survive, while the MSC appeared better placed to thrive. Curiously, however, actual experience shows the opposite occurred.

To understand why, we next examine the evolution of the FSC and the MSC in the country.

Evolution of network reactions to the FSC and the MSC

In the first decade of the twenty-first century, progress was made in certifying Canada's forests and fisheries to the FSC and MSC standard. Of the two schemes – and not withstanding its contested beginnings – surprisingly the FSC's is the better established today with over 35 million hectares of forests certified. In contrast, while two fisheries have been certified to the MSC standard and four are currently under assessment, strong industry, scientific and environmental criticisms persist related to the complexity, rigour and transparency of MSC certification arrangements. To understand the different paths taken by the FSC and the MSC in Canada, we again investigate changing ecological values and shifts in commodity chains and natural resource discourses.

In the forestry sector, such shifts in ecology, market and discourse have exercised a decisive influence, creating pressure at federal and provincial levels for a more favourable evaluation of the FSC. A key driver of changed attitudes within the forest policy network to the FSC was the endorsement by the FSC-IC in 2004 of a standard for the country's Boreal forests. The FSC-Boreal Standard achieved all-chamber agreement on the meaning of SFM for Canada's Boreal forests which, as noted in Chapter 4, constitute about three-quarters of the country's forest resource. When firms scrutinised the FSC-Boreal Standard, they realised that, while rigorous, the bar it established was attainable. This resulted in many large integrated forest companies applying for FSC certification after 2004 and a fragmenting of the previously united front that had opposed the FSC through the 1990s.

The negotiation of the FSC-Boreal Standard constituted a major break-through for the FSC in Canada. It was followed a year later by the successful conclusion of the tortured negotiations over an FSC-BC Standard. While the FSC-BC Standard was recognised as very rigorous, Tembec demonstrated that it could be met by being certified to it in 2005. Later, the concept of SFM embedded in the FSC-BC Standard assisted in breaking the deadlock over the fate of about five million hectares on BC's Central Coast in the Great Bear Rainforest. To negotiate a solution for this region, a newly created, pluralistic policy network had come into being, composed of government, industry, environmentalists and First Nations. The final settlement obliged companies logging in the area to obtain FSC certification, increasing both FSC-certified area and the scheme's relevance to resolving forest disputes.

In the fisheries sector, the MSC achieved no such significant certification breakthroughs until recently. At the outset, the opposite occurred when the pragmatic decision of the British Columbian Salmon Marketing Association to seek to certify the Pacific salmon fishery resulted in a deadlock. Hoping to achieve certification relatively quickly and in accordance with the process adopted to certify the US Alaskan salmon fishery, the complexity and slowness of the Canadian salmon certification process created enormous frustration within DFO and the industry. In the absence of an early MSC breakthrough, Canadian fishers continued to experiment with alternative, local solutions including the use of alternative ecolabels. It was only towards the end of the decade, with the certification of two smallish fisheries – the Canadian Northern Prawn Trawl fishery and the Gulf of St Lawrence Shrimp fishery – that the MSC began to establish itself. These certifications, and another four in the offing, as well as the endorsement of the Canadian supermarket chain Loblaws, indicate that the MSC may be about to become more established in Canada.

The contrasting paths taken by FSC and MSC certification in Canada are partly explained by differential changes in public perceptions of the ecological value of forests and fisheries. As in Australia, Canadian public opinion concerning the value of forests was influenced by new evidence concerning global warming and climate change. Pressure grew within and outside Canada for the forest industry to demonstrate it was practising 'sustainable forest management'. It became clear that industry-backed certification schemes that did not have environmental and First Nations support struggled to provide these assurances. Moreover, as industry gained experience with the FSC, it realised that it could meet the various regional standards with only a modest impact on harvest levels in most instances. By obtaining FSC certification, therefore, companies addressed growing public concern over forest management while simultaneously obtaining a powerful marketing advantage over domestic and international rivals. In the fisheries sector, the link between global warming and fisheries management was not so straightforward. While standing forests make a contribution to carbon sequestration strengthening the preservationist discourse, no such equivalent contribution occurs in the fisheries sector. Instead, and as a consequence of the legacy of the cod fishery collapse, disagreement continued over how the country's fisheries should be managed. While there has been a general re-orientation towards ecosystem-based fisheries management via a 'harvest strategy' approach, a lack of consensus exists on how best to implement it. This disagreement is constitutional (whether responsibility should remain with DFO or devolve to more regional bodies), political (who should participate in the process) and substantive (how rigorous should the requirements be for implementing the harvest strategy approach). It is unclear if the MSC is well positioned to mediate these divisions because its 'technical' approach to certifying fisheries keeps parties at arms length whereas the FSC's brings them together.

The practical experience of FSC certification in Canada demonstrated its feasibility; it also led to increased volumes of FSC-labelled products in domestic and overseas markets. It thus also contributed to a transformation in Canada's commodity chain for forest products. A major problem experienced by purchasers in the 1990s was an absence of sufficient quantities of FSC-certified goods, which even committed companies found difficult to source. The dramatic increase in FSC-certified products after 2004 removed a major impediment to demand. In contrast, relatively few MSC-certified products have been available in North American stores. Demand is low because US retailers do not source certified fish products; in Canada, only Loblaws has committed to stocking MSC-certified product. In the absence of substantial natural demand, the MSC needs a strong marketing strategy to accompany its efforts to certify fisheries. While other groups are working in this area – notably the ECSO-led Sustainable Seafood Canada group and its 'Best Choice' list – they are unimpressed with the rigour of MSC's certifications to date and thus are competing rather than collaborating to build the market, potentially leading to consumer confusion.

The impact of ecological, market and natural resource management considerations has substantially shifted the forest policy networks' evaluation of the FSC. From a position of unified opposition, today the FSC receives active support in the form of widespread acceptance of its certification scheme. As of 2010, almost 35 million hectares were certified to FSC standards, with another eight million hectares under assessment. In 2008, both the Manitoba and Ontario governments introduced policies that required the procurement of FSC-certified paper. In the same year, the Nova Scotia government awarded funds to the Nova Scotia Landowners and Forest Fibre Producers Association to assist woodlotters to develop management plans to meet the FSC Maritimes Regional standard. And in 2009, the BC government agreed to make FSC certification a requirement for the management of lands in the Great Bear Rainforest. These and other developments demonstrate the sea change that has occurred in the network's response to the FSC. They are suggestive of a 'pluralisation' of Canada's forest policy network, an ambition recognised in the CCFM's recent release of a *Vision for Canada's Forests: 2008 and Beyond*. Building explicitly on an ecosystem-based approach to forest management, the document states:

> To be effective, the Vision needs to be embraced by Canada's entire forest sector including the future players. Governments, forest products companies, Aboriginals, private woodlot owners, forest communities, professional associations, governments, researchers and educators, the environmental community, non-traditional partners (including energy, chemical, and pharmaceutical industries), and the public all have unique and critical roles to play in advancing the sustainable management of forests. Partnerships will be key as innovations and change lead to new players in the sector.
> (CCFM 2008, 11)

There is nothing equivalent to this vision of multi-constituency partnerships in fisheries in Canada, somewhat surprising given the Federal Government's constitutional responsibility for fisheries management. In *Our Waters Our Future*, DFO discusses adopting an ecosystem-based management approach to fisheries (DFO 2005). However, implementation of this approach is discussed largely in technical terms related to the adoption of a precautionary approach, and the notion of broad-based multi-constituency partnerships is not referenced. In contrast to the forestry policy network, therefore, the evolution of Canada's fisheries policy network has been modest in the past decade. Actors appear locked into a contest over authority (DFO versus regions), over the meaning of sustainable fisheries management (sustained-yield versus narrow and broad interpretations of ESBM) and over mechanisms for its achievement (including competing ecolabelling schemes). The MSC appears poorly positioned to resolve these conflicts because its technical approach to fisheries certification does not mediate fisheries interests either federally or at the individual fisheries level. While it is clear that there is 'relatively substantial support for private regulation... the emergence and spread of the MSC [in Canada] cannot be explained straightforwardly as private, non-governmental and voluntary' (Foley 2010, 2).

UK's response to forestry and fisheries certification

Of the three case studies considered in this book, the UK's is the most positive in terms of the state's response to the FSC and the MSC. However, in the case of the FSC, the positive reaction that led to the negotiation of the UKWAS only occurred after vigorous debate and opposition within the forest policy network and in the face of a decisive shift in the structure of the commodity chain. In the case of the MSC, early strong support was forthcoming from the elite, while the location of the MSC in London gave the organisation a strong British character and enabled it to be very active in its own 'backyard'. Despite the endorsement that the FSC and the MSC ultimately received from their respective policy networks, of the two the MSC's appears to be the more problematic. This is because, while a number of generally small-scale fisheries have been certified to the MSC standard, the major UK industry association, Seafish, remains critical and unsupportive. A summary of our analysis of responses to forestry and fisheries certification in the UK is set out in Table 8.3.

Structure of the forestry and fisheries policy networks

At the outset of forest certification in the 1990s, the UK forest policy network was firmly clientelistic, centred on the Forestry Commission and with partners located in the Country Landowners Association, the Timber Growers' Association and the Timber Trade Federation. Despite a history of gradual accommodation in the 1970s and 1980s to recreational, environmental and amenity interests and the official adoption of 'multiple purpose' forestry, the network remained closed to external groups and strongly

Table 8.3 Policy network evaluation – UK

National jurisdiction	Sub-national jurisdiction	Policy network type at outset of certification	Crucial elements influencing policy network at outset	Initial status	Shifts in crucial elements influencing policy network over past decade	Induced shifts in policy network type	Current status
UK – FSC		Clientelist forest policy network linking business (CLA, TGA and TTF) to state (Forestry Commission);	EC: SSSIs; Ancient Woodlands; Amenity; CC: Small profit margins; dependence on imports; active retail sector interested in certification; MD: SFM as 'multiple use' forestry; sustained-yield plus with accommodation of recreational, amenity and limited environmental interests.	FSC strongly resisted by FSC and industry as costs imposed viewed as unnecessary and severe.	EC: SSSIs; Ancient Woodlands; Amenity; CC: Imports of certified forest products from Sweden/Poland after 1996; defection of timber suppliers to FSC; MD: involvement in EU's MCPFE; enhanced 'multiple use' forestry; Negotiation of UKWAS compromise in 1998–9.	Clientelist forest policy network disintegrated after 1997–8; replaced by more pluralist policy network involving environmental and amenity groups; opening up of RACs to include wider range of interests; SFM interpreted as 'multiple use' forestry meets national requirements.	FSC certification dominates and all public and most industrial private land now certified to FSC standard.

UK – MSC	Participatory pluralistic policy network led by DEFRA; increasing formalisation of interests of sub-national governments and stakeholders (e.g. Scotland).	EC: food scares in beef and prawn sectors give rise to public health safety discourse; CC: retailers supporting domestic demand for MSC-certified products (Sainsbury's, Tescos); significant role of supermarkets in fish trade. Commitment by British elites to MSC; location of MSC in London; MD: Influenced by EU Common Fisheries Policy.	MSC supported by DEFRA; elite networks engage industry to seek certification of specific fisheries; very early uptake of MSC; fisheries in UK part of first phase of MSC certifications (small-scale operations). ECSOs such as WWF interested in fishing and marine environment.	EC: food scares; CC: public demand for certification and labelling continues to grow; retailers remain interested in sourcing MSC products: problems with rigour of MSC certifications limits environmental support; industry remains critical of MSC; MD: influenced by EU Common Fisheries Policy; conventional sustained-yield plus approach pursued.	Participatory pluralistic network still in operation; ongoing MSC governance reforms to address criticisms; industry ambivalence with respect to MSC remains; benefits to catch sector not realised in all cases; criticism of costs of certification.	Large number of small fisheries certified to MSC standard; South Georgian Patagonian toothfish and North East Atlantic mackerel fishery are exceptions; government supports certification of fisheries in policy statements but does not endorse MSC as only approach; industry organisations support 'sustainability agenda' and certification but does not endorse MSC; development of alternative scheme based on ISO standard; MSC working with retailers and restaurants to publicise sustainability of fisheries.

focused on protecting the rights of landowners, the interests of tree growers and the structure of the UK forest industry. While some within the Forestry Commission recognised that the new forestry implied serving a wider range of clients, the institution had not yet reached this understanding. Consequently, when the CLA, TGA and TTF ran aggressive individual campaigns against certification and the FSC, the Forestry Commission sided with them and against the new governance approach. As we saw in Chapter 7, both the CLA and TGA campaigned against the FSC, the former concerned about the property rights of landowners and the latter about the costs imposed on timber growers. Meanwhile, the TTF – which had been extensively involved in debating certification at the ITTO in relation to tropical deforestation – argued it was unworkable because of the demands of CoC monitoring.

The FSC's outsider status with respect to the UK forest policy network forms a stark contrast with the MSC's insider status within the fisheries policy network. The fisheries policy network in the latter half of the 1990s in the UK can best be described as a form of participatory pluralism. While industry, environmental and recreational interests all had access to the Department, DEFRA restricted its role to participating with these interests rather than attempting to foster all-constituency agreement over policy and practices. In considering the MSC, DEFRA officials were impressed by its 'science-based' approach and its elite board of directors, where John Gummer, a former minister, was in charge. Unlike the FSC, the MSC had easy access to senior DEFRA officials and – being based in London – its staff were able to network with the bureaucracy to obtain desired outcomes. In performing its role as participant, DEFRA chose to ignore the views of the industry's major peak body, Seafish. In an early 1998 report on the MSC, Seafish argued that it would be costly to set up, that there was a lack of retail demand for certified fish and that industry representatives had not been involved in developing the standard to be applied. Despite hearing these complaints, DEFRA engaged with the MSC and supported its efforts to certify UK fisheries. The contrast between its behaviour and the Forestry Commission could hardly be greater.

Initial network reactions to the FSC and the MSC

The clientelistic structure of the forest policy network in the early 1990s in the UK meant that it responded vigorously and with hostility to the FSC. With small profit margins, landowners and timber growers united to object to the potential abridgement of private property rights and the imposition of unnecessary costs on the industry. Despite the Forestry Commission's emerging commitment to SFM and 'multiple forest management', its institutional structure remained firmly clientelistic and it responded to the representations made by the CLA, TGA and TTF by ignoring the FSC and considering strategies for sidestepping its requirements.

As a consequence of this hostility from the UK forest policy network, the FSC developed as an exclusively non-state initiative that was driven by a business–environmental partnership within the UK and bolstered by support from the FSC-IC from without. In 1994, shortly after Tim Synnott was appointed FSC executive director, he nominated Hannah Scrase to be the FSC's first-ever contact person. Scrase immediately set about establishing an FSC working group, a task made easier by the early involvement of FoE-UK and WWF-UK in promoting the idea of forest certification and the setting up of the WWF 95 and WWF 95+ Buyers Groups. The support the FSC received internally and externally enabled the nascent organisation to employ a consultant to prepare an early draft of an FSC-Great Britain standard, a process well underway by 1995. Half way through the decade, there was every indication that the path that certification would take in the UK would be similar to that which emerged later in Canada and Australia: a 'certification war' between the FSC and a domestically negotiated competitor standard. As progress was made on developing an FSC-Great Britain standard, the Forestry Commission began to prepare the ground to produce an UK Forestry Standard based on a series of guidelines that had been put in place from the mid-1980s onwards. The animosity of the forest industry to the FSC at this time cannot be underestimated as witnessed by the head of the TGA, Peter Wilson, comparing a meeting organised by the WWF to discuss FSC certification to a 'Nazi rally'.

DEFRA's response to the MSC was much more positive. Officials at DEFRA with responsibility for fisheries management were interested in experimenting with MSC certification, notwithstanding the negative perspective of the major industry peak body, Seafish. These officials were able to ignore Seafish's objections because they did not stand in a clientelistic relationship to the industry. DEFRA was a multi-industry department (environment, fisheries, rural affairs) which was lobbied not only by industry but by environmental interests. Groups like WWF-UK had good, direct access to DEFRA while the MSC's elite structure meant that it too established good working relations with the department. The appointment of Carl Christian-Schmidt of the OECD as executive director meant that he, too, was able to use his networks of influence to secure the MSC's broader objectives.

Evolution of network reactions to the FSC and the MSC

The initial hostility of the UK forestry policy network disintegrated in the late 1990s following the negotiation of the UKWAS standard. In contrast, the MSC has failed to win over the UK's fishing industry, which continues to view it, at best, as a pragmatic contributor to potential marketing problems. The different trajectories followed by the FSC and the MSC reflect the fundamental transformation of the clientelistic forest policy network in the direction of arbiter pluralism and the relative stagnation of the fisheries policy network's participatory pluralism.

The decisive event triggering a fundamental change in the UK's forest policy network was the shock announcement in 1996 that Poland's national forest estate has been certified to the FSC standard. This startling news was soon followed by another disturbing announcement: that FSC Sweden had agreed a national standard which was expected to be endorsed by the FSC-IC. These developments created a split within the previously united industry position on the FSC. Those engaged in the sawmilling and processing sectors announced that if their customers demanded certified timber they were obliged to supply it and would have to source it from elsewhere if such timber was not being produced in the UK. As noted in Chapter 4, the UK's heavy dependence on timber imported from abroad meant that the domestic industry would lose market share unless it too was able to provide certified timber products. The significance of the Poland announcement, therefore, was that it forced the UK forest policy network to confront the issue of certification. However, in early 1997, many still hoped that a UK Forestry Standard would be negotiated within the industry that could compete with the FSC.

It is at this point that the incremental shifts in the UK's discourse on forestry that had occurred from the 1970s onwards began to play a role. While the commission as an institution was opposed to the FSC, some officials within it were more sympathetic. These officials recognised that if the UK could reach a compromise with the FSC on the standard to apply, it would be to the industry's competitive advantage. The Forestry Commission thus sought to shape rather than merely respond to its client's interests by commissioning a report on the gaps between the draft UK Forestry standard and the draft FSC-Great Britain standard. Once the consultants reported that there were only modest differences that were not unbridgeable, the stage was set for a potential compromise between parties over the forest standard to apply in the UK.

That the gap between the emerging FSC and UK government standards was bridgeable reflected, in part, the millennia of settlement, clearance and agriculture within the UK. As noted in Chapter 4, the UK had no large tracts of old-growth native forests remaining, its biodiversity had been severely compromised by the clearing and planting of crops and exotic monocultures and the environmental movement was fighting a rearguard action to protect a range of species, very few of which were endemic or iconic. In short, the ecological context in which forestry was practised in the UK was significantly different to that in Canada and Australia – in comparison to these regions, the ecological stakes were low in the UK and recreational, amenity and site-specific environmental issues came to the fore. 'Multiple purpose' forestry was thus much more adapted to the UK context, whereas as we have seen, ecosystem-based forestry approaches are required to manage forests in the far more important and biodiversity-rich regions of Canada and Australia.

The negotiation of UKWAS resulted in an immediate and dramatic shift in the UK forest network towards the FSC. For almost all network members, UKWAS represented an acceptable compromise and the Forestry Commission and the private sector immediately moved to obtain FSC certification under the UKWAS standard. The only holdouts were the large country landowners represented by the CLA. Members of this group continued to oppose the FSC and sought to establish a PEFC presence in the UK after 1999. The problem they confronted, however, was that while timber growers could certify to the UKWAS standard by contracting with either an FSC or PEFC registered certifier, the costs of certification would be the same – both directly in terms of paying for the certification and indirectly in terms of the changes required to forest-management practices. Since the FSC was recognised as the 'Gold Standard' in the market, since many British retailers had committed to purchasing the FSC, since PEFC was as yet in its infancy, and there was as yet no low cost, dual certification options, there was no demand for PEFC certification. Consequently, in the UK, all forests were until very recently certified only to the FSC standard.

The dramatic transformation of the UK forest policy network and the subsequent take-off of the FSC in the UK contrasts with the relative stagnation of the MSC after a promising start. Although the MSC has been able to certify a large number of UK fisheries, these have mostly been very small operations and it has not yet convinced many in the industry of the value of its scheme. Significant resistance remains to the MSC in the industry's peak body, Seafish, which promotes alternative certification approaches including ISO 14000. The structure of the fisheries policy network appears unchanged in the past 15 years since the MSC was launched. DEFRA appears to still reside as a participant in a pluralistic network and to encourage the adoption of the MSC without being able to mediate a fundamental compromise between environmental and industry interests. Although it supports the MSC – and has in recent years backed up that support with a tangible financial contribution by supporting its Fish and Kids programme – it does not endorse it as the only approach to certification.

Market signals remain mixed, which constitute another significant difference from the FSC-UK forestry case. While retailers continue to back the MSC and seek to stock products certified to the scheme in their stores, the wider industry remains unconvinced that the benefits of such market access, in the absence of price premiums, are worth the costs. The MSC's efforts to build the market too have been somewhat compromised by lukewarm support from the environmental movement. While WWF-UK remains strongly supportive, Greenpeace and other ECSOs are unimpressed. The compromises that the MSC makes in certifying a fishery – and the absence of a thoroughgoing ecosystem-based approach to fisheries management – have made it difficult for the movement to offer wholehearted support. Thus the MSC does not get the strong backing that has accompanied the FSC from the broader environmental movement. This lack of backing results in the public receiving a rather mixed message when it comes to purchasing MSC-certified fish products.

Responses to the FSC and the MSC

Our analysis shows that the FSC and MSC standards have had influence and impact in each of our selected countries, although to varying and changing degrees. The UK is the outlier with regard to the FSC. As we have noted, FSC certification dominates and all public and most industrial private land are now certified to FSC standard in the UK. In Australia and Canada network dynamics have led to significant change over time. The increased responsiveness of the Australian forestry sector in 2009–10 is a good example of these dynamics. Having major Australian interests join the FSC has increased its legitimacy in the eyes of former critics. While this shift has been influenced by demands from buyers of Australian (and Tasmanian) wood products, these developments may foreshadow the end of the conflict over standards and open hostility to the FSC that has bedevilled Australian forestry politics and policy. In Canada too the FSC is increasingly accepted across the country; large area of forests are certified to the FSC, especially in the provinces of BC, Quebec and Ontario.

The response to the MSC is more similar across our comparator countries, although again it has the most penetration and impact in the UK. Even in the UK, however, the MSC is not widely endorsed by peak industry bodies. We have noted the pragmatic response to the MSC in Canada and Australia where certification is directed at fisheries where direct tangible benefits are delivered. Despite a decade of activity, industry remains suspicious of the MSC's agenda. The Australian and Canadian governments have taken a hands-off approach, although government-funded catch data and management information which is critical in pursuing certification has been supplied. Successful certifications have been publicised and acknowledged by these governments who have used them as validations of management processes. In contrast the UK government has directly funded the MSC, but not on a recurrent basis. ECSOs, while supporting the focus on sustainability, have been critical of the MSC's approach to certification, arguing that the standard is not being maintained, that insufficient attention is being paid to the need for corrective actions or re-certification of fisheries.

From the discussion in the preceding chapters it is clear, however, that divergent responses to the FSC and the MSC have occurred within each state. One explanation of this divergence is the fundamentally different institutional forms of the FSC and the MSC. The differences in organisational structures also shape interactions with the broader policy network, with national FSC members engaged directly in this interaction. The MSC, in contrast, is not directly involved in such national network relationships, with its staff first providing broad-based support to actors with the MSC process and second, undertaking higher-level lobbying. This does not mean that the MSC is not influential but, lacking a membership base and a role in determining the standard to be applied in a specific fishery, it often focuses on certification in general rather than the implementation of the MSC standard in particular.

In contrast, looking at responses to the FSC in Australia and Canada, we see interaction centring on the FSC standard in relation to opposition or competing standards. This reflects in part the highly politicised nature of forestry policy and management, yet it is clear too that there are different responses within each state. Recent moves towards increased support for the FSC standard in Australia from the native forest sector reflect external pressures from Asian buyers and potential investors. With regard to fisheries, early support for the MSC in Australia within the Western Rock lobster fishery – then the most valuable fishery in the country – was driven by key industry figures and an industry keen to increase market penetration into Europe. In this example and in others in forestry as well, we find that buyer-driven commodity chains in each sector are a major factor in relation to certification. Export markets appear to be a determining factor in relation to the strength of take-up of the FSC and MSC standards, with demand for certified products a major pull factor. On the other hand suppliers seeking to locate products in markets with high environmental sensitivity (e.g. Western Europe) are likely to push for certification as a way of overcoming what may be a soft trade or market barrier.

While influencing the shape and form of global and national commodity governance, interaction over certification and labelling in each sector has also led to the development of the FSC and the MSC. While the basic political, institutional and regulatory arrangements of each council remain (e.g. the FSC as a direct membership organisation in contrast to the MSC's foundation model), each has adapted over time in response to changing demands and needs. For example, the FSC has adopted and adapted the new ecosystem-based forest management in the preparation of national and regional forest-management standards. It has also reached out to social chamber members to build up this underdeveloped constituency. And it is currently engaged in consultations on far-reaching revisions to its ten principles and 56 criteria. The MSC has adapted too in the face of considerable criticism over the implementation of its standard, criticisms over specific fishery assessments and the scores assigned by assessors with regard to the principles and criteria, and concerns that insufficient attention has been given to monitoring of corrective actions in fisheries. The MSC's focus on fisheries rather than stocks raises interesting dilemmas with regard to development of a broader ecosystem-based approach. We have noted the central challenges to the MSC from ECSOs over the lack of broad-based stakeholder involvement; from scientists, most notably associated with the University of

British Columbia's Fisheries Centre, with the transparency of the assessment process; and from civil society organisations concerned over the governance framework of the MSC.

Conclusion

States have responded differentially to FSC and MSC certification in our three comparator countries. These responses reflect differences in the governance arrangements of the FSC and the MSC as well as the structure of the policy networks in place when the FSC and the MSC were introduced into the country. The structure of the policy networks has been shaped by the natural ecology of the resource, the structure of the commodity chain and the prevailing management discourse. Broadly, the FSC received a more hostile reception in each of three countries at the outset – virulently so in Australia and the UK with Canada moving more quickly to a position of 'neutrality' among schemes. However, the FSC has managed to transcend this hostility and is becoming well established in all three jurisdictions. It has 'converted' not merely the sceptics but in many cases those who were fundamentally opposed to its vision of sustainable forest management embedded in its ten principles and 56 criteria.

Ironically, almost the opposite trajectory appears to have befallen the MSC. It received considerably more support than the FSC at the outset in all three jurisdictions, with financial support offered from the UK government and technical and data support from all three governments. However, as a certification system, the compromises arrived at within the MSC create an elite organisation that appears to be neither rigorous enough to be endorsed wholeheartedly by the broad environmental movement nor industry-friendly enough to be fully endorsed by the fishing industry. While it has avoided the FSC's fate of being excoriated by its associated policy network and has not had to struggle against industry and government-sponsored competitor schemes, it has failed to convince the broader environmental movement that MSC-certified fisheries are in fact sustainable and this has compromised the message sent to retailers and consumers. Short of a significant further restructuring of the MSC's governance arrangements, it seems difficult to envisage how the organisation might overcome this fundamental dilemma.

9
Conclusion

The preceding chapters have examined how the development and institutionalisation of the FSC and the MSC have contributed to global commodity governance. While trade in commodities, including forest and fisheries products, is longstanding, late twentieth-century globalisation and concomitant development of natural resources have provided significant opportunities for growth and expansion. At the same time, increasing demand for commodities has heightened concern over the depletion of natural resources, the degradation of associated environments, the negative impacts on workers and indigenous peoples and the responsibility of conventional approaches to management for these negative externalities. This enhanced environmental and social awareness of the state of the world's natural resources and how they are managed is another artefact of globalisation. In response states have acted individually to implement policy and legislative reforms that centre on improving the social and environmental management of resources and worked with each other in intergovernmental forums and institutions to address these concerns. We note the impact and consequences of the most significant of these initiatives: the World Commission on Environment and Development, whose report was released in 1987; the United Nations Conference on Environment and Development of 1992; and the World Summit on Sustainable Development held in 2002.

The Brundtland Commission's focus on sustainable development drove increased interest in natural resource management by states and embedded the sustainability agenda in a variety of national initiatives as well as placing it on the agenda of bilateral, regional and international forums. Despite these developments the world's forests and fisheries remain imperilled, with continued, widespread deforestation, forest degradation and 'illegal' logging and illegal, unreported and unregulated fishing. Business and civil society actors have also fully engaged with the post-Brundtland agenda including addressing the meaning of, and means to implement, sustainable development. One outcome of these efforts has been increased scrutiny of governmental actions and outcomes. The two stewardship councils, while significantly different in structure and operation, were developed from a similar standpoint: both were responses to the perceived limitations of traditional state-based actions and intergovernmentalism in achieving sustainable commodity governance in a globalising world.

As we have noted, unpacking the forms of global commodity governance conceptualised as institutionalised systems of rule provides useful insights into the limits of intergovernmentalism, most graphically depicted in the regular reports of the FAO of ongoing declines in the quantity and quality of the world's forests and fisheries despite a broadening of relevant international instruments and regimes. This has led to interest in new approaches, shifting emphasis away from intergovernmentalism to alternative forms of commodity governance, most especially in certification and labelling schemes. These business- and civil society-led efforts have, in turn, encouraged states to respond. This response has variously led to attempts to incorporate certification and labelling into standard regulatory arrangements, engage in co-regulation, or regard such processes and practices as new political and, potentially challenging, organisational forms.

Forests and fisheries have traditionally been managed by formal rules set out in state-based legislation and regulatory instruments to control exploitation, initially based on the allocation of rights to what were either state-controlled or common-property resources. More recently, market-type instruments have been introduced by states with the development of rights-based management and a focus on output rather than input controls on the resource. In fisheries this has led to the development of tradeable rights and the creation of quasi-market approaches and has increased the use of trade and market instruments as management tools. This has allowed fish products to be traced more accurately and resulted in improved compliance with management rules. In forestry, too, tenure reform to provide long-term security of access to forests has been advocated as a mechanism to improve corporate incentives to practice more sustainable forest management often coupled with codes of conduct with regard to forest practices. While these developments matched the broadening of the 'neoliberal' economic and political agenda that emerged in the last quarter of the twentieth century, they also reflected a desire to experiment with innovative approaches to resource management associated with the emerging, new 'smart regulation' paradigm.

The new paradigm did not negate the role of the state but matched it to increasing opportunities provided by market and civil society initiatives. It also incorporated the increased scope and salience of voluntary environmental certification. The development of external, third party, certification and labelling schemes to assess the sustainability of forestry and fisheries operations was targeted directly at market actors – producers, retailers and consumers. Consumer demands for food safety and quality have emerged as major drivers in fisheries markets, opening up opportunities for product certification. It is clear that food safety and quality have driven the growth of Type I and Type II labelling in fisheries as well as ensuring broad-based implementation of HACCP concepts and quality assurance systems according to ISO 9000. These schemes focus on processing facilities and identification and labelling of products. Likewise, consumer demand for sustainable wood products has increased as awareness has grown of tropical, temperate and boreal deforestation and degradation, the importance of forests in mitigating greenhouse gas emissions and the widespread existence in the marketplace of forest products made from illegal timber.

We have adopted a critical, ecological political economy approach that builds on the policy network, commodity chain and sustainability literatures. While we were interested in examining the development of the FSC and the MSC as parallel examples of civil-society led commodity governance, our major focus has been on differences. As we note in Chapter 1, we focus on differences between schemes, differences between governments and differential governmental, industry and societal responses to these two stewardship schemes. While we have compared the FSC and the MSC across the three dimensions of governance (political, institutional and regulatory), we have also examined the impact of the FSC and the MSC in three selected countries: Australia, Canada and the UK. This 'double comparison' of schemes and states offers a rich analysis of the ecological political economy of these 'new governance' arrangements. We hope too, that it will contribute to greater understanding of the opportunities and constraints in implementing certification and labelling in general, to forestry and fisheries schemes in particular, and to making scheme proponents, business, civil society and governments more self-reflective about their actions.

To analyse state responses, we adopted a disaggregated approach to the state which behaves differently depending on the structure of the policy networks in operation in a given sector. This approach enabled us to examine how the structure of a policy network (bureaucratic, clientelistic, triadic, pluralistic) shaped the state's response to specific certification schemes. From the sustainability literature, we employed the view that the problems posed by forestry and fisheries management are not merely technical but paradigmatic with emerging 'ecosystem-based' approaches to management challenging the earlier focus on the sustained yield of single stock or species management. Developments in ecosystem-based management bring new actors as well as new approaches to the policy network. A focus on sustainability introduces a broader management paradigm where the social and ecological dimensions of management are as important as the economic. Finally, we also included insights from the commodity chain literature, which examines the relative power of actors at different locations along the product chain. Both fisheries and forestry products chains are 'buyer-driven' chains, placing considerable power in the hands of retailers and consumers.

While there has been much debate over the efficacy of the 'triple bottom line' or 'three legged stool' approaches, they nonetheless provide opportunities for business and civil society to engage in debates over natural resource management. In addition such debates can broaden a policy network away from bureaucratic, clientelist or triadic arrangements. As networks broaden, so too do opportunities for civil society actors to increase input into policy proposals and management decisions. Our analysis has shown, however, that pluralistic policy networks are the exception rather than the rule and that state officials, locked into non-pluralistic networks, viewed the emergence of both the FSC and the MSC with some suspicion and/or concern. In such circumstances, state agencies in conjunction with their clients may reassert their responsibilities for management while incorporating initiatives emerging from the new sustainability agenda. Canada and Australia's establishment of competitor schemes to the FSC's are good examples of such action in the forestry sector as state officials were heavily involved in the development of CSA and AFS. In Australia, the Australian government's introduction of external strategic assessment of fisheries on sustainability criteria (drawn from MSC principles) and the introduction of a harvest strategy approach focusing on biomass targets also illustrate the phenomenon.

The FSC and the MSC

We have noted that despite the obvious links – with the MSC being modelled on the earlier FSC and both organisations carrying the 'stewardship' title – their institutional structures differ markedly (Chapter 3). Elite criticism of the FSC saw the MSC's proponents seek to streamline the new body. The FSC's chamber-based, membership association model contrasts directly with the MSC's stakeholder, foundation model; its governing body, the General Assembly, is likewise very different from the MSC's Board of Trustees. As we note, despite, or perhaps because of, the difficulties emerging from the FSC's more complicated structure the legitimacy of its decision-making processes has rarely been called into question. While the MSC operates more 'efficiently', it has encountered ongoing criticism concerning the adequacy of its standard and the rigour and transparency of its implementation.

It is important to note too that both organisations have evolved to maintain their 'fitness for purpose'. The FSC has incorporated more ecosystem-based science into its deliberations over national and regional standards while the MSC has broadened its stakeholder representation, including by establishing a Stakeholder Council. While both organisations have evolved in terms of their organisational structures, the initial criticisms of the FSC's somewhat cumbersome membership-based operational arrangements have given way to broad recognition of the legitimacy that these arrangements play. In contrast the more 'efficient' MSC has found itself undertaking much more significant changes to cope with ongoing criticism from environmental and social stakeholders which nonetheless persist. We recognise that despite these differences both the FSC and the MSC have developed into robust organisations (albeit despite or in response to critiques) but that the MSC, given its organisational form and structure – with a head office and a small number of regional offices – is seen as an elite and distant presence when compared with the membership base and national committees of the FSC.

The model of 'proponent driven' certification underpinning the operation of the FSC and the MSC has been criticised as contributing to 'green-' or 'blue-washing' of products. This concern is increasing in relation to the MSC where fishers seek certification of 'their' fishery. The narrowness of certifications that may not necessarily reflect the health of the broader stocks, or address impacts of fishing practices or outcomes outside those that have been undertaken by the group seeking certification, appear to be the antithesis of an ecosystem-based approach. The consequence appears to be, as some analyst have noted, that consumer-driven changes to fisheries sustainability has been well below the aspirations set for the approach a decade or more ago. This forms a sharp contrast to perceptions of the FSC's impact in boreal and temperate forests where it has been very successful, although concerns regarding its relevance to tropical forests remain.

Networks and certification

Business responses to certification, while influenced by increasing scrutiny by civil society organisations, are still shaped by national governmental policy frameworks and settings that influence policy networks. We placed interaction over forestry and fishery management at the centre of this analysis and have utilised insights from the literature on policy networks to explore the dynamics of these interactions. In addition our study of civil society global governance initiatives provides interesting insights on the modelling of networks. As has been noted in preceding chapters, the FSC and the MSC have provided interesting challenges to traditional state-centred resource management and governance. This leads to both new, hybrid forms of interaction over forests and fisheries (among a range of issue areas) and new pressures on old forms of governance – in this case traditional state-centred, clientelistic or triadic networks. Both the FSC and the MSC were developed in response to perceptions of state and intergovernmental failures to sustainably manage forests and fisheries. The turn to market and civil society governance using retailer and consumer pressure to encourage change was and is a direct challenge to the state and to business/industry practices. At the same time the new governance fostered by the FSC and the MSC was situated in a system of global commodity production that relied and continues to rely on state action.

In Chapter 1 we outlined the focal point of our research. We are interested in the normative and empirical questions arising from the state's interactions with the FSC and the MSC, and market- and civil-society driven forms of global commodity governance more generally. What are the factors shaping a policy networks response? What factors encourage change in the network? What lessons can be learnt? We see 'the state' as both an actor and a framework; state agencies become powerful actors within and also help shape and maintain the policy network(s) that interact over forestry and fisheries policy and management. While the state can facilitate or constrain the new global commodity governance, it cannot ignore it. Our cross-sectoral/cross-national case studies show that even when the state is ambivalent over certification and labelling the actions of others have great impact on the policy network. In the fisheries sector, for example, the nexus between the MSC's initial foundation partners WWF and Unilever worried industry – first in terms of the WWF's agenda and second over the potential market power of Unilever. Announcements of support for the MSC by Sainsbury's, a major UK food retailer, and subsequently by other UK retailers Tesco's and Safeway, indicated a potential change to business-as-usual arrangements. Whether the state operated alone or was enmeshed in bureaucratic or clientelistic policy networks, it could not by any means ignore the expressed concerns of the sector.

We have noted how differences in natural ecologies, policy contexts, management discourses and commodity chains shaped interaction within the policy network over forestry or fisheries certification. It is clear that the relative natural ecologies and concomitant commodity chains are major factors in shaping responses to the FSC or the MSC. The ecological significance of the resource, concerns over the sustainability of extraction and/or demands from export markets are factors encouraging certification. In each state fisheries certification is driven by the influence of European and North American markets. In contrast North Asian markets do not currently demand MSC-type certification, so products exported to these markets are less influenced by developments in third-party certification. The same situation pertained in forestry until recently. The initial engagement of Australian, Canadian and UK producers with certification was in response to developments in North America and Europe. However, in the past two years, ECSO pressure on Japanese buyers intersected with a significant drop in demand for forest products as a consequence of the GFC to induce a change in direction. Suddenly, Japanese buyers of forest products began demanding FSC-certified woodchips, inducing wrenching changes in Australia's and particularly Tasmania's forest industry.

Our analysis has reinforced the point that a network's attitude to certification and/or the FSC/MSC is not static. We have noted evidence of change within forestry and fishery networks in each country that have been influenced by both internal and external factors. The 'certification wars' in forestry in Australia and Canada illustrate the role of internal factors. The struggle here was not over the concept of certification per se but over how it should be institutionalised and what its political, institutional and regulatory arrangements should be. Recent decisions in the native forest sector to implement the FSC in Australia are a result of external pressures. The slow take-up of MSC certification in Canada reflects concerns at external factors – the apparent differential treatment of Canadian Pacific salmon certification as opposed to the process undertaken by similar US fisheries. While the networks differ across sectors and countries, as noted in Chapter 8, we see a general trend in the opening up of traditional bureaucratic, clientelistic or triadic networks to incorporate environmental and social civil society actors. It is important however not to underestimate the strength and resilience of state actors, who still retain influential roles. This latter comment is most clearly relevant in the UK in relation to forestry and Canada in relation to fisheries as single agencies with major policy and management responsibilities.

Hybrid governance

Our premise for this study of the stewardship councils was that they provide examples of emerging new patterns of market- and civil-society led global commodity governance. In exploring the impacts and influences of this form of commodity governance we have emphasised that the stewardship councils have had both direct and indirect effects. They have been responsible for developing standards and establishing ecolabels, but also have contributed to increased awareness of the broader question of sustainability in each sector. Not surprisingly this form of governance both engages with and challenges the state at the same time. Developed in response to ECSO perceptions of failure of intergovernmentalism, the FSC and the MSC nonetheless are becoming embedded in state practices related to forestry and fisheries management and marketing.

In place of state- or market-led governance we are seeing the emergence of new hybrid forms of state/market/civil society governance. These hybrid forms of governance render increasingly problematic the traditional public/private dichotomy that has been a foundational concept of modernity. Hybrid arrangements appear better adapted to the complex reality of the twenty-first century, where agreeing how to manage forests, fisheries and the wider range of natural resources more generally is the collective responsibility of affected constituencies at multiple levels and across multiple sectors. Conceptualising the state as the embodiment of this collective responsibility has always been problematic in theory, while outcomes have been seriously suboptimal in practice. Hybrid governance arrangements, of the kind embedded in the FSC and the MSC, provided a vehicle for transcending outmoded public/private divisions by refocusing attention on commodity sectors and commodity chains and on the enterprise of sustainable production literally from 'vessel to plate' and from 'tree to book'.

Notes

1 Commodity Governance in a Globalising World

1. The group of 20 nations that formed around trade policy issues at the WTO was composed of developing countries including China, Brazil and India. It is not to be confused with the G-20 nations meeting on the global financial system which consists mainly of industrialised countries.
2. Following intensive negotiations, the UK delegation agreed to reformulate the project proposal to focus on 'incentives' for sustainable forest management. The Friends of the Earth International delegate, Simon Counsell, was outraged at the change and refused to endorse the new proposal (Gale 1998).
3. The UK Forestry Standard was used for internal assessment purposes following the negotiation of UKWAS. See Chapter 6 on the UK Case.

2 Global Commodities, Sustainable Governance: An Analytic Framework

1. A word of caution, however, is in order. Despite the enormous efforts undertaken by the FAO to ensure that its data is of high quality, it relies on countries to compile and submit information to it. While the data contained in its reports is the most globally authoritative, there are numerous gaps.
2. There is debate over the validity of China's official fisheries and aquaculture production, see R. Watson and D. Pauly, 'Systematic Distortions in Fisheries Catch Trends', *Nature*, 29 November 2001: 534–6. The 2006 SOFIA report noted that 'there are continued indications that capture fisheries and aquaculture production statistics for China may be too high' (FAO 2007).
3. The ITTO was an atypical commodity agreement because it did not utilise a buffer stock, it being agreed that it was preferable to have them standing, and growing in the forest.
4. The final ECSO statement to the ITTO Plenary Meeting on the Sarawak Mission stated: 'This must be a day of shame for the ITTC [International Tropical Timber Council, the ITTO's governing body]. It is the culmination of a week of vacillation, compromise and lack of determination to tackle issues central to the organization's mandate. ... The resolution before the Council is unacceptable. It is limited to the weak responses proposed by the Sarawak and Malaysian governments. It fails to mention several key issues raised in the body of the report' (NGO Statement 1990).
5. The full title of the Forest Principles was: 'Non-legally binding authoritative statement of principles for a global consensus on the management, conservation and sustainable development of all types of forests.' They are available at http://www.un.org/documents/ga/conf151/aconf15126-3annex3.htm.
6. For a detailed review of theories of the state, see Carnoy 1984.

4   Forest and Fisheries Management in Comparative Perspective

1. The FAO defines a forest as 'Land spanning more than 0.5 hectares with trees higher than 5 meters and a canopy cover of more than 10 percent, or trees able to reach these thresholds in situ. It does not include land that is predominantly under agricultural or urban land use.' See Ridder 2007. Dargavel uses a 'conventional' foresters' definition of a forest as the 'capacity to produce timber' to identify 43.2 million hectares of forest in Australia (Dargavel 1995).
2. There are over 4000 SSSIs in England alone, with the more important designated Special Areas of Conservation (SACs), Special Protection Areas (SPAs) or sites under the Ramsar Convention (Natural England 2009a). SSSIs may also be National Nature Reserves (NNRs) or Local Nature Reserves (LNRs). With respect to forests, Natural England lists 33 SSSIs in its online database (Natural England 2009c). These include Windsor Forest and Great Park, and Hatfield, Lune and Naddle Forests in Essex, Durham and Cumbria respectively and, the jewel in the crown, The New Forest. The New Forest is mainly Crown land under the management of the Forestry Commission, but parts are also owned by the Hampshire County Council, by private owners, and by civil society organisations such as the Hampshire Wildlife Trust (Natural England 2009b).
3. The Resource Assessment Commission was set up in 1989 by the Hawke Labor Government and between then and 1993 it launched inquiries in three issue areas – forests, mining (in the Kakadu National Park) and coastal zone management (Stewart and McColl 1994).
4. All states and territories bar Tasmania signed the National Forestry Policy Statement in 1992, with Tasmania holding out until 1995.
5. The conflict had a long history, rooted in a 1979 decision by the Bennett Social Credit government to allow logging on Meares Island in the traditional territories of the Nuu-chah-nulth (Parai and Esakin 2003). While that conflict concluded in 1985 when the Nuu-chah-nulth were granted an injunction by the British Columbia Supreme Court banning logging pending the settlement of outstanding land claims, it prompted forest industry and government officials to develop plans to log adjacent lands in the region, notably in the old-growth temperate rainforests of Clayoquot Sound, which were also subject to a land claim by the Nuu-chah-nulth.
6. The historic owners of the land – the Teme-Augama Anishnabai had been in dispute with the provincial government and development interests for decades. Following the election of the NDP government in 1990, negotiations on the Temagami dispute eventually resulted in the establishment of a co-management arrangement through the Wendaban Stewardship Authority.
7. There were two periods during which FRDAs were signed. FRDA I lasted from 1985 to 1990 and FRDA II lasted from 1991 to 1996. FRDA programmes followed on from a range of similar initiatives in the 1950s and 1960s under the Mackenzie King and Diefenbaker administrations under the Canada Forestry Act 1949 (Howlett 2001).
8. Environmental and conservation civil society organisations signing the accords include the Canadian Nature Federation (now Nature Canada) (Accord I), Canadian Wildlife Federation (Accords I, II and III), Ducks Unlimited Canada (Accords II and III), Prince Edward Island Nature Trust (Accords II and III), Friends of Oldman River (Accord III) and Ontario Federation of Anglers and Hunters (Accord III).

9. The Oceans Act expanded the role of the Department of Fisheries and Oceans, placing an increased emphasis on ocean governance, with the Department designated as lead agency with responsibility for assuring effective coordination and integration of activities carried out by many agencies – federal, provincial, territorial and local in relation to management of Canada's large ocean estate. See Haward and Vince 2008.
10. Fisheries management and policy in the UK has been embedded within the frameworks, policies and guidance of the European Commission since the UK joined what was then the European Economic Community on 1 January 1973. Prior to the implementation of the Common Fisheries Policy (CFP) – the premier fisheries instrument of the EU in 1983 – fisheries had been considered as part of the agricultural policy framework embedded in the EEC's foundation document, the 1957 Treaty of Rome, with specific regulations first developed for fisheries in 1970. During the 1970s developments in customary international law of the sea saw states proclaim over extended areas of fisheries jurisdiction, and extend their territorial seas (Haward and Vince 2008). As noted earlier the UK became embroiled in the cod wars, one of the most notable conflicts over these shifts. The first moves towards the CFP began in 1976 and involved significant negotiations that led to the conclusion of the CFP in 1983. The CFP has been subject to regular revisions, with the most recent revision in 2002 being the most significant, with detailed information on background to the revisions, key components and implementation plan included in what is termed the 'Roadmap' (see Haward and Vince 2008).
11. In the mid-1990s, in response to major changes in the DFO's approach to fishery management in Atlantic Canada, inshore 'fixed gear' (hook and line, gillnet) fishers began developing an innovative mechanism for creating greater local control over fishing arrangements, through quotas allocated by government to communities rather than to individuals or companies (see Charles 2001) so the available catch is managed locally. This approach was pioneered in the small community of Sambro, near Halifax, Nova Scotia and has since spread throughout the small-boat fixed-gear fishery in Scotia-Fundy region. Information provided by Anthony Charles, see Haward et al. 2003.

5 Forest and Fisheries Certification in Australia

1. WWF-Australia has a marine campaign, with most of its attention being focused on Southern Ocean and IUU fishing.
2. Doyle's criticism was taken further with the publication of *Taming the Panda*, a critique of WWF-Australia by The Australian Institute in 2004 (Hamilton and Macintosh 2004). These authors argued that WWF-Australia was losing its independence and 'capacity to make dispassionate assessments of what is in the interests of the environment' (2004, ix) due to increased political and financial ties to the coalition government of then Prime Minister John Howard.
3. ACF has a marine campaign that has focused on Australia's oceans policy framework and issues with sustainable fisheries.
4. Attendees included Jag Maini (Head, IPF), Stephen Bass (IIED, UK), Mike Fullerton (Canadian Forest Service, Ottawa), Peter Kanowski (ANU, Canberra), Jim Bourke (FAO, Rome) and Tim Synnott (FSC, Mexico).

6 Forest and Fisheries Certification in Canada

1. These alliances with government and industry create friction with the broader environment movement, which views WWF-Canada as a captured agency. Corporate donations constitute a substantial component of WWF-Canada's resources. In 2008, for example, it lists CanWest Mediaworks, The Toronto Star, Coca-Cola Canada, the Forest Products Association of Canada, Rio Tinto Alcan, Bell Canada, Canada Life, Hewlett-Packard (Canada) Co., Hudson's Bay Company and Wal-Mart Canada Corp as corporate donors that have contributed above $50,000 to the organisation (WWF-Canada 2008, 44–6). WWF-Canada also receives significant support from government agencies. In the 2008 Annual Report, donations are listed from the Canadian International Development Agency (CIDA), Environment Canada, Canadian Environmental Assessment Agency (CEAA), Department of Indian and Northern Affairs (DIAND), Fisheries and Oceans Canada, Natural Resources Canada (NRCan) and the Ontario Ministry of the Environment.

2. In a rather unsympathetic account of Greenpeace's organisational arrangements, Harter argues that Greenpeace is run by a 'professional managerial class' that has no difficulties in exploiting the working class either indirectly as victims of their campaigns or directly as employees within the larger Greenpeace system. He highlights the internal tensions that emerged in 1993 when workers at Greenpeace's Toronto office attempted to form a union, and argues that Greenpeace adopted the conventional union busting activities of corporate capitalists. According to Harter, Greenpeace 'hired an anti-union law firm, Mathews, Dinsdale, and Clark, best known as a defender of corporate polluters' to bust the union and paid them $100,000 to secure a victory. Tactics included laying off unionised workers, fighting attempts to negotiate an enterprise bargaining agreement and opposing staff actions at the Ontario labour relation's board (Harter 2004).

3. Although the Sierra Club Canada claims to be a membership organisation, there is a caveat. The membership body is separate from its fund-raising body, the Sierra Club Canada Foundation. In a note to the accounts, the auditors state 'The Sierra Club of Canada Foundation and the Sierra Club of BC Foundation are registered charitable organizations governed by their own Boards of Directors. The Foundations' missions are to advance the preservation and protection of the natural environment through charitable projects of the Sierra Club of Canada and its chapters' (Sierra Club Canada 2008, 19). There were six directors of the SCCF in early 2009, with no overlapping directors between SCCF and SCC. It thus appears that while the members run SCC, they may or may not get funding for their activities depending on the willingness of the SCCF to fund them. As noted above, SCCF and SCC are funded from a variety of sources. The 2007 accounts record contributions from SCCF and SCBCF of $1.263 million and $0.549 million respectively to SCC. Government contracts ($0.428 million) and donations and memberships ($0.467 million) made up the bulk of the remaining contributions.

4. The official acronym for the Canadian Environmental Network is RCEN, where the R stands for 'Réseau', the French for 'network'.

5. The essay, written by Michael Schellenberger and Ted Nordhaus, was released at an October 2004 meeting of the US Environmental Grantmakers Association and provoked considerable controversy. It is available online at http://oracle.cas.muohio.edu/ies/students 2005/DeathOfEnvironmentalism.pdf.

6. According to Thomson, the BCEN was a much more effective organisation in the early 1990s, but declined markedly through the decade, both in terms of

membership and in terms of funding. Thomson notes, 'At its peak around 1993, membership within the BCEN had reached 302 groups, including associates. This has changed dramatically over the past couple of years. As of April 2005 only 36 members had paid their dues' (Thomson 2006, 8). Relationships between many environmental groups deteriorated through the 1990s and became dysfunctional in the early 2000s as large, well-resourced, professionalised and strategic ECSOs overpowered grassroots groups. According to Thomson, the Great Bear Rainforest campaign, which brought together Greenpeace, Sierra Club of BC and ForestEthics, was a lightning rod for movement discontent between the well-resourced, urban 'insiders' and the poorly resourced, rural 'outsiders' (Thomson 2006, 5).

7. For a more detailed account of Hammond's ecosystem-based approach to forestry, see Hammond 1993.

7 Forest and Fisheries Certification in the UK

1. FoE-UK was intended to be more centrally run at the outset, but as local groups were established they increasingly demanded a greater say in how the organisation should be managed and on the policy priorities and campaigns in which it should engage (Lamb 1996).

2. There are now two organisational entities that make up FoE-UK: FoE-UK Limited and FoE-UK Trust. FoE-UK Trust is the charitable arm of the organisation and is formally run by a board of 14 trustees who work closely with FoE-UK Limited. The majority of FoE-UK Trust trustees are active members of FoE-UK local groups, but the major role of the trustees is to oversee the financial management and strategic direction of the organisation. It meets four times a year to 'determine mission, policy and strategy; to monitor the organisation's performance; to appoint and manage the CEO; and to manage the governance process' (FoE-UK 2008b, 2). A critical feature of FoE-UK Trust is its relative autonomy from governmental and corporate funds. Almost four-fifths (79%) of FoE-UK Trust 2008 income, which was up slightly on the previous year, derived from individual donations from members and supporters. In contrast, there was a decline in income in FoE-UK Limited, mainly because it now encourages 'new supporters to join Friends of the Earth Trust, a registered charity, so they can make their gift go further by taking advantage of Gift Aid' (FoE-UK 2008a, 6).

3. Following the Forestry Commission's establishment, they reputedly vied with each other for the honour of being the first person to plant a tree for the Forest Commission, the race being won by Lord Clinton, who was able to get to his lands in Devon quicker than Lord Lovat who had to take the overnight train to Scotland (James 1981, 216–17).

4. The CLA was founded in 1907 as the Central Land Association, changing its name to the Central Landowners' Association in 1918, to the Country Landowners Association in 1949 and to the Country Land and Business Association in 2001.

5. The CLA remains an association with a conservative ethos, however, with many members concerned to defend their private property rights and economic development options against perceived threats from socialism and, more recently, environmentalism. Frances Beatty, the CLA's West Midlands Director, states in a recent speech that 'The CLA is the premier organisation safeguarding the interests of those responsible for land, property and business throughout rural England and Wales'. She goes on to observe that 'the CLA's original 1907 mission to protect private property rights remains fundamental to us' (CLA 2009).

6. ConFor is the name for the current umbrella groups that represents forest and timber interests in the UK. Its predecessors were the Timber Growers Organisation (TGO), which was established in 1958 as a consequence of recommendations of the government's Watson Committee and which represented woodland owners in England and Wales. A year later the Scottish Woodland Owners' Association (SWOA) was formed to represent timber-growing interests there (Crawford 1965). The two bodies merged in 1983 to form Timber Growers UK (Winter 1996, 290), with that body and the Association of Professional Foresters merging to form the Forestry and Timber Association (FTA) in 2002 to bring together 'woodland owners, managers, contractors, suppliers to the industry and others interested in trees and woodland' (Small Woods Association 2009). Subsequently, however, the FTA merged with the UK Forest Products Association and other groups to form ConFor in 2004, which now represents the interests subsumed within it.

7. At this juncture, the FSC-UK was writing a standard for Great Britain due to the lack of resources and representation from Northern Ireland. It was only later, and as a result of the UKWAS process, that the scope of the standard was extended to cover all regions of the UK.

8. Given the fairly limited options available to PEFC in the UK, it might be thought that there would be no local presence, but this is not the case. PEFC-UK was established in 2001 by private forest interests and endorsed the following year by the international PEFC Council (UKWAS 2006). The driver behind this initiative was Len Yull, the former chairman of the TGUK. Yull appears to have been personally affronted by the FSC initiative when it was originally mooted in the early 1990s. He saw the PEFC initiative as an opportunity to offer forest managers in the UK a low-cost certification option that would break the looming FSC monopoly and challenge the right of environmentalists to set the forestry agenda.

Bibliography

AAD (Australian Antarctic Division). 2008. 'Makerel Icefish'. Online at http://www. aad.gov.au/default.asp?casid=29301, accessed April 2010.

ABARE (Australian Bureau of Agricultural and Resource Economics). 2009a. Australian forest and wood product statistics: March and June Quarters 2009. November. Bureau of Regional Statistics, ABARE.

——. 2009b. *Australian Fisheries Statistics 2008*, Canberra, ABARE.

——. 2007a. *Australian Forest and Wood Product Statistics, September and December Quarters 2006*. Canberra, ACT: Department of Agriculture, Fisheries and Forestry. Online at http://www.abareconomics.com/publications_html/afwps/afwps_07/ afwps_may07.pdf, accessed December 2007.

——. 2007b. *Australian Fisheries Statistics 2006*, Canberra, ABARE.

ABC (Australian Broadcasting Corporation). 2010a. 'Managed Investment Schemes Still Falling Over'. Inside Business, 18 April. Online at http://www.abc.net.au/ insidebusiness/content/2010/s2875821.htm, accessed April 2010.

——. 2010b. 'Gunns Seeks Certification for its Timber Products'. Online at http:// www.abc.net.au/pm/content/2010/s2807061.htm, accessed March 2010.

——. 2005. 'Green Groups Oppose Lobster Fishery Approval'. 10 January 2005. Online at http://www.abc.net.au/news/stories/2005/01/10/1279255.htm, accessed March 2010.

ACF (Australian Conservation Foundation). 2008a. 'The Birth of ACF'. Online at http://www.acfonline.org.au/articles/news.asp?news_id=860, accessed March 2008.

——. 2008b. 'The Successes'. Online at http://www.acfonline.org.au/articles/news. asp?news_id=862, accessed March 2008.

AFS (Australian Forestry Standard Limited). 2006. 'A Response to Incredible Claims by The Wilderness Society'. Canberra, ACT: AFS Limited. Online at http://www. forestrystandard.org.au/files/News_Feb06Response.pdf, accessed March 2008.

Agnew, D., C. Grieve, P. Orr, G. Parkes and N. Barker. 2006. *Environmental Benefits Resulting From Certification Against MSC Principles and Criteria for Sustainable Fishing*, Final Report. London, MRAG/MSC London.

Ajani, J. 2007. *The Forest Wars*. Carlton, VIC: Melbourne University Press.

Akerlof, G. 1970. 'The Market for Lemons: Quality, Uncertainty and the Market Mechanism'. *Quarterly Journal of Economics* 84, 3 (August): 488–500.

Arts, B. 2003. 'Non-State Actors in Global Environmental Governance: New Arrangements Beyond the State'. Online at http://www.essex.ac.uk/ECPR/events/ jointsessions/paperarchive/edinburgh/ws11/Arts.pdf, accessed April 2010.

ASOC (Antarctic and Southern Ocean Coalition). 2001. 'Antarctic and Southern Ocean Coalition Submission on the Moody Marine Assessment of the Suitability of the South Georgia and South Sandwich Islands Patagonian Toothfish Fishery for Certification by the Marine Stewardship Council'. Washington, Antarctic and Southern Ocean Coalition.

ASP (Association of Seafood Producers) 'Welcome to the Association of Seafood Producers'. Online at www.seafoodproducers.org, accessed 17 October 2010.

Atkinson, M. and W. Coleman. 1989. 'Strong State and Weak States: Sectoral Policy Networks in Advanced Capitalist Economies'. *British Journal of Political Science* 19, 1 (January): 47–67.

Bache, S., M. Haward and S. Dovers. 2001. *Economic, Trade and Environmental Instruments: Their Impact on Australian Fisheries Policy and Management*. Fisheries Resources Research Fund, Centre for Maritime Policy University of Wollongong.

Baily, M., R. Litan and M. Johnson. 2008. *The Origins of the Financial Crisis*. Washington, DC: The Brookings Institute.

Baker, D. and B. Pierce (eds). 1998. *Environmental Management Plan of the Southern Fishermen's Association for the Lakes and Coorong Fishery*. Adelaide: Southern Fisherman's Association.

Balaam, D. and M. Veseth. 2008. *Introduction to International Political Economy, Fourth Edition*. Upper Saddle River, NJ: Pearson Education International.

Bartley, T. 2003. 'Certifying Forests and Factories: States, Social Movements, and the Rise of Private Regulation in the Apparel and Forest Products Fields'. *Politics & Society* 31, 3 (September): 433–64.

——. Forthcoming. 'Transnational Private Regulation in Practice: The Limits of Forest and Labour Standards Certification in Indonesia'. Online at http://www.indiana. edu/~tbsoc/indo.pdf, accessed October 2010.

BC Ministry of the Environment, 2008a. 'BC Species and Ecosystem Explorer'. Online at http://www.env.gov.bc.ca/atrisk/toolintro.html, accessed November 2008.

BCNDP (British Columbia New Democratic Party). 2009. 'Labour, the Environment and Politics: An Evening with United Steelworkers International President Leo Gerard and Sierra Club Executive Director Carl Pope, Leading Members of The Blue-Green Alliance and Apollo Alliance'. Online at http://www.usw.ca/UserFiles/File/D3%20documents/ Leo_Gerard_Carl_Pope_NDP_Fundraiser.pdf, accessed March 2009.

BCSMC (British Columbia Salmon Marketing Council). 2008. 'BC Salmon Marketing Council'. Online at www.bcsalmon.ca/files/bsmc.html, accessed 10 December 2008.

Beder, Sharon. 2004. 'Moulding and Manipulating the News'. Faculty of Arts Papers, University of Wollongong. Online at http://ro.uow.edu.au/cgi/viewcontent.cgi? article=1043&context=artspapers, accessed October 2010.

——. 1991. 'Activism versus Negotiation: Strategies for the Environmental Movement'. *Social Alternatives* 10, 4: 53–6.

Beddington, J. R., D. J. Agnew and C. W. Clark. 2007. 'Current Problems in the Management of Marine Fisheries'. *Science* 316, 5832: 1713–16.

Beeson, M. 1999. *Competing Capitalisms: Australia, Japan and Economic Competition in the Asia-Pacific*. London: Macmillan.

Beeton R., K. Buckley, G. Jones, D. Morgan, R. Reichelt and D. Trewin. 2006. *Australia State of the Environment 2006*. Independent Report to the Australian Government. Canberra, ACT: Minister for the Environment and Heritage, Department of the Environment and Heritage. Online at http://www.environment.gov.au/soe/2006/ publications/report/pubs/soe-2006-report.pdf, accessed March 2010.

Bell, S. 1997. *Ungoverning the Economy: The Political Economy of Australian Economic Policy*. Oxford: Oxford University Press.

Bergin, A. 1993. *Aboriginal and Torres Strait Islander Interests in the Great Barrier Reef Marine Park*. Townsville: GBRMPA.

——. 1991. 'Aboriginal Sea Claims in the Northern Territory of Australia'. *Ocean and Shoreline Management* 15, 3: 171–204.

Bergin, A. and M. Haward. 2000. *Australia and International Fisheries Management Issues – Future Directions*. Canberra: Fisheries Resources Research Fund.

——. 1999. 'Australia's New Oceans Policy'. *International Journal of Marine and Coastal Law* 14, 3: 387–98.

——. 1995. 'Australia's Approach to High Seas Fishing'. *International Journal of Ocean and Coastal Law* 10, 3 (August 1995): 349–67.

——. 1994. 'The Southern Bluefin Tuna Fishery; Recent Developments in International Management'. *Marine Policy* 18, 3: 289–309.

Bernstein, S. and B. Cashore. 2001. 'The International-Domestic Nexus: The Effects of International Trade and Environmental Politics on the Canadian Forest Sector'. In *Canadian Forest Policy: Adapting to Change*, edited by M. Howlett, pp. 65–93. Toronto: University of Toronto Press.

Beyers, J. and L. A. Sandberg. 1998. 'Canadian Federal Forest Policy: Present Initiatives and Historical Constraints'. In *Sustainability the Challenge: People, Power, and the Environment*, edited by L. Sandberg and S. Sverker, pp. 99–107. Montreal: Black Rose Books.

Bickerton, G. and J. Stinson. 2008. 'Challenges Facing the Canadian Labour Movement in the Context of Globalisation, Unemployment and the Casualisation of Labour'. In *Labour and the Challenges of Globalisation: What Prospects for Transnational Solidarity?* edited by Andreas Bieler, Ingemar Lindberg and Devan Pillay. London: Pluto Press.

Bills, D. 2001. 'Comment: The UK Government and Certification'. *International Forestry Review* 3, 4 (December): 1–5.

Birnie, P. W. and A. Boyle. 1995. *Basic Documents on International Law and the Environment*. Oxford: Clarendon Press.

Boehm, F. 2007. 'Regulatory Capture Revisited – Lessons from Economics of Corruption'. Research Centre in Political Economy, Universidad Externado de Colombia, Colombia. Online at http://www.icgg.org/downloads/Boehm%20-%20 Regulatory%20Capture%20Revisited.pdf, accessed February 2010.

Borg, J. 2005. 'Added Value in Seafood and Fish Farming'. Speech to the Conference of Fisheries Ministers, Iceland, 8 September.

Brewer, C. 1999. 'Certifying the Woodlot of Rod Blake: Here's How it Unfolded'. *Ecoforestry* 14, 3 (Fall): 26–31.

British North America Act, 1867. Online at http://www.solon.org/Constitutions/ Canada/English/ca_1867.html, accessed November 2008.

Brown, B. 2004. 'Forestry: Timber Communities Australia'. Canberra, ACT: Hansard, pp. 19916–19. Online at http://www.aph.gov.au/HANSARD/senate/dailys/ds110204. pdf, accessed March 2008.

Brown, D. 2005. *Salmon Wars: The Battle for the West Coast Salmon Fishery*. Madeira Park, British Columbia: Harbour Publishing.

Brueckner, M. 2007. 'The Western Australian Regional Forest Agreement: Economic Rationalism and the Normalisation of Political Closure'. *Australian Journal of Public Administration* 66, 2: 148–58.

Brueckner, M. and P. Horwitz. 2005. 'The Use of Science in Environmental Policy: a Case Study of the Regional Forest Agreement Process in Western Australia'. *Sustainability: Science, Practice & Policy* 1, 2 (Fall): 14–24.

Bun, Y. and I. Bewang. 2006. 'Papua New Guinea Case Study'. In *Confronting Sustainability: Forest Certification in Developing and Transitioning Countries*, edited by B. Cashore, F. Gale, E. Meidinger and D. Newsom. New Haven, CT: Faculty of Forestry and Environmental Studies Publication Series, Yale University.

Bureau of Rural Sciences. 2007. *Australia's Forests at a Glance*. Canberra, ACT: Department of Agriculture, Fisheries and Forestry. Online at http://affashop.gov. au/PdfFiles/forest_at_a_glance_reduced.pdf, visited November 2007.

Burton, B. 1997. 'Wise Guys Down Under: PR's Eco-Front Moves on Australia'. *PR Watch Newsletter* 4, 4. Online at http://www.prwatch.org/prwissues/1997Q4/wise. html, accessed March 2008.

Cadman, T. 2009. 'Quality, Legitimacy and Global Governance: A Comparative Analysis of Four Forest Institutions'. PhD Dissertation, School of Government, University of Tasmania, Australia.

——. 2003. 'Progress of the Forest Stewardship Council in Australia'. Online http://www.aela.org.au/conference/Conference%20Proceedings/1_31_ForestryCertificat ionPanel/1_TimCadman_FSC.pdf, accessed April 2008.

Cairns, R. 1992. 'Natural Resources and Canadian Federalism: Decentralization, Recurring Conflict, and Resolution'. *Publius: The Journal of Federalism* 22 (Winter): 55–70.

Canadian Intergovernmental Conference Secretariat. 1999. 'Fisheries Ministers Agree in Principle on a Framework for Cooperation'. 12 April. Online at http://www.scics.gc.ca/cinfo99/ 83064409_e.html.

Carnoy, M. 1984. *The State and Political Theory*. Princeton, NJ: Princeton University Press.

Cartwright, J. 2003. 'Environmental groups, Ontario's Lands for Life process and the Forest Accord'. *Environmental Politics* 12, 2 (Summer): 115–32.

Cashore, B., F. Gale, E. Meidinger and D. Newsom. 2006. 'Forest Certification in Developing and Transitioning Countries: Part of a Sustainable Future?' *Environment* 48, 9 (November 2006), pp. 6–25.

—— (eds). 2006. *Confronting Sustainability: Forest Certification in Developing Countries and Countries in Transition*. New Haven, CT: Yale School of Forestry & Environmental Studies Publication Series.

Cashore, B., G. Auld and D. Newsom. 2004. *Governing Through Markets: Forest Certification and the Emergence of Non-State Authority*. New Haven, CT: Yale University Press.

Cashore, B., G. Auld, J. Lawson and D. Newsom. 2008. 'The Future of Non-State Authority in Canadian Staples Industries: Assessing the Emergence of Forest Certification'. In *Canada's Resource Economy: The Past, Present, and Future of the Canadian Staples Industries*, edited by M. Howlett and K. Brownsey, pp. 209–29. Toronto: Edmond Montgomery Publications.

Cashore, B. and J. Lawson. 2003. 'Private Policy Networks and Sustainable Forestry Policy: Comparing Forest Certification Experiences in the US Northeast and the Canadian Maritimes'. Yale Program on Forest Certification, Yale School of Forestry and Environmental Studies, March 16.

CCFM (Canadian Council of Forest Ministers). 2008. 'About the CCFM'. Online at http://www.ccmf.org/main/about_e.php, accessed November 2008.

——. 1992. *Sustainable Forests: A Canadian Commitment*. Ottawa: CCFM.

CCAMLR (Commission for the Conservation of Antarctic Marine Living Resources) 2008. Report of the Working Group on Fish Stock Assessment, Hobart, 13–24 October.

CCWA (Conservation Council of Western Australia). 2010. 'Marine Stewardship Council Credibility is on the Line over Certification of Rock Lobsters'. NumbatNews, 15 January. Online at http://nevillenumbat.wordpress.com/2010/01/15/marine-stewardship-council-credibility-is-on-the-line-over-rock-lobsters/, accessed October 2010.

CFA (Canadian Forestry Association). 2009a. 'About the CFA'. Online at http://www.canadianforestry.com/html/about_cfa/about_cfa_e.html, accessed February 2009.

——. 2008. 'CFA Timeline'. Online at http://www.canadianforestry.com/html/about_cfa/cfa_timeline_e.html, accessed November 2008.

CFMEU (Construction Forestry Mining Energy Union). 2008. 'History in Brief'. Online at http://www.cfmeu-forestdivision.com.au/history.html, accessed March 2008.

Chaffee, C., D. Leadbitter and A. Aalders. 2003. 'Seafood Evaluation, Certification and Consumer Information'. In *Ecolabelling in Fisheries: What is it all About?* edited by B. Phillips, T. Ward and C. Chaffee, 4–13. Oxford: Blackwell Science.

Charles, A. T. 2005. Personal communication. 10 November 2005.

——. 2001. *Sustainable Fishery Systems*. Oxford: Blackwell Science.

——. 1995. 'The Atlantic Canadian Groundfishery: Roots of a Collapse'. *Dalhousie Law Journal* 18: 65–83.

——. 1992. 'Fishery Conflicts: A Unified Framework'. *Marine Policy*, 16: 379.

Chimni, B. 1986. *International Commodity Agreements: A Legal Study*. Beckenham, Kent: Croom Helm.

CLA (Country Land and Business Association). 2009. 'West Midlands Regional News'. Online at www.cla.org.uk/In_Your_Area/West_Midlands/Regional_News_Archive/CLA/CLA/6219.htm, accessed September 2009.

Clancy, P. 1998. 'Some Political Aspects of Certification for Sustainable Forest Management in Canada'. In *The Challenge of Sustainability: Interdisciplinary and Cross-Cultural Perspectives on Environmentalism*, edited by Anders Sandberg and Sverker Sorlin. Montreal: Black Rose Books.

Clarke, J. 2001. 'Forest Industry is Lost in the Glare of High Tech'. *Logging and Sawmilling Journal* (Feb).

CMFN (Canadian Model Forest Network). 2008. 'Canadian Model Forests'. Online at http://www.modelforest.net/cmfn/en/forests/, accessed November 2008.

Coast Forest Products Association. 2008. *2007 Annual Report*. Vancouver, BC: Coast Forest Products Association. Online at www.coastforest.org/media_pdf/2007ar_website.pdf, accessed March 2009.

Cohn, T. 2009. *Global Political Economy: Theory and Practice*, 5th edn. New York: Longman.

Colchester, M. 1990. 'The International Tropical Timber Organization: Kill or Cure for the Rainforests'. *The Ecologist* 20, 5 (November): 38–45.

Confederation of Paper Industries. 2009. 'Forest Certification: Fact Sheet'. August. Online at http://www.paper.org.uk/information/factsheets /forest_certification.pdf, accessed September 2009.

Constance, D. H. and A. Bonnano. 2000. 'Regulating the Global Fisheries: The World Wildlife Fund, Unilever, and the Marine Stewardship Council'. *Agriculture and Human Values* 17: 125–39.

ContractJournal.Com. 2004. 'PEFC Slams Government's Timber Buying Guidelines'. 14 December. Online at http://www.contractjournal.com/Articles/2004/12/14/44386/pefc-slams-governments-timber-buying-guidelines.html, accessed September 2009.

CPET (Central Point of Expertise on Timber). 2010. 'Announcement: Sustainability Definition now Includes Social Criteria'. Online at http://www.cpet.org.uk/activities-and-news/news%20stories/announcement-sustainability-definition-now-includes-social-criteria, accessed April 2010.

——. 2009. 'Current Reviews and Consultations'. Online at http://www.proforest.net/cpet/review-comments-1/current-reviews-and-consultations/, accessed September 2009.

——. 2004. 'Assessment of Forest Certification Schemes'. UK Government Timber Procurement Policy. CPET Phase 1 Final Report, November. Online at http://www.proforest.net/cpet/files/Phase%201%20Final%20Report%20-%202004%20archive.pdf/view, accessed September 2009.

Crawford, D. B. 1965. 'A Woodland Owners' Association'. *Unasylva* 79, 19 (Issue 4).

Cullen, R. 1992. 'Constitutional Federalism and Natural Resources'. In *Comparative Political Studies: Australia and Canada*, edited by M. Alexander and B. Galligan, pp. 119–32. Melbourne: Longman.

Cummine, A. and G. Wettenhall. 2002. 'Introduction of Forest Stewardship Council is Only Just Moving'. *Australian Forest Grower* (Winter): 9.

Cummins, A. 2004. 'The Marine Stewardship Council: A Multi-Stakeholder Approach to Sustainable Fishing'. *Corporate Social Responsibility and Environmental Management*, 11, 2: 85–94.

DAFF (Department of Agriculture, Fisheries and Forestry). 2010a. 'What is a Forest?' Online at http://www.daff.gov.au/brs/forest-veg/nfi/forest-info/what-is, accessed March 2010.

——. 2010b. 'Plantations for Australia: The 2020 Vision'. Online at http://www.daff.gov.au/forestry/plantation-farm-forestry/2020, accessed March 2010.

——. 2007. 'RFA History'. Online at http://www.daff.gov.au/rfa/about/history, accessed December 2007.

——. 2000. 'Forest and Wood Futures: An Action Agenda to Pursue the Vision for Australia's Forest and Wood Products Industry'. Online at http://www.daff.gov.au/_data/assets/pdf_file/0007/37861/action_agenda.pdf, accessed March 2008.

Dale, S. 1996. *McLuhan's Children: The Greenpeace Message and the Media*. Toronto: Between the Lines Press, 1996.

Daly, H. 1973. *Toward a Steady-State Economy*. New York, NY: W. H. Freeman.

Daly, H. and J. Cobb. 1994. *For the Common Good: Redirecting the Economy toward Community, the Environment, and a Sustainable Future*. Boston, MA: Beacon Press.

Dargavel, J. 1998. 'Politics, Policy and Process in the Forests'. *Australian Journal of Environmental Management* 5 (March): pp. 25–30.

——. 1995. *Fashioning Australia's Forests*. Melbourne: Oxford University Press.

Dauvergne, P. 2008. *The Shadows of Consumption: Consequence for the Global Environment*. Cambridge, MA: MIT Press.

Dauvergne, P. and J. Clapp. 2005. *Paths to a Green World: The Political Economy of the Global Environment*. Boston, MA: MIT Press.

Daw, T. and T. Gray. 2005. 'Fisheries Science and Sustainability in International Policy: A Study of Failure in the European Union's Common Fisheries Policy', *Marine Policy* 29: 189–97.

DeMarco, J. 2005. 'Ideas for a More Effective Environmental Movement in Canada: A Discussion Paper Prepared for The J. W. McConnell Family Foundation'. 8 June. Online at http://www.cen-rce.org/eng/publications/DeMarco%20Discussion%20Paper% 20June8%202005.pdf, accessed January 2009.

Deere, C. 1999. *Eco-Labeling and Sustainable Fisheries*. Washington: International Union for the Conservation of Nature and FAO.

De Freitas, A. 2008a. 'Business Report'. Report by the FSC Executive Director to the FSC General Assembly 2008, Cape Town, South Africa. Online at http://www.fsc.org/fileadmin/web-data/public/document_center/GA_2008 _English/Presentations/Freitas_ED_Business_Report_GA_2008.pdf, accessed October 2009.

——. 2008b. 'Update to the Membership'. Report by the FSC Executive Director to the FSC General Assembly 2008, Cape Town, South Africa. Online at http://www.fsc.org/fileadmin/web-data/public/document_center/GA_2008_English/Presentations/Forest_Conference/Freitas_Update_to_the_membership_2008-11-04.pdf, accessed October 2009.

Descarries, R. 2001. 'CPPA becomes the Forest Products Association of Canada (FPAC)'. *The Forestry Chronicle* 77, 3 (May/June): 410.

David Suzuki Foundation. 2010. 'SeaChoice: Canada's Sustainable Seafood Program'. Online at www.davidsuzuki.org/issues/oceans/science/sustainable-fisheries-and-aquaculture/seachoice-canadas-sustainable-seafood-program/index.php.

DEFRA (Department of Environment, Food and Rural Affairs). 2009. *UK Sea Fisheries Statistics 2008*. London: Marine and Fisheries Agency DEFRA.

——. 2008. *Securing the Benefits: The Joint UK Response to the Prime Minister's Strategy Unit Net Benefits Report on the Future of the Fishing Industry in the UK*. London: DEFRA.

——. 2007. *Fisheries 2027: A Long Term Vision for Sustainable Fisheries*. London: DEFRA.

DEWHA (Department of Environment, Water, Heritage and the Arts). 2008. 'Fisheries and the Environment'. Online at http://www.environment.gov.au/coasts/fisheries/index.html, accessed 27 November 2008.

——. 2007. *Guidelines for the Ecologically Sustainable Management of Fisheries – 2007* Sustainable Fisheries Section, Department of Environment and Water, Canberra, ACT: 6–8.

DEWR (Commonwealth Department of Environment and Water Resources). 2007. 'EPBC Act List of Threatened Species'. Canberra, ACT: Commonwealth Government. Online at http://www.environment.gov.au/cgi-bin/sprat/public/publicthreatened list.pl?wanted=fauna#MAMMALS_ENDANGERED, visited November 2007.

——. 2006. *State of the Environment Report 2006*. Canberra, ACT: Government of Australia. Online at www.environment.gov.au/soe/2006/index.html, visited December 2007.

DFAIT (Department of Foreign Affairs and International Trade, Government of Canada). 2006. 'The US-Canada Softwood Lumber Agreement'. October 17. Online at http://www.dfait-maeci.gc.ca/eicb/softwood/SLA-backgrounder-en.asp?format= print, accessed November 2008.

DFID (Department for International Development). 2009. 'Project Record: Responsible Forestry Programme'. Online at http://www.research4development. info/SearchResearchDatabase.asp?ProjectID=1178, accessed September 2009.

DFO (Department of Fisheries and Oceans Canada). 2008a. 'Canada's First Marine Stewardship Council Certified Fishery'. Press Release, 19 August 2008, Vancouver, DFO.

——. 2008b. *Canadian Fisheries Statistics 2006*. Ottawa: Fisheries and Oceans Canada.

——. 2007. 'Canada's New Government Announces New Activities to Protect the Health of the Oceans'. Press Release, 9 October, Vancouver, DFO.

——. 2005. *Our Waters, Our Future: 2005–2010 Strategic Plan*. Online at www.dfo-mpo. gc.ca/dfo-mpo/plan-eng.htm#5a3, accessed 21 October 2010.

——. 1993. 'Charting a New Course: Towards the Fishery of the Future: Report of the Task Force on Incomes and Adjustment in the Atlantic Fishery'. Ottawa, ONT: Task Force on Incomes and Adjustment in the Atlantic Fishery, Communications Directorate, Department of Fisheries and Oceans.

Dicken, P. 2003. *Global Shift: Reshaping the Global Economic Map in the 21st Century*. 4th edn. London: Sage.

Dimitrov, R. 2010. 'Inside Copenhagen: The State of Climate Governance'. *Global Environmental Politics* 10, 2 (May): 18–24.

Djelic, M-L. and K. Sahlin-Andersson (eds). 2006. *Transnational Governance: Institutional Dynamics of Regulation*. Cambridge: Cambridge University Press.

Doelle, M. 2010. 'The Legacy of the Climate Talks in Copenhagen: Hopenhagen or Brokenhagen? *The Carbon & Climate Law Review* 4, 1: 86–100.

Dossa, S. 2007. 'Slicing Up "Development": Colonialism, Political Theory, Ethics'. *Third World Quarterly* 28, 5: 887–99.

Doyle, T. 2000. *Green Power: The Environment Movement in Australia*. Sydney: UNSW Press.

Doyle, T. and A. Kellow. 1995. *Environmental Politics and Policy Making in Australia*. South Melbourne: Macmillan Education Australia.

DPC (Department of Premier and Cabinet). 2007. *Sustainability Indicators for Tasmanian Forests 2001–2006*. Prepared by the Tasmanian and Australian Governments for the 2007 Ten Year Review of the Tasmanian Regional Forest Agreement. Online at http://www.dpac.tas.gov.au/divisions/policy/rfa/documents/2007%20INDICATORS%20REPORT_FINAL%2024%20May%202007.pdf, accessed December 2007.

DPIE (Department of Primary Industries and Energy). 1996. *Proceedings of the International Conference on Certification and Labelling of Products from Sustainably Managed Forests, 26–31 May 1996, Brisbane, Australia*. Canberra, ACT: Commonwealth of Australia.

DPIW (Department of Primary Industries, Parks, Water and Environment). 2010. 'Native Plants and Animals: Tasmanian Tiger'. Hobart, Tasmania. Tasmanian Government. Online at http://www.dpiw.tas.gov.au/inter.nsf/WebPages/BHAN-54F4ED?open, accessed October 2010.

Drushka, K. 1999. *In the Bight: The BC Forest Industry Today*. Madeira Park, BC: Harbour Publishing.

Earth Trends. 2009. 'Forests, Grasslands and Drylands – Trade in Forest Products: Exports'. Online at http://earthtrends.wri.org/searchable_db/results.php?years=all&variable_ID=857&theme=9&country_ID=all&country_classification_ID=all, accessed December 2009.

Eba'a Atyi, R. 2006. 'Gabon Case Study'. In *Confronting Sustainability: Forest Certification in Developing and Transitioning Countries*, 443–76, edited by B. Cashore, F. Gale, E. Meidinger and D. Newsom. New Haven: Yale School of Forestry & Environmental Studies Press.

Economou, N. 1996. 'Australian Environmental Policy Making in Transition: The Rise and Fall of the Resource Assessment Commission'. *Australian Journal of Public Administration* 55, 1 (March): 12–22.

Eden, S. 2009. 'The Work of Environmental Governance Networks: Traceability, Credibility and Certification by the Forest Stewardship Council'. Online at www.hull.ac.uk/geog/research/credibility/pdf/pubs-3.pdf, accessed September 2009.

Elliott, C. 1999. *Forest Certification: Analysis from a Policy Network Perspective*. PhD thesis, Départment de Génie Rural, École Polytechnique Fédérale de Lausanne, Lausanne, Switzerland.

Elliott, C. and R. Schlaepfer. 2001. 'The Advocacy Coalition Framework: Application to the Policy Process for the Development of Forest Certification in Sweden'. *Journal of European Public Policy* 8, 4 (August): 642–61.

Environment Canada. 2008. 'All Endangered, Threatened, and Special Concern Species with a Map'. Online at http://www.sis.ec.gc.ca/ec_species/ec_species_e.phtml, accessed November 2008.

ESDWG (Ecologically Sustainable Development Working Groups). 1991. *Ecologically Sustainable Development Working Groups: Final Report – Executive Summaries*. Canberra, ACT: Australian Government Publishing Service (November).

Esty, D. 2002. 'The World Trade Organization's Legitimacy Crisis'. *World Trade Review* 1, 1: 7–22.

Europa. 2006. 'Common Fisheries Policy'. Online at http://ec.europa.eu/fisheries/cfp_en.htm, accessed 19 May 2008.

European Commission. 2002. Communication from the Commission on Reform of the Common Fisheries Policy-Roadmap. Luxembourg: European Commission.

Exel, M. 2006. Personal communication, October 2006.

Exel, M. 1994. 'Australian Fisheries Management – Resource Allocation and Traditional Rights' *Outlook 94 Vol. 2* ABARE Canberra: 231–7.

FAO (United Nations Food and Agriculture Organization). 2010. 'Background: Code of Conduct for Sustainable Fisheries'. Rome: FAO. Online at http://www.fao.org/docrep/x2410e/x2410e02.htm, accessed April 2010.

——. 2009a. *Forest Yearbook 2007*. Rome: FAO. Online at ftp://ftp.fao.org/docrep/fao/012/i0750m/i0750m00.pdf, accessed December 2009.

——. 2009b. *State of World Fisheries and Aquaculture (SOFIA) 2008*. Rome: FAO.

——. 2008. 'Fishery and Aquaculture Country Profiles: United Kingdom'. Online at www.fao.org/fishery/countrysector/FI-CP_GB/en, accessed 21 October 2010.

——. 2007. *State of World Fisheries and Aquaculture (SOFIA) 2006*. Rome: FAO.

——. 2006. *Global Forest Resources Assessment 2005*. Rome: FAO. Online at www.fao.org/forestry/fra2005, accessed December 2009.

——. 2004. 'Outcome of Expert Consultation on Eco-Labeling'. Committee on Fisheries, Sub-Committee on Fish Trade, Ninth Session. Rome: FAO.

——. 2003. *Expert Consultation on the Development of International Guidelines for Eco-Labeling of Fish and Fishery Products from Marine Capture Fisheries*. Rome: FAO.

——. 2000. *State of World Fisheries and Aquaculture (SOFIA)*. Rome: FAO.

Ferguson, M. 2006. 'Speech to the House of Representatives, Forty-First Parliament, First Session, Seventh Period'. House of Representatives. Official Hansard, No. 10, Thursday, 10 August, 142–7. Online at http://www.aph.gov.au/HANSARD/reps/dailys/dr100806.pdf, accessed March 2010.

FERN. 2004. *Footprints in the Forest: Current Practice and Future Challenges in Forest Certification*. Moreton-on-Marsh, Glocestershire: FERN. Online at http://www.fern.org/pubs/reports/footprints.pdf, accessed March 2008.

FERN, Greenpeace, FoE-UK and WWF-UK. 2006. 'UK Government Decision Undermines its Own Timber Policy'. London, 20 December. Online at http://www.foe.co.uk/resource/press_releases/uk_government_decision_und_21122006.html, accessed September 2009.

FFIC (Forests and Forest Industry Council of Tasmania). 2010. 'About FFIC'. Online at http://www.ffic.com.au/index.php?option=com_content&view=article&id=139&Itemid=79, accessed March 2010.

Fischer, C., F. Aguilar, P. Jawahar, and R. Sedjo. 2005. *Forest Certification: Towards Common Standards?* Washington, DC: Resources for the Future. Online at http://www.rff.org/documents/RFF-DP-05-10.pdf, accessed March 2010.

FoE-Australia (Friends of the Earth-Australia). 2008a. 'About us'. Online: http://www.foe.org.au/about-us, accessed March 2008.

——. 2008b. 'FOEA Indigenous Land and Rights Policy'. Online: http://www.foe.org.au/indigenous/policy-position/foea-indigenous-land-and-rights-policy, accessed March 2008.

——. 2006. *Annual Report 2005–2006*. Melbourne: FoE-Australia. http://www.foe.org.au/resources/organisational/annualreports/FoEA%20annual%20report%202006.pdf, accessed March 2008.

FoE-Melbourne (Friends of the Earth-Melbourne). 2008. 'Barma-Millewah Collective: The Issues'. Online at http://www.melbourne.foe.org.au/campaigns/barmah/issues.htm, accessed March 2008.

FoE-UK (Friends of the Earth UK). 2008a. 'Friends of the Earth Limited Report and Accounts for the Year Ended 31 May 2008'. London: Friends of the Earth. Online at www.foe.co.uk/resource/reports/limited_accounts_08.pdf, accessed September 2009.

——. 2008b. 'Shaping Our Future: Friends of the Earth's Annual Review 2007–08. London: Friends of the Earth'. Online at www.foe.co.uk/resource/reports/annual_review_0708.pdf, accessed September 2009.

——. 1993a. 'Notes on the Meeting with Overseas Development Administration, 18 January 1993'. London: Friends of the Earth. Mimeo.

——. 1993b. Annual Review 1993. London: Friends of the Earth: mimeo.

——. Circa 1990a. 'Special Briefing: The International Tropical Timber Organisation – A Call for Urgent Reforms'. London: Friends of the Earth. Mimeo.

——. Circa 1990b. 'Press Release: Governments Fail to Agree on Conservation Initiative for Rainforests'. London: Friends of the Earth. Mimeo.

——. Circa 1988. 'Tropical Forest Conservation and the Timber Trade: A Call to Action'. London: Friends of the Earth: Mimeo.

FOCS (Friends of Clayoquot Sound). 2009a. 'Who we are'. Online at http://www.focs.ca/about/index.asp, accessed January 2009.

——. 2009b. 'Our History'. Online at http://www.focs.ca/about/history.asp, accessed January 2009.

Foley, P. 2010. 'The Global Political Economy of Public Support for Private Regulation: The Impacts of International Eco-certification and Labeling on Canadian Fisheries Management', paper presented at Canadian Political Science Association 2010 Annual Conference, Concordia University, Montreal 1–3 June.

Forest Practices Board, Tasmania. 2000. *Forest Practices Code 2000*. Hobart, Tasmania: Forest Practices Board. Online at http://www.fpa.tas.gov.au/fileadmin/user_upload/PDFs/Admin/FPC2000_Complete.pdf, accessed December 2007.

Forestry Chronicle. 1999. 'Canadian Forestry Association designates May 2 to 8 National Forest Week in 1999'. Volume 75, No. 3 (May/June 1999): 333.

Forestry Commission. 2010. 'FRA 2010 – Country Report United Kingdom of Great Britain and Northern Ireland'. Online at http://www.forestry.gov.uk/pdf/Table1UKReport.pdf/$FILE/Table1UKReport.pdf, accessed March 2010.

——. 2009. 'Forestry Statistics 2009'. Online at http://www.forestry.gov.uk/website/forstats2009.nsf/0/733EDABDF0EA86B780257360003953A3, accessed September 2009.

Foster, E. and M. Haward. 2003. 'Integrated Management Councils: A Conceptual Model for Ocean Policy Conflict Management in Australia'. *Ocean and Coastal Management* 46: 547–63.

Fowler, P. 1998. 'Learning from the Marine Stewardship Council: A Business-NGO Partnership for Sustainable Marine Fisheries'. *Greener Management International* (Winter): 77–90.

FPAC (Forest Products Association of Canada). 2009. 'Member Companies'. Online at http://www.fpac.ca/en/who_we_are/member_companies/index.php, accessed February 2009.

——. 2007. 'FPAC Market Acceptance: Customer Briefing Note: Forest Certification'. January. Online at www.fpac.ca, accessed March 2009.

François, D. 1984. 'Lessons for the Future – A Look at Commonwealth/State Relations in Fisheries'. *The Australian Fishing Industry – Today and Tomorrow Seminar*. Launceston: Australian Maritime College.

FRDC (Fisheries Research and Development Corporation). 2008. 'SA Fishery Credentials Recognised'. *Fish* 16, 3: 20.

——. 2006. 'Supermarkets Rule in UK' *Fisheries R & D News* 14, 3: 37–38.

FSC-IC (Forest Stewardship Council). 2010a. 'Some History'. Online at http://www.fsc.org/history.html, accessed February 2010.

——. 2009a. 'A Stronger Scheme'. *FSC News & Notes* 7, 11 (December) 10. Online at http://www.fsc.org/fileadmin/web-data/public/document_center/publications/newsletter/newsletter_2009/FSC-PUB-20-07-11-2009-12-23.pdf, accessed April 2010.

——. 2009b. 'Global FSC Certificates: Type and Distribution'. September. Online at http://www.fsc.org/fileadmin/web-data/public/document_center/powerpoints_graphs/facts_figures/09-09-15_Global_FSC_certificates_-_type_and_distribution_-_FINAL.pdf, accessed October 2009.

——. 2009c. 'FSC Approved Forest Stewardship Standards'. September. Online at http://www.fsc.org/fileadmin/web-data/public/document_center/national_FSC_standards/FSC_Approved_FSS_2009-09-24.pdf, accessed October 2009.

——. 2009d. 'FSC Controlled Wood Risk Assessments by FSC accredited National Initiatives, National and Regional Offices'. February. Online at http://www.fsc.org/fileadmin/web-data/public/document_center/international_FSC_policies/procedures/FSC_PRO_60_002_V_2_0_EN_NI_Controlled_Wood_Risk_Assessments_.pdf, accessed October 2009.

——. 2009e. 'FSC Advice Note: Implementation of FSC Controlled Wood requirements in FSC STD-40-005 V2-1 and FSC-STD-20-011 V1-1'. Online at http://www.fsccanada.org/docs/fsc-adv-40-016%20v2-0%20en%20implementation%20of%20fsccontrolledwood.pdf, accessed October 2009.

——. 2009f. 'Overview of the FSC Principles and Criteria'. Online at http://www.fsc.org/pc.html, accessed October 2009.

——. 2008. 'FSC Governance Review Process Final Proposals'. October. Online at http://www.fsc.org/fileadmin/web-data/public/document_center/institutional_documents/FSC_Governance_Paper_2008.pdf, accessed October 2009.

——. 2007a. 'FSC Controlled Wood: A Guide for FSC Chain of Custody Certified Companies'. Online at http://www.fsc.org/fileadmin/web-data/public/document_center/international_FSC_policies/brochures/Controlled_Wood_friendly_user_guide-EN.pdf, accessed October 2009.

——. 2007b. 'FSC Standard for Chain of Custody Certification'. November. Online at http://www.fsc.org/fileadmin/web-data/public/document_center/international_FSC_policies/standards/FSC_STD_40_004_V2_0_EN_Standard_for_CoC_Certification_2008_01.pdf, accessed October 2009.

——. 2006. 'FSC Controlled Wood Standard for Forest Management Enterprises'. November. Online at http://www.fsc.org/fileadmin/web-data/public/document_center/international_FSC_policies/standards/FSC_STD_30_010_V2_0_EN_Controlled_Wood_standard_for_FM_enterprises.pdf, accessed October 2009.

——. 2004a. 'FSC Standard for Forest Management Enterprises Supplying Non-FSC Certified Controlled Wood'. FSC-STD-30-010 (Version 1-0)'. Online at http://fsccontrolledwood.org/Documents/FSC_STD_30_010_V1_0_EN_controlled_wood_for_forest_managers.pdf, accessed October 2009.

——. 2004b. 'Looking to the Future …: Ten Years of FSC 1993–2003'. Online at http://www.fsc.org/fileadmin/web-data/public/document_center/publications/annual_reports/10_Years_of_FSC_final.pdf, accessed October 2009.

——. 2002. 'FSC Principles and Criteria for Forest Stewardship'. Online at http://www.fsc.org/fileadmin/web-data/public/document_center/international_FSC_policies/standards/FSC_STD_01_001_V4_0_EN_FSC_Principles_and_Criteria.pdf, accessed October 2009.

FSC-UK. 2008. 'Forest Stewardship Council UK Working Group Annual Report 2007–2008'. Llanidloes, Powys, Wales: FSC Working Group. Online at http://www.fsc-uk.org/wp-content/plugins/downloads-manager/upload/FSC_AnnualReport_WEB.pdf, accessed September 2009.

——. 2007. 'Forest Stewardship Council UK Working Group Annual Report 2006–2007'. Llanidloes, Powys, Wales: FSC Working Group. Online at http://www.fsc-uk.org/wp-content/plugins/downloads-manager/upload/FSC_AnnuaL_Report_06_07.pdf, accessed September 2009.

——. 2006. 'Forest Stewardship Council UK Working Group Annual Report 2005–2006'. Llanidloes, Powys, Wales: FSC Working Group. Online at http://d515375.u48.pipeten.co.uk/wp-content/plugins/downloads-manager/upload/5840_FSC%20AnnRpt%202005-2006.pdf, accessed September 2009.

——. 2005. 'Forestry Commission Re-Certified'. *FSC UK News* 8 (March): 4.

——. 2001a. 'DETR/Meacher Announcement'. *FSC UK News* 10 (May): 2.

——. 2001b. 'Certification Withdrawn from Teak Plantations'. *FSC UK News* 11 (November): 1.

FSC-Watch. 2008. 'FSC Dumps Asia Pulp and Paper – but who was to Blame?' Online at http://www.fsc-watch.org/archives/2008/01/10/FSC_dumps_Asia_Pulp_, accessed October 2009.

Gale, F. 2004. 'The Consultation Dilemma in Private Regulatory Regimes'. *Journal of Environmental Policy & Planning* 6, 1 (March 2004): pp. 57–84.

——. 2003. 'Discourse and Southeast Asian Deforestation: A Case Study of the International Tropical Timber Organization'. In *The Political Ecology of Tropical Forests in Southeast Asia: Historical Perspectives*, edited by L. Tuck-Po, W. de Jong and K. Abe. Kyoto, Japan: Kyoto University Press.

——. 2002. '*Caveat Certificatum*: The Case of Forest Certification'. In *Confronting Consumption*, edited by Ken Conca, Mike Maniates and Thomas Princen, pp. 275–99. New York: MIT Press.

——. 1998a. 'Ecosystem-Bound: How International Trade Agreements Constrain the Adoption of an Ecosystem-Based Approach to Forest Management'. In *The Wealth of Forests*, edited by Chris Tollefson. Vancouver, BC: University of British Columbia Press.

——. 1998. *The Tropical Timber Trade Regime*. Basingstoke, Hampshire: Macmillan Press.

——. 1996. 'The Mysterious Case of the Disappearing Environmentalists: The International Tropical Timber Organization'. *Capitalism Nature Socialism* 7, 3: 103–17.

Gale, F. and C. Burda. 1998. 'The Pitfalls and Potential of Eco-Certification as a Market Incentive for Sustainable Forest Management'. In *The Wealth of Forests*, edited by Chris Tollefson. Vancouver, BC: University of British Columbia Press.

Gale, F. and T. Cadman. Forthcoming. 'Whose Norms Prevail? Multilateralism, Vested Interests and Transnational Forest Politics'.

Gale, F. and M. Haward. 2004. 'Public Accountability in Private Regulation: Contrasting Models of the Forest Stewardship Council (FSC) and Marine Stewardship Council (MSC)'. Paper presented at the 2004 Australasian Political Studies Association Conference (29 September–1 October 2004), Adelaide, South Australia, Australia.

Gale, M. 2006. 'UKWAS Launch Presentations'. Online at www.ukwas.org.uk/assets/documents/UKWASlaunchpresentations.doc, accessed September 2009.

Gale, R. and F. Gale. 2006. 'Accounting for social impacts and costs in the forest industry, British Columbia'. *Environmental Impact Assessment Review* 26, 2 (March): pp. 139–55.

Gardner Pinfold Consulting Economists Limited. 2004. *Seafood Eco-Labelling; Trends, Challenges and Opportunities*. Report Prepared for Fisheries and Oceans Canada.

Gereffi, G. 2001. 'Shifting Governance Structures in Global Commodity Chains with Special Reference to the Internet'. *The American Behavioural Scientist* 44, 10 (June): 1616–37.

Global Forest Watch. 2000. *Canada's Forests at a Crossroads: an Assessment in the Year 2000*. Washington, DC: World Resources Institute.

Globe and Mail (Toronto). 2007. 15 November.

Goldsmith, E. 1993. *The Way: An Ecological World-View*. Boston, MA: Shambala Press.

Goodall, S. 2001. 'Update on the UK Woodland Assurance Scheme'. Presentation to the FAO-GTZ-ITTO Seminar on 'Building Confidence among Forest Certification Schemes and Their Supporters'. Rome, Italy, 19–20 February. Online at www.fao. org/docrep/003/X6720E/x6720e19.htm, accessed September 2009.

——. 2000. 'Forest Certification and the UK Woodland Assurance Scheme'. *Quarterly Journal of Forestry* 94, 3: 239–44.

Goodlund, R. 2002. *Eco-Labeling: Opportunities for Progress towards Sustainability*. Washington, Consumer Choice Council.

Gorte, R. and J. Grimmett. 2006. 'Softwood Lumber Imports from Canada: Issues and Events'. Congressional Research Service Report for Congress, Washington, DC. December. Online at http://ncseonline.org/NLE/CRSreports/07Jan/RL33752.pdf, accessed October 2010.

Government of Australia. 1989. *New Directions for Commonwealth Fisheries Management in the 1990s*. Canberra, ACT: Government of Australia.

Government of British Columbia. 2001. 'Framework Agreement'. Online at http:// archive.ilmb.gov.bc.ca/slrp/lrmp/nanaimo/cencoast/docs/framework_agreement. pdf, accessed April 2010.

Government of New South Wales. 1987. *Wilderness Act 1987*. Online at http://www. environment.nsw.gov.au/legislation/WildernessAct1987.htm, accessed March 2010.

Government of Nova Scotia. 2009. 'Province Funds Forestry Development in Southwestern Nova Scotia'. Online at http://www.gov.ns.ca/news/details.asp?id= 20091030002, accessed April 2010.

Government of Tasmania. 1997. *Tasmanian Regional Forest Agreement*. Hobart, Tasmania: Forestry Tasmania.

GPI Atlantic. 2008. 'A New Measure of Progress'. Online at http://www.gpiatlantic. org/ accessed December 2008.

Gramsci, A. 1988. 'Selected Writings'. In *A Gramsci Reader: Selected Writings*, edited by D. Forgacs. London: Lawrence and Wishart.

Gray, T. and J. Hatchard. 2003. 'The 2002 Reform of the Common Fisheries Policy's System of Governance – Rhetoric or Reality?' *Marine Policy* 27: 545–54.

Gray, T. S., M. J. Gray and R. A. Hague. 1999. 'Sandeels, Sandals and Suits: The Strategy of the Environment Movement in Relation to the Fishing Industry'. *Environmental Politics* 8, 3: 119–39.

Greenpeace. 2008. 'Organizational Profile'. Online at http://www.greenpeace.org/ canada/en/about-greenpeace/organizational-profile, accessed January 2009.

——. 2007. *Greenpeace Annual Report 2007*. Online at http://www.greenpeace.org/raw/ content/canada/en/documents-and-links/publications/annual-report-2007-en.pdf, accessed January 2009.

Gregersen, H., A. Contreras-Hermosilla, A. White and L. Phillips. 2004. *Forest Governance in Federal Systems: An Overview of Experiences and Implications for Decentralization*. Jakarta, Indonesia: CIFOR/Forest Trends.

Griss, P. 1993. 'Environmental Groups: Adapting to Changing Times'. National Round Table Review (Spring). Ottawa, Ontario: NRTEE.

Grumbine, R. 1994. 'What is ecosystem management?' *Conservation Biology* 8, 1: 27–38.

Guardian. 2004. 21 February.

Gulbrandsen, L. H. 2010. *Transnational Environmental Governance: The Emergence and Effects of the Certification of Forests and Fisheries*. Cheltenham, UK; Northampton, MA: Edward Elgar.

——. 2009. 'The Emergence and Effectiveness of the Marine Stewardship Council'. *Marine Policy*, 33: 654–60.

——. 2008. 'Enhancing Accountability in Non-State Governance Schemes: Control and Responsiveness'. Paper prepared for the 49th Convention of the International Studies Association, San Francisco, 26–29 March.

——. 2005. 'Mark of Sustainability? Challenges for Forestry and Fisheries Eco-Labeling'. *Environment* 47, 5 (June): 8–23.

Hall, N. and R. Taplin. 2007. 'Revolution or Inch-by-Inch? Campaign Approaches on Climate Change by Environmental Groups'. *Environmentalist* 27: 95–107.

Hall, H. R. and M. Haward. 2001. 'Enhancing Compliance with International Legislation and Agreements Mitigating Seabird Mortality on Longlines'. *Marine Ornithology*, 28: 183–90.

Haley, D. and M. K. Luckert. 1998. 'Tenures as Economic Instruments for Achieving Objectives of Public Forest Policy in British Columbia'. In *The Wealth of Forests*, edited by C. Tollefson, 123–52. Vancouver, BC: University of British Columbia Press.

Haley, D. and H. Nelson. 2007. 'Has the Time Come to Rethink Canada's Crown Tenure Forest Systems?' *The Forestry Chronicle* 83, 5 (September/October): 629–41.

Hamilton, C. and A. Macintosh. 2004. *Taming the Panda: The Relationship between WWF Australia and the Howard Government*. Canberra, ACT: The Australian Institute.

Hammond, H. 1993. 'Wholistic Forest Use'. Slocan Park, BC: Silva Forest Foundation.

Hammond, S. 1995a. 'The Silva Forest Foundation'. *International Journal of Ecoforestry* 11, 4 (Winter): 112.

——. 1995b. 'Silva Forest Foundation conducts Canada's First Wood Certification'. *International Journal of Ecoforestry* 11, 4 (Winter): 128–9.

Hasenclever, A., P. Mayer and V. Rittberger. 1997. *Theories of International Regimes*. Cambridge: Cambridge University Press.

Harris, M. 1998. *Lament for an Ocean: The Collapse of the Atlantic Cod Fishery – A True Crime Story*. Toronto: McClelland & Stewart.

Harrison, A. J. 1991. *The Commonwealth in the Administration of Australian Fisheries: A Sort of Mongrel Socialism*. National Monograph Series 6, Royal Australian Institute of Public Administration. Canberra: RAIPA.

Harter, J-H. 2004. 'Environmental Justice for Whom? Class, New Social Movements, and the Environment: A Case Study of Greenpeace Canada 1971–2000'. *Labour/Le Travail* 54.

Harvey, D. 2005. *A Brief History of Neoliberalism*. Oxford: Oxford University Press.

Haward, M. 2009. 'State, Market and Community: Managing Australian Fisheries'. *Dialogue – the Journal of the Academy of Social Sciences in Australia* 28, 1: 36–45.

——. 2006. 'Australian Aquaculture: Opportunities and Challenges'. In *Aquaculture Law and Policy: Towards Principled Access and Operations*, edited by D. VanderZwaag and G. Chao, 488–503. Abingdon, Oxon: Routledge.

——. 2004. 'IUU Fishing: Contemporary Practice'. In *Oceans Management in the 21st Century: Institutional Frameworks and Responses*, edited by A. G. Oude Elferink and D. R Rothwell, 87–106. Leiden: Martinus Nijhoff Publishers.

——. 2003. 'The Ocean and Marine Realm'. In *Managing Australia's Environment*, edited by S. Dovers and S. Wild River, 35–52. Sydney: Federation Press.

——. 1995. 'The Commonwealth in Australian Fisheries Management: 1955–1995'. *The Australasian Journal of Natural Resources Law and Policy* 2: 313–25.

——. 1989. 'The Australian Offshore Constitutional Settlement'. *Marine Policy* 13, 4: 334–8.

Haward, M., A. Bergin and H. R. Hall. 1998. 'International Legal and Political Bases to the Management of the Incidental Catch of Seabirds'. In *Albatross: Biology and Conservation*, edited by G. Robertson and R. Gales, 255–66. Chipping Norton, NSW: Surrey Beatty and Sons.

Haward, M. and B. W. Davis. 1994. 'Recent Developments in Australian Coastal Zone Management'. In *Coastal Zone Canada 94: 'Cooperation in the Coastal Zone' Conference Proceedings, Vol 1*, edited by P. G. Wells and P. J. Ricketts, 19–29. Dartmouth, NS: Coastal Zone Canada Association, Bedford Institute of Oceanography.

Haward, M., R. Dobell, A. Charles, E. Foster and T. Potts. 2003. 'Fisheries and Oceans Governance in Australia and Canada: From Sectoral Management to Integration'. *Dalhousie Law Journal* 26, 1: 6–45 [published 2005].

Haward, M. and D. VanderZwaag. 1995. 'Implementation of UNCED Agenda 21 Chapter 17 in Australia and Canada: A Comparative Analysis'. *Ocean and Coastal Management* 29, 1–3: 279–95.

Haward, M. and J. Vince. 2008. *Oceans Governance in the Twenty-First Century: Managing the Blue Planet*. Cheltenham: Edward Elgar.

Held, D, A. McGrew, D. Goldblatt and J. Perraton. 1999. *Global Transformations: Politics, Economics, Culture*. Stanford, CA: Stanford University Press.

Herr, R. A. and B. W. Davis. 1982. 'Of Federations and Fishermen: Australia, Canada and UNCLOS III'. In *Theory and Practice in Comparative Studies: Canada, Australia: Papers from the First Conference of the Australian and New Zealand Association of Canadian Studies*, edited by P. Crabb. Sydney: Australian and New Zealand Association of Canadian Studies and Macquarie University.

Hessing, M., M. Howlett and T. Summerville. 2005. *Canadian Natural Resource and Environmental Policy: Political Economy and Public Policy*, 2nd Edn. Vancouver: UBC Press.

Hollander, R. 2004. 'Changing Places? Commonwealth and State Government Performance and Regional Forest Agreements'. Refereed paper presented to the Australasian Political Studies Conference, University of Adelaide, 29 September–1 October.

Hodgins, B. 1992. 'The Temagami Dispute: A Northern Ontario Struggle Toward Co-Management'. Paper presented at the White Pine Symposium: History, Ecology, Policy and Management, Department of Forest Resources, University of Minnesota, St. Paul, Minnesota, 16–18 Sept.

Howes, R. 2005. 'Five Years On – An Update on the Marine Stewardship Council'. *FishBytes* 11, 6: 1–3.

Howlett, M. 2001. 'The Federal Role in Canadian Forest Policy: From Territorial Landowner to International and Intergovernmental Co-Ordinating Aent'. In *Canadian Forest Policy: Adapting to Change*, edited by M. Howlett, pp. 378–415. Toronto: University of Toronto Press.

Howlett, M. and J. Rayner. 2007. 'The National Forest Strategy in Comparative Perspective'. *The Forestry Chronicle* 83, 5 (Sept/Oct): 651–7.

——. 2001. 'The Business and Government Nexus: Principal Elements and Dynamics of the Canadian Forest Policy Regime'. In *Canadian Forest Policy: Adapting to Change*, edited by M. Howlett, pp. 23–62. Toronto: University of Toronto Press.

——. 1995. 'Do Ideas Matter? Policy Network Configuration and Resistance to Policy Change in the Canadian Forest Sector'. Canadian Public Administration 38, 3: 382–410.

HRH Prince Charles. 2004. 'A Speech by HRH The Prince of Wales at the Marine Stewardship Reception Gala Dinner 4 March 2004'.

Huber, J. 2000. 'Towards Industrial Ecology: Sustainable Development as a Concept of Ecological Modernization'. *Journal of Environmental Policy and Planning* 2, 4: 269–85.

Humphreys, D. 2006. *Logjam: Deforestation and the Crisis of Global Governance*. London: Earthscan.

——. 1996. *Forest Politics: The Evolution of International Cooperation*. London: Earthscan.

ICO (International Coffee Organization). 2009. 'Statistical Data'. Online at http://www.ico.org/new_historical.asp, accessed November 2009.

Industry Commission. 1992. *Cost Recovery in Australian Fisheries*, Canberra, AGPS.

IFF (Intergovernmental Forum on Forests). 1999. 'Report of the Intergovernmental Forum on Forests on its Third Session (Geneva 3–14 May 1999)'. Geneva: ECOSOC. Online at http://daccess-dds-ny.un.org/doc/UNDOC/GEN/N00/228/81/PDF/N0022881.pdf?OpenElement, accessed October 2010.

IISD (International Institute for Sustainable Development). 1996. *Global Green Standards: ISO 14000 and Sustainable Development*. Winnipeg: International Institute for Sustainable Development.

IMFN (International Model Forest Network). 2006. 'Model Forest List'. June. Online at http://www.imfn.net/en/ev-41778-201-1-DO_TOPIC.html, accessed November 2008.

IPF (Intergovernmental Panel on Forests). 1995 'Programme of Work of the Intergovernmental Panel on Forests'. First Meeting of the Ad Hoc Intergovernmental Panel on Forests, 11–15 September 1995, E/CN 17./IPF/1995/2. Commission on Sustainable Development, United Nations Economic and Social Council. Online at http://daccessdds.un.org/doc/UNDOC/GEN/N95/239/96/IMG/N9523996.pdf?OpenElement, accessed March 2008.

ITTO (International Tropical Timber Organisation). 1992. *ITTO Guidelines for the Sustainable Management of Natural Tropical Forests*. ITTO Policy Development Series 1. Yokohama, Japan: ITTO. Online at http://www.itto.or.jp/live/Live_Server/147/ps01e.doc, accessed March 2008.

IUCN (International Union for the Conservation of Nature) and WWF (World Wildlife Fund) 1998. *Creating a Sea Change: The WWF / IUCN Marine Policy*. London, UK: World Wildlife Fund and International Union for the Conservation of Nature.

James, N. D. G. 1981. *A History of English Forestry*. Oxford: Basil Blackwell.

JAS-ANZ (Joint Accreditation Systems of Australia and New Zealand). 2008a. 'About Us'. Online at http://www.jas-anz.com.au/index.php?option=com_content&task=blogcategory&id=13&Itemid=1, accessed 27 November.

——. 2008b. 'The Clean Green Australian Southern Rocklobster Certification Program'. Online at http://www.jas-anz.com.au/index.php?option=com_content&task=view&id=51&Itemid=1.

Keenan, R. 2008. 'Approaches to Providing for Multiple Values and Functions from Forests in Australia'. In *The Multifunctional Role of Forests – Policies, Methods and Case Studies*, edited by L. Cesaro, P. Gatto and D. Pettenella, 139–52. Joensu, Finland: European Forest Institute.

Kelemen, R. D. 2002. 'Regulatory Federalism: EU Environmental Regulation in Comparative Perspective'. *Journal of Public Policy* 20, 3: 133–67.

Keohane, R. 1984. *After Hegemony: Cooperation and Discord in the World Political Economy*. Princeton, NJ: Princeton University Press.

Kern, K., I. Kissling, U. Landmann and C. Mauch in collaboration with T. Löffelsend. 2001. 'Policy Convergence and Policy Diffusion by Governmental and Non-Governmental Institutions – An International Comparison of Eco-labeling Systems'. Discussion Paper FS II 01–305. Berlin: Wissenschaftszentrum Berlin für Sozialforschung.

Kirkpatrick, J. 1998. 'Nature Conservation and the Regional Forest Agreement Process'. *Australian Journal of Environmental Management* 5 (March): pp. 31–7.

Kurien, J. 1996. 'A View from the Third World'. *Samudra* 15, July: 22–5.

Lalonde, M. 2007. 'Sierra Club Sets Sights on Pig Farming in Quebec'. *Montreal Gazette*, 5 March.

Lamb, R. in collaboration with Friends of the Earth. 1996. *Promising the Earth*. London: Routledge.

Lancien, D. 2007. *Anciennes et Nouvelles Aristocraties de 1880 a nos jours*. Paris: Fondation Maison des Sciences de l'Homme: pp. 11–27.

Lane, M. 1999. 'Regional Forest Agreements: Resolving Resource Conflicts or Managing Resource Politics?' *Australian Geographical Studies* 37, 2 (July): 142–53.

Larcombe, J. and G. Begg. 2008. *Fishery Status Reports 2007: Status of Fish stocks Managed by the Australian Government*, Canberra Bureau of Rural Sciences.

Lawson, G. and G. Hemery. 2008. 'World Timber Trade and Implementing Sustainable Forest Management in the United Kingdom'. A report of the Woodland Policy Group, Land Use Policy Group. Online at http://www.lupg.org.uk/Default.aspx?page=100, accessed September 2009.

Leadbitter, D. and T. Ward. 2007. 'An Evaluation of Systems for the Integrated Assessment of Capture Fisheries'. *Marine Policy* 31, 4: 458–69.

LEEC (London Environmental Economics Centre). 1993. 'The Economics Linkages Between the International Trade in Tropical Timber and the Sustainable Management of Tropical Forests'. Main Report to the International Tropical Timber Organization. March 19. London: London Environmental Economics Centre and International Institute for Sustainable Development.

Levy, D. and P. Newell. 2005. *The Business of Global Environmental Governance*. Cambridge, MA: MIT Press.

Lindblom, C. 1982. 'The Market as Prison'. *Journal of Politics* 44: 324–36.

Lindstrom, T., E. Hansen and H. Juslin. 1999. 'Forest Certification: The View from Europe's NIPFs'. *Journal of Forestry* 97, 3 (March): 25–30.

Lincoln, R. 2005. Personal Interview, London, 3 November 2005.

Lister, J. 2009. *Co-Regulating Corporate Social Responsibility: Government Response to Forest Certification in Canada, the United States and Sweden*. PhD Dissertation, Faculty of Graduate Studies (Resource Management and Environmental Studies), University of British Columbia, Vancouver, BC, Canada. April.

——. 2005. 'Co-Regulating Self Regulation: Government Role in Forest Certification'. Submission to Tuck School of Business, Dartmouth College, Institutional Mechanisms for Self-Regulation Mini-Conference, November 18. Online at http://mba.tuck.dartmouth.edu/mechanisms/pages/Papers/Lister_paper.pdf, accessed February 2010.

Litfin, K. 1994. *Ozone Discourses: Science and Politics in Global Environmental Cooperation*. New York, NY: Columbia University Press.

MacKinnon, A. 1998. 'Biodiversity and Old-Growth Forests'. In *Conservation Biology Principles for Forested Landscapes*, edited by J. Voller and S. Harrison, 146–84. Vancouver: UBC Press.

MacMullen, P. H. 1998. *A Report to the Fish Industry Forum on the Marine Stewardship Council and Related Topics*. United Kingdom: Sea Fish Industry Authority.

Markell, D. and J. Knox. 2003. *Greening NAFTA: The North American Commission for Environmental Cooperation*. Stanford, CA: Stanford University Press.

Maser, C. 1994. *Sustainable Forestry: Philosophy, Science and Economics*. Delray Beach, FL: St. Lucie Press.

Mason, W. 2007. 'Silviculture of Scottish Forests at a Time of Change'. *Journal of Sustainable Forestry* 24, 1 (May): 41–57.

Mathews, R. 1983. *The Creation of Regional Dependency*. Toronto: University of Toronto Press.

May, B., D. Leadbitter, M. Sutton, and M. Weber. 2003. 'The Marine Stewardship Council (MSC): Background, Rationale and Challenges'. In *Eco-Labelling in Fisheries: What is it All about?* edited by B. Phillips, B., T. Ward and C. Chaffee, 14–33. Oxford: Blackwell Science.

McAfee, B. and C. Malouin. 2008. *Implementing Ecosystem-Based Management Approaches in Canada's Forests: A Science-Policy Dialogue*. Ottawa: Natural Resources Canada. Online at http://dsp-psd.pwgsc.gc.ca/collection_2008/nrcan/Fo4-21-2008E.pdf, accessed March 2010.

McAlexander, J. and E. Hansen. 1999. *Business Case Study Series: J Sainsbury plc and The Home Depot: Retailers' Impact on Sustainability*. Washington: Island Press.

McDonald, J. (1999), 'Regional Forest (Dis)Agreements: The RFA Process and Sustainable Forest Management'. *Bond Law Review* 11: pp. 295–342.

McLaren, W. 2010. 'Japan wants Worlds Largest Woodchip Exporter to Seel FSC Certified Product'. *Treehugger*. Online at http://www.treehugger.com/files/2010/03/japan-wants-worlds-largest-woodchip-exporter-to-sell-fsc-certified.php, accessed March 2010.

McLoughlin, K. 2006. *Fisheries Status Reports 2005: Status of Fish Stocks Managed by the Australian Government*. Canberra: BRS.

McNichol, J. 1999. 'Contesting Governance in the Global Marketplace: A Sociological Assessment of British Efforts to Build New Markets for NGO-Certified Sustainable Wood Products'. University of California Department of Sociology (Spring). Online at www.irle.berkeley.edu/culture/papers/McNichol.pdf, accessed September 2009.

MCPFE (Ministerial Conference on the Protection of Forests in Europe). 2008. 'First Meeting of the MCPFE Working Group on Exploring the Potential Added Value of and Possible Options for a Legally Binding Agreement on Forests in the Pan-European Region'. Minutes of Meeting of 27–28 November 2008. Athens, Greece.

Meidinger, E., C. Elliott and G. Oesten (eds). 2003. *Social and Political Dimensions of Forest Certification*. Kessel: Forestbusch.

Mercury (Hobart). 2008. 'Certified Tasmanian Seafood or Not'. 1 March.

Metafore. 2009. 'SFM Certification in Canada 1999–2009'. Metafore's Forest Certification Resource Centre. Online at www.certifiedwood.org, accessed March 2009.

Mfodwo, K. 1998. *Non-Governmental Initiatives in Global Environmental Responsibility: A Review of the WWF/Unilever Marine Stewardship Council Proposal*. International Conference on Environmental Justice and Market Mechanisms: Key Challenges for Environmental Law and Policy, Auckland.

Miles, E. 1997. 'The Approaches of UNCLOS and Agenda 21 – A Synthesis'. In *Sustainable Development and Preservation of the Oceans: The Challenge of UNCLOS and Agenda 21*, edited by M. Kusuman-Atmaja, T. A. Mensah and B. H. Oxman, 16–42. Proceedings of the Law of the Sea Institute 29th Annual Conference, Bali Indonesia, Hawaii, LSI and University of Hawaii.

Miles, E. L. and W. T. Burke. 1989. 'Pressures on the United Nations Convention on the Law of the Sea of 1982 Arising from New Fisheries Conflicts – the Problem of Straddling Stocks'. *Ocean Development and International Law* 20, 4: 343–57.

MPIG (Montreal Process Implementation Group). 2008. *Australia's State of the Forests Report: Five-Yearly Report 2008*. Canberra, ACT: Government of Australia. Online at http://adl.brs.gov.au/forestsaustralia/_pubs/sofr2008reduced.pdf, accessed March 2010.

MSC. 2009a. 'MSC Stakeholder Council'. Online at http://www.msc.org/about-us/governance/structure/msc-stakeholder-council, accessed October 2009.

——. 2009b. 'Fisheries under Assessment'. http://www.msc.org/get-certified/find-a-certifier/fisheries-assessments, accessed Oct 2009.

——. 2009c. 'DEFRA Funding puts MSC on the menu' 24 July 2009. www.msc.org/newsroom/news/defra-funding-puts-msc-on-the-menu, accessed April 2010.

——. 2008. First Canadian Fishery Earns Marine Stewardship Council Certification'. Press release, 29 August 2008.

——. 2008a. 'Sustainable Fish Project Expands to Swedish Schools' 2 December 2008. www.msc.org/newsroom/sustainable-fish-project-expands-to-swedish, accessed April 2010.

——. 2007a. 'Consultation Document: Articulating the Intent of MSC's Criteria, MSC Quality and Consistency Project'. January.

——. 2007b. 'Mackerel Icefish'. Online at http://www.msc.org/cook-eat-enjoy/fish-to-eat/mackerel-icefish, accessed April 2010.

——. 2007c. 'MSC Response to Marks and Spencer Announcement on Sustainable Fish' 15 January 2007. www.msc.org/newsroom/news/msc-response-to-marks-spencer-announcement-on, accessed April 2010.

——. 2006. *MSC Annual Report: 2005–06* London, MSC.

——. 2004a. 'Marine Stewardship Council'. Online at http://www.msc.org, accessed September 2004.

——. 2004b. 'Minutes from the 4th Stakeholder Council Meeting 26–27th May 2004'. London: Marine Stewardship Council.

——. 2003. 'Lessons Learned in Fisheries Certification – The First Four Years of the Independent Marine Stewardship Council'. London: Marine Stewardship Council.

——. 2002a. *MSC Annual Report: 2002.* London: Marine Stewardship Council.

——. 2002b. 'Using the AHP and Expert Choice to Support the MSC Certification Process'. MSC and Hartley McMaster Ltd. London: Marine Stewardship Council.

——. 2001a. 'MSC Certification Methodology, Issue 3'. London: Marine Stewardship Council.

——. 2001b. 'The MSC Program: Certifying Seafood on its Journey from the Fishery to the Consumer'. London: Marine Stewardship Council.

——. 2000. 'Conclusions of the MSC Governance Review'. London, MSC: 6pp.

——. 1998. 'Principles and Criteria for Sustainable Fishing (Airlie House Draft)'. Washington: Marine Stewardship Council, October.

Mol, A. 2003. *Globalization and Environmental Reform: the Ecological Modernization of the Global Economy.* Boston: MIT Press.

Molenaar, E. 2000. 'The Concept of "Real Interest" and Other Aspects of Co-Operation Through Regional Fisheries Management Mechanisms'. *International Journal of Marine and Coastal Law* 15: 475–531.

Moody Marine. 2007. *Association of Seafood Producers – Canadian Northern Prawn Trawl Fishery MSC Certification: Nomination for Assessment Team Membership.* Derby, Moody Marine. Online at www.msc.org/track-a-fishery/certified/north-west-atlantic/Canada-northern-prawn/copy_of_assessment-downloads-1/assessment-downloads/Team_Nominees.pdf, accessed 21 October 2010.

Musselwhite, G. and G. Herath. 2005. 'Australia's Regional Forest Agreement Process: Analysis of the Potential and Problems'. *Forest Policy and Economics* 7, 4: 579–88.

NAFA (National Aboriginal Forestry Association). 2003. *Aboriginal-Held Forest Tenures in Canada 2002–2003.* Online at http://www.nafaforestry.org/forest_home/documents/AboriginalTenures2003.pdf, accessed November 2008.

NAFI (National Association of Forest Industries). 2008. 'About NAFI'. Online at: http://www.nafi.com.au/about/?printfriendly=1, accessed July 2010.

——. 2006. 'NAFI Welcomes Review of Forest Certification'. NAFI Media Release, 23 October. Online at http://www.nafi.com.au/media/view.php3?id=442, accessed April 2008.

——. 1999. 'Certification 'not a magic wand' – Lang'. Report of the NAFI 1999 Annual Conference, Forest Futures Australia. Online at http://archive.nafi.com.au/newsletter/forest.pdf, accessed March 2008.

——. 1998. 'Forest Industry Certification: A Given for all Internationally Traded Wood Products within 2 Years'. *NAFI Newsletter*, May. Online at http://archive.nafi.com.au/newsletter/1998-05/section1.html, accessed March 2008.

——. 1996. 'Brisbane Conference on Certification'. *NAFI News*, July. Online at http://archive.nafi.com.au/newsletter/july/sect2.html#9, accessed March 2008.

NAFI Newsletter. 1997. November. Online at http://archive.nafi.com.au/newsletter/97november/section2.html, accessed October 2010.

Natural England. 2009a. 'How are SSSIs Designated'. Online at www.naturalengland.org.uk/conservation/designatedareas/SSSI/designation.aspx, accessed September 2009.

——. 2009b. 'The New Forest'. Online at www.english-nature.org.uk/special/sssi/sssi_details.cfm?sssi_id=1003036, accessed September 2009.

——. 2008. *State of the Natural Environment Report. Chapter 3, Biodiversity*. Sheffield: Natural England.

Newell, P. 2008. 'The Political Economy of Global Environmental Governance'. *Review of International Studies* 34: 507–29.

NFFO (National Federation of Fishermen's Organisations). 2010. 'Sustainability Initiatives in Fishing'. Online at www.nffo.org.uk/sustainability.html, accessed 21 October 2010.

NFPS (National Forest Policy Statement). 1992. *National Forest Policy Statement: A New Focus for Australia's Forests*. Canberra, ACT: Commonwealth of Australia 1992/1995.

NFCS (National Forest Coalition Strategy). 2003a. 'Canada Forest Accord 2003–2008.' Online at http://csfn.foret.ca/accords/accord3.html, accessed November 2008.

——. 2003b. 'Canada Forest Accord 2003–2008'. Online at http://csnf.forest.ca/accords/accord3.html, accessed November 2003.

——. 2003c. 'Canada Forest Accord 1998–2003'. Online at http://csnf.forest.ca/accords/accord2.html, accessed November 2003.

——. 2003d. 'Canada Forest Accord 1992–1997'. Online at http://csnf.forest.ca/accords/accord1.html, accessed November 2003.

NGO Statement. 1990. 'NGO Statement to the Final Plenary, 9th Meeting of the International Tropical Timber Council'. Yokohama, Japan. 23 November.

NRCan (Natural Resources Canada). 2010a. 'Forest'. Online at http://canadaforests.nrcan.gc.ca/glossary/f, accessed March 2010.

——. 2010b. 'The Atlas of Canada: Boreal Forest'. Online at http://atlas.nrcan.gc.ca/site/english/learningresources/theme_modules/borealforest/index.html, accessed October 2010.

——. 2008a. 'Quick Facts: Table 2: Ownership of Forest Lands'. Online at http://www.pfc.forestry.ca/monitoring/inventory/canfi/facts/table2_e.html, accessed November 2008.

——. 2008b. 'Forested Ecozones'. Online at http://atlas.nrcan.gc.ca/site/english/maps/environment/forest/forestcanada/forestedecozones/1, accessed November 2008.

——. 2008c. 'Hitting Home – Impacts of the US Housing Crisis'. Online at http://foretscanada.rncan.gc.ca/articletopic/184, accessed November 2008.

——. 2008d. *Canada's National Forest Inventory*. Ottawa: NRCan. Online at http://www. pfc.forestry.ca/monitoring/inventory/canfi/data/area-class-small_e.html, accessed November 2008.

——. 2008e. 'Annual Economic Review and Outlook for the Canadian Forest Sector 2008–09'. September. Online at http://warehouse.pfc.forestry.ca/HQ/29039.pdf, accessed October 2010.

——. 2007. *The State of Canada's Forests: Annual Report 2007*. Ottawa: NRCan. Online at canadaforests.nrcan.gc.ca, accessed November 2007.

O'Connor, J. 1973. *The Fiscal Crisis of the State*. New York, NY: St. Martin's Press.

ODA (Overseas Development Administration). 1993. 'Overseas Development Administration Forestry Policy'. December. Mimeo.

OMNR (Ontario Ministry of Natural Resources). 2008a. 'Ontario's Tenure and Licensing System'. Online at http://www.mnr.gov.on.ca/en/Business/Forests/2ColumnSubPage/STEL02_167460.html, accessed November 2008.

——. 2008b. *Canada's National Forest Inventory*. Ottawa: NRCan. Online at http://www. pfc.forestry.ca/monitoring/inventory/canfi/data/area-class-small_e.html, accessed November 2008.

——. 2006. 'Ontario's State of the Forest Report 2006'. Peterborough, ON: OMNR. Online at http://www.mnr.gov.on.ca/en/Business/Forests/2ColumnSubPage/STEL02_179267.html, accessed November 2008.

Oosthoek, Jan-Willem. 2000. 'The Logic of British Forest Policy, 1919–1970.' Paper presented at the 3rd Conference of the European Society for Ecological Economics 'Transitions Towards a Sustainable Europe. Ecology – Economy – Policy'. Vienna, Austria (May 3–6).

Parai, B. and T. Esakin. 2003. 'Beyond Conflict in Clayoquot Sound: The Future of Sustainable Forestry'. In *Natural Resource Conflict Management Case Studies: An Analysis of Power, Participation and Protected Areas*, edited by A. P. Castro and E. Nielsen, 163–82. Rome: United Nations Food and Agricultural Organization. Online at http://www.fao.org/docrep/005/y4503e/y4503e00.htm.

Paschalis-Jakubowicz, P. 2006. 'Poland'. In B. Cashore, F. Gale, E. Meidinger and D. Newsom *Confronting Sustainability: Forest Certification in Developing and Transitioning Countries*, edited by B. Cashore, F. Gale, E. Meidinger and D. Newsom. New Haven, CT: Yale Faculty of Environmental Studies Publication Series.

Pattberg, P. 2007. *Private Institutions and Global Governance: The New Politics of Environmental Sustainability*. Northhampton: Edward Elgar.

Peet, R. and E. Hartwick. 1999. *Theories of Development*. Oxford: The Guildford Press.

Pellew, R. and A. Burgmans. 1996. 'Statement of Intent'. Reprinted in *Samudra* 15: 16.

Payne, R. and J. Nasser. 2009. *Politics and Culture in the Developing World: The Impact of Globalisation*, 4th edn. New York, NY: Longman.

Phillips, B., T. Ward and C. Chaffee. 2003. 'Case Study 1: The Western Rock Lobster: A The Fishery and its assessment'. In *Eco-Labelling in Fisheries: What is it all About?* edited by B. Phillips, T. Ward and C. Chaffee, 94–102. Oxford: Blackwell Science.

—— (eds). 2003. *Eco-Labeling in Fisheries: What Is It All About?* Oxford: Blackwell Science.

PHMA (Pacific Halibut Management Association). 2009. About us. www.phmana.org/about.htm

Pierre, J. and B. G. Peters. 2000. *Governance, Politics and the State*. Basingstoke: Macmillan.

Pinkerton, E. 2009. 'Partnerships in Management'. In *A Fishery Manager's Handbook*, edited by Kevern L. Cochrane, 159–73. FAO Fisheries Technical Paper 424. Rome: Food and Agriculture Organisation.

——. 1999. 'Factors in Overcoming Barriers to Implementing Co-Management in British Columbia Salmon Fisheries'. *Conservation Ecology*, 3, 2: 2.

Pitcher, T. 2007. Personal communication (e-mail). 11 September 2007.

——. 2001. 'The FC and MSC'. *FishBytes* 7, 2 (March/April): 1.

Pollock, K. 2009. '70 Sleeps'. United Steelworkers of America. Online at http://www.usw.ca/UserFiles/File/D3%20documents/70_sleeps_March_3_09.pdf, accessed March 2009.

——. 1996. 'The IWA Speaks Out on Job Loss'. Letter to the editor. *BC Environmental Reporter* 7: 30.

Poore, D. 2003. *Changing Landscapes: The Development of the International Tropical Timber Organization and its Influence on Tropical Forest Management*. London: Earthscan.

Potts, T. 2006. 'A Framework for the Analysis of Sustainability Indicator Systems in Fisheries'. *Ocean and Coastal Management* 49: 259–80.

Potts, T. and M. Haward. 2007. 'International Trade, Ecolabelling, and Sustainable Fisheries – Recent Concepts and Practices'. *Environment, Development and Sustainability* 9: 91–106.

——. 2001. 'Sustainability Indicator Systems and Australian Fisheries Management'. *Maritime Studies* 117: 1–11.

Poulain, F. 2003. *Recent Trends in Environmental Labelling and Certification in Fisheries, Forestry and Organic Agriculture: Information Paper Prepared for the Expert Consultation on the development of International Guidelines for Ecolabelling of Fish and Fish Products from Marine Capture Fisheries*. Rome Italy 14–17 October.

PriceWaterhouseCoopers. 2000. *State of the BC Seafood Industry Report*. Vancouver: PriceWaterhouseCoopers.

Priest, J. 2003. 'Private Native Forestry Law in Tasmania'. In *The Forgotten Forests: The Environmental Regulation of Forestry on Private Land in New South Wales Between 1997 and 2002*. PhD Thesis, Centre of Natural Resources Law and Policy, Faculty of Law, University of Wollongong (December). Online at http://www.library.uow.edu.au/adt-NWU/uploads/approved/adt-NWU20050922.121033/public/12Chapter11.pdf, visited December 2007.

Prime Minister's Strategy Unit. 2004. *Net Benefits: A Sustainable and Profitable Future for UK Fishing*. London: Cabinet Office.

Princen, T. 2002. 'Consumption and Its Externalities'. In *Confronting Consumption*, edited by T. Princen, M. Maniates and K. Conca. Boston, MA: The MIT Press.

Princen, T., M. Maniates and K. Conca. 2002. *Confronting Consumption*. Boston, MA: MIT Press.

Pross, A. P. 1992. *Group Politics and Public Policy*. 2nd edn. Toronto: Oxford University Press.

Pross, A. P. and S. McCorquodale. 1987. *Economic Resurgence and the Constitutional Agenda*, Kingston, Institute of Intergovernmental Relations, Queen's University.

Raban, J. 1986. *Coasting*, London: Picador.

Raynolds, L. 2000. 'Re-Embedding Global Agriculture: The International Organic and Fair Trade Movements'. *Agriculture and Human Values* 17: 297–309.

RCEN (Réseau Canadian Environment Network). 2008a. 'Background'. Online at ww.cen-rce.org/eng/about_us.html, accessed August 2008.

——. 2008b. *RCEN 2007–2008 Annual Report*. Online at http://www.cen-rce.org/eng/publications/RCEN%202007-2008%20Annual%20Report.pdf, accessed January 2009.

Read, M. 1991. 'An Assessment of Claims of "Sustainability" Applied to Tropical Wood Products and Timber Retailed in the UK, July 1990–January 1991'. Godalming, Surrey: World Wide Fund for Nature-UK.

Redclift, M. 1987. *Sustainable Development: Exploring the Contradictions*. London: Routledge.

RESOLVE. 2007. 'About Resolve Consulting Group'. Online at http://www.resolve consulting.co.uk/index.html, accessed April 2010.

Resource Assessment Commission. 1992. *Forest and Timber Inquiry: Final Report: Volume 1*. Canberra, ACT: Australian Government Publishing Service (March).

Restino, C. 1993. 'The Cape Breton Island Spruce Budworm Infestation: A Retrospective Analysis'. *Alternatives* 19, 4 (August): 28–36.

Rhodes, R. 1997. *Understanding Governance: Policy Networks, Governance, Reflexivity and Accountability*. Buckingham: Open University Press.

Ridder, R. 2007. 'Global Forest Resources Assessment 2010: Options and Recommendations for a Global Remote Sensing Survey of Forests'. Forest Resources Assessment Program. Working Paper 141, March. Rome: FAO.

——. 1994. 'The Hollowing Out of the State'. *Political Quarterly* 65: 138–51.

Roberts, C. 2007. *The Unnatural History of the Sea: The Past and Future of Humanity and Fishing*. London: Gaia Thinking/Octopus Publishing.

Robertson, R. 1992. *Globalization: Social Theory and Global Culture*. London: Sage.

Rogers P., R. Gould and B. McCallum. 2003. 'Case Study 1: The Western Rock Lobster: B What Certifcation has Meant to the Department of Fisheries and the Industry'. In *Eco-Labelling in Fisheries: What is it all About?* edited by B. Phillips, T. Ward and C. Chaffee, 103–8. Oxford: Blackwell Science.

Rollinson, T. 2003. 'Changing Needs – Changing Forests: The UK Experience'. Presentation to the UNFF Intersessional Experts Meeting on the Role of Planted Forests in Sustainable Forest Management, 24–30 March 2003, New Zealand.

Rootes, C. 2007. 'Nature Protection Organizations in England'. Working Paper 1/2007. Centre for the Study of Social and Political Movements, School of Social Policy, Sociology and Social Research, University of Ken at Canterbury.

——. 2006. 'Facing South? British Environmental Movement Organisations and the Challenge of Globalisation'. *Environmental Politics* 15, 5 (November): 768–86.

Rootes, C., B. Seel and D. Adams. 2000a. 'The Old, the New and the Old New: British Environmental Organisations from Conservation to Radical Ecologism'. Paper presented to the workshop 'Environmental Organisations in Comparative Perspective,' ECPR Joint Sessions, Copenhagen, 14–19 April.

Rootes, C. and A. Miller. 2000b. 'The British Environmental Movement: Organisational Field and Network of Organisations'. Paper presented to the workshop 'Environmental Organisations in Comparative Perspective', ECPR Joint Sessions, Copenhagen, 14–19 April.

Ross, M. 1997. 'A History of Forest Legislation in Canada 1867–1996'. CIRL Occasional Paper #2 (March). Calgary, Alta.: Canadian Institute of Resources Law, University of Calgary.

Russell, S. 2003. 'Northeast NSW: The Community and the Forests'. Community Forestry Forum, Forest and Ecosystems Science Institute (FESI), University of Melbourne, Creswick Campus (25/26 October).

Sabatier, P. 2007. *Theories of the Policy Process*. Boulder, CO: Westview Press.

Salzman, J. 1991. *Environmental Labeling in OECD Countries: OECD Report No. 12*. Paris, OECD.

Sandberg, L. A. and P. Clancy. 2002. 'Politics, Science and the Spruce Budworm in New Brunswick and Nova Scotia'. *Journal of Canadian Studies* 37, 2 (Summer): 164–91.

Santiago Declaration. 1995. Online at http://www.rinya.maff.go.jp/mpci/rep-pub/1995/santiago_e.html, accessed November 2008.

Sayer, J. 1989. 'Memorandum by the International Union for Conservation of Nature and Natural Resources: Examination of Witness'. House of Lords Inquiry into Tropical Deforestation Chaired by the Earl of Cranbrook, House of Lords, Tuesday 16 January.

Schellenberg, M. and T. Nordhaus. 2005. 'The Death of Environmentalism: Global Warming Politics in a Post-Environmental World'. *Grist Magazine*, 13 January.

Schirmer, J. 2005. *Achieving Successful Change in Conflict over Afforestation: A Comparative Analysis*. PhD dissertation, School of Resources, Environment and Society, Australian National University, Canberra, ACT, Australia.

Scholte, J-A. 2001. 'The Globalization of Politics'. In *The Globalization of World Politics*, edited by J. Baylis and S. Smith. Oxford: Oxford University Press.

Scottish Fishermen's Federation. 2010. Online at www.sff.co.uk/, accessed 21 October 2010.

SCS (Scientific Certification Systems). 2010. Western Australian Rock Lobster Fishery Maintains 'Sustainable Seafood' Certification. Media release, 13 January.

——. 2003. *MSC Evaluation of BC Salmon Fisheries: Units of Certification, Performance Indicators and Scoring Guideposts*. Emeryville, CA: SCS.

SeaChoice. 2008. "Not Green Yet' – Sustainability Certification for Atlantic Canada's Northern Shrimp Trawl Fishery Leaves Conservation Concerns Unaddressed'. 21 August 2008.

Seafish. 2010. 'About us'. www.seafish.org/about/default.asp?p=ba, accessed 21 October 2010.

Seafood Services Australia. 2008. SeaQual Food Safety Guidelines. www.seafood services.com.au/quality/foodsafety.php.

Segura, G. 2004. *Forest Certification and Governments: The Real and Potential Influence on Regulatory Frameworks and Forest Policies*. Washington, DC: ForestTrends.

Searle, R., S. Colby and K. Milway. 2004. *Moving Eco-Certification Mainstream*. Boston and San Francisco: Bridgespan Group.

SFF (Silva Forest Foundation). 1999. *Ecosystem-Based Certification*. Slocan Park, BC: Silva Forest Foundation.

Shelton, P. A. and A. F. Sinclair. 2008. 'Its Time to Sharpen Our Definition of Sustainable Fisheries Management'. *Canadian Journal of Fisheries and Aquatic Science*, 62: 2305–14.

Shotton, R. and M. Haward. 2005. 'Requirements for Managing Deep-Sea Fisheries'. In *Deep Sea 2003: Conference on the Governance and Management of Deep-sea Fisheries, Queenstown New Zealand 27–29 November 2003, FAO Fisheries Proceedings No 3/1*, edited by R. Shotton, 686–710. Rome FAO.

Short, K. 2003. 'New Zealand Hoki – the WWF Perspective'. In *Eco-Labelling in Fisheries: What is it all About?* edited by B. Phillips, T. Ward and C. Chaffee. Oxford: Blackwell Science.

Sierra Club of BC. 2009. 'Promise to Protect Great Bear Rainforest becomes a Reality – Five-Year Plan in Place'. Online at http://www.savethegreatbear.org/mediacentre/gbr_release_mar31, accessed April 2010.

——. 2008. 'Our Roots: A Brief History of the Sierra Club of BC'. Online at http://www.sierraclub.bc.ca/quick-links/about/our-roots, accessed January 2009.

Sierra Club Canada. 2009. 'Mission'. Online at http://www.sierraclub.ca/national/aboutus/mission.html, accessed January 2009.

——. 2007. *Sierra Club Canada 2007 Annual Report*. Ottawa, ON: Sierra Club Canada. Online at http://www.sierraclub.ca/national/aboutus/annual-reports/annual-report-2007-web.pdf, accessed January 2009.

——. 2006. 'National Forest Strategy 2003–2008: An Assessment in 2006 – is it Making a Difference?' Ottawa, ON: Sierra Club Canada.

——. 2005. 'Audited Financial Statements: Sierra Club of Canada'. Ottawa: Sierra Club Canada. 31 December.

Small Woods Association. 2009. 'Forestry and Timber Association'. Online at http://initiatives.smallwoods.org.uk/?link=directory.php&id=2039, accessed September 2009.

Smith, W. 2006. 'Regulating Timber Commodity Chains: Timber Commodity Chains Linking Cameroon and Europe'. Paper presented at the 11th biennial Congress of the International Association for the Study of Common Property, Bali, Indonesia.

——. 1995. 'What is the Pacific Certification Council?' *International Journal of Ecoforestry* 11, 4 (Winter): 105–7.

Southern Fishermen's Association. 1998. *Wild Fisheries with a Future: Environmental Management Plan of the Southern Fishermen's Association*. Adelaide: Southern Fishermen's Association.

SSCISTTCI (Senate Standing Committee on Industry, Science, Technology, Transport, Communications and Infrastructure). 1993. Fisheries Reviewed. Canberra, ACT: Parliament of Australia.

Standards Australia. 2003. *The Australian Forestry Standard: AS 4708(Int)*. Canberra, ACT: Australian Forestry Standard Steering Committee and Australian Forestry Standard Project Office.

Stewart, D. and G. McColl. 1994. 'The Resource Assessment Commission: An Inside Assessment'. *Australasian Journal of Environmental Management* 1, 1: 12–23.

Stewart, L. 2009. 'Defra Funding Puts MSC on the Menu'. Marine Stewardship Council. Online at http://www.msc.org/newsroom/news/defra-funding-puts-msc-on-the-menu, accessed April 2010.

Stokes, D. 2009. 'The War Gamble: Understanding US interests in Iraq'. *Globalizations* 6, 1 (2009): 105–10.

Sumaila, R. T. Pitcher and D. Pauly 2005. 'On Eco-Labeling, the MSc and Us', *FishBytes*, 11, 6: 4–5.

Sun, C., L. Chen, L. Chen, L. Han and S. Bass. 2008. *Global Forest Product Chains: Identifying Challenges and Opportunities for China through a Global Commodity Chain Sustainability Analysis*. Winnipeg, Manitoba: IISD.

Sutton, D. 2003. 'An Unsatisfactory Encounter with the MSC – a Conservation Perspective'. In *Eco-Labelling in Fisheries: What is it all About?* edited by B. Phillips, T. Ward and C. Chaffee. Oxford: Blackwell Science.

Symes, D. and J. Phillipson. 2009. 'Whatever Became of Social Objectives in Fisheries Policy?' *Fisheries Research* 95: 1–5.

Synnott, T. 2005. 'Some Notes on the Early Years of FSC'. November. Online at http://www.fsc.org/fileadmin/web-data/public/document_center/publications/ Notes_on_the_early_years_of_FSC_by_Tim_Synnott.pdf, accessed October 2009.

Talbot, J. 2004. *Grounds for Agreement: The Political Economy of the Coffee Commodity Chain*. Lanham, MD: Rowman & Littlefield.

Tanzer, J. 1995. 'New Management Arrangements for Queensland's Fisheries'. Paper presented at the Outlook 95 Conference, 7–9 February, 1995, Canberra.

TAVEL (TAVEL Certification Inc). 2008. 'Advisory – Transition of BC Sockeye Salmon Certification Assessment from SCS to TAVEL Certification'. 8 May 2008.

Taylor, I. 2003. 'The United Nations Conference on Trade and Development'. *New Political Economy* 8, 3 (2003): pp. 409–18.

TCA (Timber Communities Australia). 2008. 'About TCA'. Online at http://www.tca.org.au/abouttca/index.shtml, accessed April 2010.

Teisl, M. F., B. Roe and R. L. Hicks. 2002. 'Can Eco-Labels Tune a Market? Evidence from Dolphin-Safe Labeling'. *Journal of Environmental Economics and Management* 43: 339–59.

Thomson, S. 2006. 'Supporting Grassroots Environmental Groups in British Columbia: An Evaluation of the Role of the BCEN: Summary Report'. Report prepared for the Canadian Environmental Network, Ottawa, Ontario, Canada. 16 January.

Timber Trade Journal. 1989a. 'TTF Tells Members to Ignore Sticker System'. 30 September: 1.

——. 1989b. 'Government to Study Sustainable Incentives'. 11 November: 1.

timbertrends. 2007. 'Measuring Timber Certification: The UK Timber Industry: All Sector Report'. Alicante, Spain: timbertrends.

Tollefson, C., F. Gale, J. Rayner and A. Zito. 2010. 'Three Dimensions of Governance: An Analytic Framework'. Paper presented to the New Governance and Natural Resources Management Workshop, Victoria, BC, Canada, 3–4 March.

Tollefson, C., F. Gale and D. Haley. 2008. *Setting the Standard: Certification, Governance and the Forest Stewardship Council.* Vancouver, BC: UBC Press.

TTF (Timber Trade Federation). 2009a. 'Frequently Asked Questions'. Online at http://www.ttf.co.uk/About_TTF/Frequently_asked_questions.aspx, accessed April 2010.

——. 2009b. 'Responsible Sourcing'. Online at http://www.ttf.co.uk/Environment/Responsible_Sourcing.aspx, accessed April 2010.

——. 1993. 'Forests Forever: A Campaign for Wood: 14th Session of the ITTC and 12th Session of the Permanent Committees: Kuala Lumpur 11–19 May 1993: Forests Forever Statement: Sustainability of Timber in Trade'. Mimeo.

——. 1992. 'Forests Forever: A Campaign for Wood. UK Forests Forever: Items for ITTO Consideration: Yaounde, Cameroon, May 1992'. Mimeo.

——. 1991. 'Forests Forever: A Campaign for Wood: Eleventh Session of the International Tropical Timber Council and Ninth Session of the Permanent Committees, Yokohama 28 November–4 December 1991: UK Forests Forever'. Mimeo.

Treib, O., O. Bähr and G. Falkner. 2007. 'Modes of Governance: Towards a Conceptual Clarification'. *Journal of European Public Policy* 14, 1 (January): 1–20.

——. 2005. 'Modes of Governance: Towards a Conceptual Clarification'. In *European Governance Papers (EUROGOV)* No. N-05-02. Online at http://www.connex-network.org/eurogov/pdf/egp-newgov-N-05-02.pdf.

Tsouvalis, J. 2000. *A Critical Geography of Britain's State Forests.* Oxford: Oxford University Press.

Tully, S. 2004. 'Access to Justice within the Sustainable Development Self Government Model'. Discussion paper 21, June 2004, ESRC Centre for the Analysis of Risk and Regulation. London: London School of Economics and Political Science.

TWS (The Wilderness Society). 2008a. 'From People Power Came Victory for the River'. http://wenlock.wilderness.org.au/articles/about-us-history-from-people-power-came-victory/?searchterm=%20people%20power, accessed April 2010.

——. 2008b. 'History of the Franklin River Campaign 1976–1983'. Online at http://wenlock.wilderness.org.au/campaigns/wild-rivers/franklin/?searchterm=%20history%20franklin%20river, accessed March 2008.

——. 2007. *Annual Report for the Year Ended 2007.* Online at http://wenlock.wilderness.org.au/articles/tws-annual-report-06-07/?searchterm=%20annual%20report, accessed March 2008.

——. 2005a. *WildCountry: A New Vision for Nature.* Hobart: TWS. Online at http://wenlock.wilderness.org.au/files/WildCountry-a_new_vision_for_nature2005.pdf/view?searchterm=%20wildcountry, accessed March 2008.

——. 2005b. *Certifying the Incredible, The Australian Forestry Standard: Barely Legal and Not Sustainable.* Hobart, Tasmania: TWS. Online at http://wenlock.wilderness.org.au/pdf/TWS%2C%20Certifying%20the%20Incredible-responseto%20AFSLtdV

1%2C%20Feb%202006.pdf/view?searchterm=%20certifying%20the%20incredible, accessed March 2008.

UBCFC (University of British Columbia Fisheries Centre). 2004. Fisheries Centre Research Report 12, 2.

UCMP (University of California Museum of Paleontology). 2009. 'The Forest Biome'. Online at http://www.ucmp.berkeley.edu/exhibits/biomes/forests.php, accessed November 2009.

UKRCEP (Royal Commission on Environmental Pollution). 2004. *Turning the Tide: Addressing the Impact of Fisheries on the Marine Environment.* 25th Report, London Cm 6392.

UK Tropical Forest Forum, ITTO Working Group. 1992. 'Minutes of Meeting Held on September 4 1992 at the WWF offices, Beauchamp Place, 10 am–1 pm'. Godalming, Surrey: WWF-UK. Mimeo.

UKWAS. 2006. 'UKWAS (Second Edition) – Summary of Main Changes'. September. Online at http://www.ukwas.org.uk/standard/background_information/index.html, accessed September 2009.

——. 2000. 'Certification Standard for the UK Woodland Assurance Scheme'. Edinburgh, Scotland: UKWAS Support Unit, Forestry Commission. Online at http://www.ukwas.org.uk/standard/ukwas_archive/index.html, accessed September 2009.

UNECE (United Nations Economic Commission for Europe). 2009. *Forest Products Annual Market Review 2008–2009.* New York and Geneva 2009. Online at http://timber.unece.org/index.php?id=208, accessed October 2009.

UNEI (United Nations Environmental Indicators). 2010. 'Forests'. Online at http://unstats.un.org/unsd/environment/indicators.htm, accessed April 2010.

UNGA (United Nations General Assembly). 1992. 'Non-Legally Binding Authoritative Statement of Principles for a Global Consensus on the Management, Conservation and Sustainable Development of all Types of Forests'. Online at http://www.un.org/documents/ga/conf151/aconf15126-3annex3.htm, accessed December 2009.

UPM Tilhill. 2007. 'UKWAS Woodland Certification – UPM Tilhill Breaks through 100,000 Hectares'. News release. October. Online at http://www.upm-tilhill.com/NetsiteCMS/pageid/25/offset/0/pressreleaseid/584/Press/Press.html, accessed October 2010.

USWA (United Steelworkers of America). 2009. 'Home Page'. Online at http://www.usw.org/, http://www.uswa.ca/program/content/index.php?lanaccessed March 2009.

——. 2008. 'Cambell's Legacy of Mismanagement and Steelworkers' 10 Point Plan for the BC Forest Sector'. United Steelworkers District 3 & USW Wood Council. Online at http://www.usw.ca/UserFiles/File/D3%20Images/BC_March08.pdf, accessed March 2009.

Vallejo, N. and P. Hauselmann. 2004. *Governance and Multistakeholder Processes.* May. Winnipeg, Manitoba: International Institute for Sustainable Development. Online at http://www.iisd.org/pdf/2004/sci_governance.pdf, accessed October 2009.

Van der Goot, C. 2008. 'Financial Report 2005–2007'. Presentation to the 2008 General Assembly, Cape Town, South Africa.

VanderZwaag, D. (ed.). 1992. *Canadian Ocean Law and Policy,* Toronto, Butterworths.

——. 1983. 'Canadian Fisheries Management: A Legal and Administrative Overview'. *Ocean Development and International Law,* 13, 2: 171–211.

VanderZwaag, David L. and G. Chao (eds.) 2006. *Aquaculture Law and Policy: Towards Principled Access and Operations.* Abingdon: Routledge.

Vince, J. 2007. 'Policy Responses to IUU Fishing in Northern Australian Waters'. *Ocean & Coastal Management* 50, 8: 683–98.

Vitalis, V. 2002. *Roundtable on Sustainable Development – Private Voluntary Eco-Labels: Trade Distorting, Discriminatory and Environmentally Disappointing*. Paris, OECD.

Voller, J. and S. Harrison. 1998. *Conservation Biology Principles for Forested Landscapes*. Vancouver: UBC.

Waack, R. 2008. 'Board Report of the FSC Board of Directors 2005–2008'. Presentation to the 2008 General Assembly, Cape Town, South Africa.

Wade, R. 2009. 'From Global Imbalances to Global Reorganisations'. *Cambridge Journal of Economics* 33: 539–62.

Walker, K. 1999. 'Statist Developmentalism in Australia'. In *Australian Environmental Policy 2: Studies in Decline & Devolution*, edited by K. Walker and K. Crowley, 22–44. Sydney, NSW: NSW Press.

Warhurst, J. 1994. 'The Australian Conservation Foundation: The Development of a Modern Environmental Interest Group'. *Environmental Politics* 3, 1 (Spring): 68–90.

WCED (World Commission on Environment and Development). 1987. *Our Common Future*. Oxford: Oxford University Press.

WCL (Wildlife and Countryside Link). 2009a. 'Who we are'. Online at http://www.wcl.org.uk/who-we-are.asp, accessed September 2009.

——. 2009b. 'About us'. Online at http://www.wcl.org.uk/about-us.asp, accessed September 2009.

——. 2009c. 'Current Work Areas'. Online at http://www.wcl.org.uk/currentwork areas.asp, accessed September 2009.

——. 2009d. 'Recent Publications'. Online at http://www.wcl.org.uk/recent publications.asp, accessed September 2009.

——. 2007. 'Woodlands in England: Wildlife and Countryside Link's Manifesto'. June. Online at http://www.wcl.org.uk/docs/Link_Woodland_Manifesto_21Jun07.pdf, accessed September 2009.

WCWC (Western Canada Wilderness Committee). 2008. 'Western Canada Wilderness Committee Annual Report for the Year Ended 30 April 2008'. Vancouver, BC: WCWC.

——. 2006. 'Logging in Clayoquot Sound's Pristine Valleys and Islands? No Way!' Wilderness Committee, Victoria, British Columbia Chapter. Victoria, BC: WCWC.

——. 1992. '1992 Official WCWC AGM Business'. *Wilderness Committee Education Report* 11, 10 (Winter).

Weatherby, J. 2010. 'The Old and the New: Colonialism, Neo-Colonialism, and Nationalism'. In *The Other World*, edited by J. Weatherby, C. Arcenaux, E. Evans, D. Long, I. Reed and O. Carter. New York, NY: Longman.

Weber, G. 2005. 'Report of the Dispute Resolution Workshop'. Manaus, Brazil: FSC General Assembly.

Weldon, Sue. 2004. 'Social Science in Forestry: Public Participation and Partnership: A Review of Forestry Commission Practice and Governance in a Changing Political and Economic Context'. Edinburgh: Forestry Commission. Online at http://www.forestry.gov.uk/pdf/fcrp007.pdf/$FILE/fcrp007.pdf, accessed September 2009.

Wessells, C. R., K. Cochrane, C. Deere, P. Wallis and R. Willmann. 2001. *Product Certification and Ecolabelling for Fisheries Sustainability*. FAO Fisheries Technical Paper, No. 422. Rome: FAO.

Wettenhall, R. 1985. 'Intergovernmental Agencies: Lubricating a Federal System'. *Current Affairs Bulletin* April: 28–35.

Wildhavens. 2004. *An Independent Assessment of the Marine Stewardship Council*. Report prepared for The Homeland Foundation, the Oak Foundation and the Pew Charitable Trusts.

Willman, R., K. Cochran and W. Emerson. 2008. 'FAO Guidelines for Ecolabelling'. In *Seafood Ecolabelling: Principles and Practice*, edited by T. Ward and B. Phillips. Oxford: Blackwell.

Wilson, J. 1998. *Talk and Log: Wilderness Politics in British Columbia*. Vancouver: UBC Press.

Wilson, D., R. Curtotti, G. Begg and K. Phillips (eds) 2009. *Fisheries Status Reports 2008: Status of Fish Stocks and Fisheries Managed by the Australian Government*. Canberra, ACT: BRS and ABARE.

Winter, M. 1996. *Rural Politics: Policies for Agriculture, Forestry and the Environment*. London: Routledge.

Wood, P. 2000. A Comparative Analysis of Selected International Forestry Certification Schemes. Report Prepared for the Government of British Columbia, Victoria, British Columbia. Online at www.for.gov.bc.ca/het/certification/WoodReportOct00.PDF, accessed April 2010.

World Bank. 2009. 'Country Classifications'. Online at http://data.worldbank.org/about/country-classifications, accessed December 2009.

WWF-Australia. 2008a. 'Governors'. Online at http://wwf.org.au/about/structure/governors/, accessed March 2008.

——. 2008b. 'How we Work'. Online at http://www.wwf.org.au/about/howwework/, accessed March 2008.

WWF-Canada. 2008. *World Wildlife Fund Canada Annual Report 2008*. Toronto: WWF Canada 2008. Online at http://assets.wwf.ca/downloads/wwfcanada_annual report2008.pdf, accessed January 2009.

——. 2003. 'Groundbreaking Forest Certification Effort in Canada'. November. Online at http://gftn.panda.org/newsroom/?9783/Groundbreaking-forest-certification-effort-in-Canada, accessed April 2010.

——. 2001. 'World Wildlife Fund and Tembec Inc. Reach Historic Accord to Promote Long-Term Sustainability of Canadian Forestry'. Online at http://wwf.ca/newsroom/?1131, accessed April 2010.

WWF-UK. 2009. 'About us'. Online at http://www.wwf.org.uk/what_we_do/about_us/, accessed September 2009.

——. 2008. 'Report and Financial Statements'. Godalming, Surrey: WWF-UK. Online at http://assets.wwf.org.uk/downloads/2009_wwf_annual_report.pdf, accessed September 2009.

——. 2001. 'Environment Minister Reaffirms Government Position on Responsible Timber'. 23 November.

WWF/World Bank Alliance. 2002a. 'Case Study No. 8 – United Kingdom: FSC UK Working Group'. July. Online at http://www.piec.org/mswg_toolkit/mswg_toolkit/data/casestudies/8_FSC_UK.doc, accessed September 2009.

——. 2002b. 'Case Study No. 9 – The Development of the United Kingdom's Woodland Assurance Standard (UKWAS) Steering Group'. July. Online at http://www.piec.org/mswg_toolkit/mswg_toolkit/data/casestudies/9_UKWAS.doc, accessed September 2009.

Yearsley, G., P. R. Last and R. D. Ward. 1999. *Australian Seafood Handbook: A Guide to Domestic Species*. Hobart: CSRIOR Marine Research.

Young, O. 1989. *International Cooperation: Building Regimes for Natural Resources and the Environment*. New York, NY: Cornell University Press.

Young, V. and T. Cadman. 2001. '18th National Forest Summit Communique'. May.

Zelco, F. 2004. 'Making Greenpeace: The Development of Direct Action Environmentalism in British Columbia'. *BC Studies* 142/143 (Summer 2004): 197–240.

Index